# POLARIZED ION SOURCES AND POLARIZED GAS TARGETS

# AIP
# CONFERENCE
# PROCEEDINGS 293

# POLARIZED ION
# SOURCES AND
# POLARIZED GAS TARGETS

MADISON, WI 1993

EDITORS:
**L.W. ANDERSON**
**WILLY HAEBERLI**
UNIVERSITY OF WISCONSIN–MADISON

**American Institute of Physics**          New York

L.C. Catalog Card No. 93-074102
ISBN 1-56396-220-9
DOE CONF-930544

Printed in the United States of America.

SPONSORS


U.S. National Science Foundation

International Committee for High Energy Symposia

# CONTENTS

## II. ATOMIC BEAM POLARIZED ION SOURCES

## III. OPTICALLY PUMPED POLARIZED TARGETS AND ION SOURCES
## (EXCEPT $^3$He)

## IV. POLARIZED $^3$He TARGETS

## V. MACHINE–TARGET INTERACTIONS

L. W. Anderson
Department of Physics
University of Wisconsin
1150 University Avenue
Madison, WI 53706
Phone: (608) 262-8962
e-mail: LWANDERSON@
UWNUCO.PHYS.EDU
FAX: (608) 265-2334

A. S. Belov
Institute for Nuclear Research
Russian Academy of Sciences
Prospect, 7a
Moscow, 117312
Russia
Phone: (095)-334-09-62
e-mail: BELOV@INR.MSK.SU

I. D. Boyd
Cornell University
Department of Mechanical and
Aerospace Engineering
246 Upson Hall
Ithaca, NY 14853
Phone: (607) 255-4563
e-mail: boyd@coriolis.tn.cornell.edu
FAX: (607) 255-1222

G. D. Cates
Department of Physics
Jadwin Hall
Princeton University
P. O. Box 708
Princeton, NJ 08544
Phone: (609) 258-4414

T. E. Chupp
University of Michigan
Randall Lab. of Physics
500 E. University
Ann Arbor, MI 48109
Phone: (313) 747-2514
FAX: (313) 763-9694

G. Clausnitzer
University of Giessen
Institut für Kernphysik
Strahlenzentrum
6300 Giessen, Germany
Phone: 0641-702-2655
FAX: 0641-702-2672

T. B. Clegg
Department of Physics and Astronomy
University of North Carolina
Phillips Hall Room 278
Chapel Hill, NC 27599-3255
Phone: (919) 962-2079
e-mail: clegg@tunl.bitnet
FAX: (919) 962-0480

A. Converse
Department of Physics
University of Wisconsin
1150 University Avenue
Madison, WI 53706
Phone: (608) 263-3091
e-mail: CONVERSE@WISCNUC
FAX: (608) 262-3598

G. Court
Physics Department
Liverpool University
P. O. Box 147
Liverpool L69 3BX
United Kingdom
Phone: 051-794-3383
e-mail: GRC@UK.AC.LIV.PHYS.IA
FAX: 051-794-3444

E. R. Crosson
Triangle Universities Nuclear Lab.
Duke University
Box 90305
Durham, NC 27708
Phone: (919) 660-2660

C. W. de Jager
NIKHEF-K
P. O. Box 41882
1009 DB Amsterdam
The Netherlands
Phone: 31-20-592-2143
e-mail: kees@paramount.
nikhefk.nikhef.nl
FAX: 31-20-592-2165


V. P. Derenchuk
IUCF
2401 Milo B. Sampson Lane
Bloomington, IN 47408
Phone: (812) 855-9130
e-mail: LADDIE@IUCF.BITNET
FAX: (812) 855-6645


W. A. DeZarn
IUCF
2401 Milo B. Sampson Lane
Bloomington, IN 47405
Phone: (812) 855-3613
e-mail: DEZARNWA@IUCF
FAX: (812) 855-6645


J. Doskow
IUCF
2401 Milo B. Sampson Lane
Bloomington, IN 47405
Phone: (812) 855-5258
e-mail: DOSKOW@IUCF
FAX: (812) 855-6645


P. D. Eversheim
Institut für Kernphysik
University of Bonn
Nussallee 14-16
W-5300 Bonn 1, Germany
Phone: 0049-228-735299
e-mail: Evershei@
servax.iskp.uni-bonn.de
FAX: 0049-228-733728


E. A. George
Department of Physics
University of Wisconsin
1150 University Avenue
Madison, WI 53706
Phone: (608) 263-2263
e-mail: GEORGE@WISCNUC
FAX: (608) 262-3598


G. Graw
Sektion Physik der LMU
Am Coulombwall 1
W-8046 Garching, Germany
Phone: 49 89 3209-4155
e-mail: Gerhard.Graw@
Physik.Uni-Muenchen.DE
Uni-Muenchen.DE
FAX: 49 89 3209-4103


W. Haeberli
Department of Physics
University of Wisconsin
1150 University Avenue
Madison, WI 53706
Phone: (608) 262-0009
e-mail: WHAEBERLI@
WISCNUC
FAX: (608)262-3598


O. Häusser
TRIUMF
4004 Westbrook Mall
Vancouver B.C. V6T 2A3
Canada
Phone: (604) 222-1047
e-mail: HAUSSER@
TRIUMFCL.BITNET
FAX: (604) 222-1074


W. Heil
Institut für Physik
EXAKT
D-55099 Mainz
Germany
Phone: 6131-39-5918
FAX: 6131-39-2991

A. Hershcovitch
Building 911B
Brookhaven National Lab.
Upton, New York 11973
Phone: (516) 282-4531


F. W. Hersman
Department of Physics
University of New Hampshire
66 Bucks Hill Road
Durham, NH 03824
Phone: (603) 862-3512
e-mail: hersman@unh.edu
FAX: (603) 862-2998


R. J. Holt
Physics Division
Argonne National Laboratory
9700 South Cass Avenue
Argonne, IL 60439-4843
Phone: (708) 252-4012
e-mail: HOLT@ANLPHY
FAX: (708) 252-3903


A. Honig
Physics Department
Syracuse University
Syracuse, New York 13244
Phone: (315) 443-3888
e-mail: Honig@Suhep.
PHY.SYR.EDU
FAX: (315) 443-9103


K. Ikegami
Cyclotron Laboratory
RIKEN Institute
Wako-shi, Saitama 351-01
Japan
Phone: 81-48-462-1111
e-mail: IKEGAMI@
RIKVAX.RIKEN.GO.JP
FAX: 81-48-462-4642

C. E. Jones
Physics Division, Bldg. 203
Argonne National Lab.
Argonne, IL 60439
Phone: (708) 252-7267
e-mail: cjones@anlphy.bitnet
FAX: (708) 253-3903


W. A. Kaufman
University of Michigan
Randall Laboratory of Physics
Ann Arbor, MI 48109
Phone: (313) 764-5111
e-mail: Kaufman@UMIPHYS
MICH::KAUFMAN
FAX: (313) 936-0794


E. R. Kinney
Nuclear Physics Laboratory
University of Colorado
Campus Box 446
Boulder, CO 80309-0446
Phone: (303) 492-3662
e-mail: kinney@dilsey.colorado.edu
FAX: (303) 492-7486


L. D. Knutson
University of Wisconsin
Department of Physics
1150 University Avenue
Madison, WI 53706
Phone: (608) 262-3096
e-mail: KNUTSON@WISCNUC
FAX: (608) 262-3598


W. Korsch
MIT, LNS, 26-456
77 Massachusetts Avenue
Cambridge, MA 02139
Phone: (617) 253-6997
e-mail: KORSCH@MITBATES
FAX: (617) 258-5440

W. Kretschmer
Physikalisches Institut
Erwin-Rommel-Str. 1
D-8520 Erlangen
Germany
Phone: 0049 9131-857075
FAX: 0049 9131-15249


A. D. Krisch
Randall Lab. of Physics
The University of Michigan
Ann Arbor, MI 48109-1120
Phone: (313) 936-1027
FAX: (313) 763-9694


W. Kubischta
CERN-EP-Division
CH-1211, Geneva 23
Switzerland
e-mail: KUB@CERNVM.cern.ch
FAX: (41 22) 782 18 20


M. Leduc
Laboratoire de Physique
Ecole Normale Superieure
24, rue Lhomond
75231 Paris Cedex 05
France
Phone: 33 1 4331 6959
e-mail: LEDUC@
MERLIN.ENS.FR
FAX: 33 1 4535 0076


J.-L. Lemaire
Laboratoire National Saturne, EN
Bat. 130, SMLNS
91191 Gif-sur-Yvette Cedex
France
Phone: 33-1-6908 3625
FAX: 33-1-6908 4858


S. Lemaître
Institut für Kernphysik
University of Köln
Zulpicherstr. 77
D-5000 Köln 41
Germany
e-mail: LEMAITRE@
IKP.UNI-KOELN.DE
FAX: 49-221-470-5168


C. D. P. Levy
TRIUMF
4004 Westbrook Mall
Vancouver B.C.V6T 2A3
Canada
Phone: (604) 222-1047
e-mail: LEVY@TRIUMFER
FAX: (604) 222-1074


V. G. Luppov
Randall Lab. of Physics
University of Michigan
Ann Arbor, MI 48109-1120
Phone: (313) 764-5114
e-mail: MICH::LUPPOV
FAX: (313) 936-0794


K. Maehata
Department of Nuclear Engineering
Kyushu University
6-10-1 Hakozaki, Highashiku
Fukuoka 812, Japan
Phone: 092-641-1101
FAX: 092-641-7098


C. Martin
Department of Physics
University of Wisconsin
1150 University Avenue
Madison, WI 53706
Phone: (608) 263-3628
FAX: (608) 265-2334

J. E. McAninch
Department of Physics
University of Wisconsin
1150 University Avenue
Madison, WI 53706
Phone: (608) 262-3091/2-5
e-mail: MCANINCH@WISCNUC
FAX: (608) 262-3598


H. O. Meyer
Physics Department
Indiana University
Swain Hall West
Bloomington, IN 47405
Phone: (812) 855-2883
e-mail: MEYER@IUCF
FAX: (812) 855-6645


H. Middleton
Department of Physics
Princeton University
P. O. Box 708
Princeton, NJ 08544
Phone: (609) 258-4647
e-mail: HMIDDLETON@
PUPGG.Princeton.edu


M. Midzor
Department of Physics
University of Wisconsin
1150 University Avenue
Madison, WI 53706
Phone: (608) 263-2263
e-mail: MIDZOR@WISCNUC
FAX: (608) 262-3598


M. A. Miller
Department of Physics
University of Wisconsin
1150 University Avenue
Madison, WI 53706
Phone: (608) 263-2263
e-mail: MILLER@WISCNUC
FAX: (608) 262-3598


R. G. Milner
MIT, 26-447, Department of Physics
77 Massachusetts Avenue
Cambridge, MA 02139
Phone: (617) 258-5439
e-mail: MILNER@MITLNS
FAX: (617) 258-5440


Y. Mori
KEK—National Lab. for High
Energy Physics
1-1 Oho, Tsukuba-shi,
Ibaraki-ken, 305
Japan
Phone: 81-298-64-5209
e-mail: MORIY@JPNKEKVX
FAX: 81-298-64-3182


E. G. Myers
Department of Physics
Florida State University B-159
315 Keen Bldg.
Tallahassee, FL 32306
Phone: (904) 644-4040
e-mail: MYERS@
FSUNUC.PHYSICS.FSU.EDU


H. Okamura
Department of Physics
University of Tokyo
Hongo 7-3-1, Bunkyo
Tokyo 113
Japan
e-mail: okamura@rikvax.riken.go.jp
FAX: 81-3-3811-0960


H. Paetz gen. Schieck
Institut für Kernphysik
University of Köln
Zulpicherstr. 77
D-5000 Köln 41
Germany
Phone: 0049-221-2796
e-mail: SCHIECK@
IKP.UNI-KOELN.DE
FAX: 0049-221-470-5168

H. R. Petri
IUCF
2401 Milo B. Sampson Lane
Bloomington, IN 47405
Phone: (812) 855-5198
e-mail: PETRI@
VENUS.IUCF.INDIANA.EDU
FAX: (812) 855-6645

M. Poelker
Argonne National Lab.
Bldg. 203
9700 S. Cass Avenue
Argonne, IL 60439
Phone: (708) 252-3619
e-mail: POELKER@ANLPHY
FAX: (708) 252-3903

S. Popov
Institute for Nuclear Physics
Novosibirsk 63009
Russia

J. S. Price
Randall Lab. of Physics
University of Michigan
Ann Arbor, MI 48109
Phone: (313) 936-1027
e-mail: PRICE@MICH
FAX: (313) 936-0794

P. A. Quin
University of Wisconsin
Physics Department
1150 University Avenue
Madison, WI 53706
Phone: (608) 262-8739
e-mail: QUIN@WISCNUC
FAX: (608) 262-3598

R. S. Raymond
Randall Lab. of Physics
University of Michigan
Ann Arbor, MI 48109
Phone: (313) 936-1027
e-mail: RAYMOND@UMIPHYS
FAX: (313) 936-0794

R. Reckenfelderbäumer
Institute of Nuclear Physics
Zülpicherstr. 77
D-5000 Köln 41, Germany
Phone: 0049-221-470-3459
e-mail: baeumer@ikp.uni-koeln.de
FAX: 0049-221-470-5168

A. D. Roberts
Department of Physics
University of Wisconsin
1150 University Avenue
Madison, WI 53706
Phone: (608) 262-3091
e-mail: ROBERTS@WISCNUC
FAX: (608) 262-3598

M. A. Ross
Department of Physics
University of Wisconsin
1150 University Avenue
Madison, WI 53706
Phone: (608) 263-7089
e-mail: ROSS@WISCNUC
FAX: (608) 262-3598

N. Sakamoto
Department of Physics
University of Tokyo
Hongo 7-3-1, Bunkyo
Tokyo 113, Japan
Phone: 81-3-5689-7343
e-mail: nsakamoto@
rikvax.riken.go.jp

J. E. Schewe
Department of Physics
University of Wisconsin
1150 University Avenue
Madison, WI 53706
Phone: (608) 262-3091
e-mail: UWNUCO::SCHEWE
FAX: (608) 262-3598

P. A. Schmelzbach
Paul Scherrer Institute
CH-5232 Villigen-PSI
Switzerland
Phone: 0041 0 56 99 3111
e-mail: SCHMELZBACH@CAGEI
FAX: 0041 0 56 99 33 8

P. W. Schmor
TRIUMF
4004 Westbrook Mall
Vancouver B.C. V6T 2A3
Canada
Phone: (604) 222-1047
e-mail: SCHMOR@TRIUMFCL
FAX: (604) 222-1074

P. Schwandt
IUCF
2401 Milo B. Sampson Lane
Bloomington, IN 47405
Phone: (812) 855-9365
e-mail: SCHWANDT@IUCF
FAX: (812) 855-6645

N. Severijns
Institut de Physique Nucleaire
Universite Catholique de Louvain
Chemin du Cyclotron 2
B-1348 Louvain-la-Neuve
Belgium
Phone: (32 10) 473273
e-mail: NATHAL@IKS.KULEUV
FAX: (32 10) 452183

M. Smith
Department of Physics
University of Wisconsin
1150 University Avenue
Madison, WI 53706
Phone: (608) 262-2263
e-mail: MKSMITH@WISCNUC

T. Smith
University of Michigan
Randall Lab. of Physics
500 East University
Ann Arbor, MI 48109
Phone: (313) 763-5981
FAX: (313) 763-9694

J. Sowinski
IUCF
2401 Milo B. Sampson Lane
Bloomington, IN 47405
Phone: (812) 855-9365
e-mail: SOWINSKI@IUCF
FAX: (812) 855-6645

F. Sperisen
IUCF
2401 Milo B. Sampson Lane
Bloomington, IN 47405
Phone: (812) 855-2948
e-mail: SPERISEN@IUCF.BITNET
FAX: (812) 855-6645

E. Steffens
MPI für Kernphysik
P. O. Box 103980
D-6900 Heidelberg 1
Germany
Phone: 0049 6221-516408
e-mail: EST@UXNHD3.MPI-
HD.MPG.DE
FAX: 0049 6221-516540

J. Stewart
Randall Laboratory of Physics
University of Michigan
Ann Arbor, MI 48109
Phone: (313) 764-5112
FAX: (313) 936-0794

F. Stock
MPI für Kernphysik
P. O. Box 103980
D-6900 Heidelberg 1
Germany
Phone: 0049-6221-516-339
e-mail: FRS@UXNHD3.MPI-
HD.MPG.DE
FAX: 0049-6221-516-540


D. R. Swenson
Los Alamos National Lab.
MP-5 H838
Los Alamos, NM 87545
Phone: (505) 665-2944
e-mail: SWENSON@
LAMPF.BITNET
FAX: (505) 665-5688


M. Tanaka
Kobe Tokiwa Jr. College
Ohtani-cho 2-6-2, Nagata
Kobe 653
Japan
Phone: 078-611-1821
e-mail: TANAKAM@
JPNRCNPF.BITNET
FAX: 078-643-4361


J. Theunissen
NIKHEF-K
P. O. Box 41882
Amsterdam 1009 DB
The Netherlands
FAX: 31-20-592-2165


D. K. Toporkov
Budker Institute of Nuclear Physics
Novosibirsk 90
Russia
Phone: 383 2 359026
e-mail: TDM@INP.NSK.SU
FAX: 383 2 352163


D. Tupa
Los Alamos National Lab.
MS H838, MP-5
Los Alamos, NM 87545
Phone: (505) 665-1820
FAX: (505) 667-1712


J. F. J. van den Brand
University of Wisconsin
Department of Physics
1150 University Avenue
Madison, WI 53706
Phone: (608) 262-9107
e-mail: JOELLE@WISCNUC
FAX: (608) 262-3598


B. v. Przewoski
IUCF
2401 Milo B. Sampson Lane
Bloomington, IN 47405
Phone: (812) 855-2882
e-mail: PRZEWOSKI@IUCF
FAX: (812) 855-6645


P. A. Voytas
University of Wisconsin
Department of Physics
1150 University Avenue
Madison, WI 53706
Phone: (608) 262-3091
e-mail: VOYTAS@WISCNUC
FAX: (608) 262-3598


T. Waites
Uppsala University
The Svedberg Lab.
Box 533
S-751 21 Uppsala
Sweden
Phone: 0046-1818-3874
FAX: 0046-1818-3833

Y. Wakuta
Department of Nuclear Engineering
Kyushu University
6-10-1 Hakozaki, Higashi-ku
Fukuoka-shi, 812
Japan
Phone: 092-641-1101
FAX: 092-641-7098

T. Walker
Department of Physics
University of Wisconsin
1150 University Ave.
Madison, WI 53706
Phone: (608) 262-4093
e-mail: WALKER@
WISCNUC.BITNET
FAX: (608) 265-2334

R. Weidmann
Physikalisches Institut, Abt. IV
Erwin-Rommel-Str. 1
8520 Erlangen
Germany
Phone: 09131/85-7700
e-mail: RAINER:WEIDMANN@
CNVE:UNI-ERLANGEN.DE
FAX: 09131/15249

D. Wessman
Uppsala University
The Svedberg Lab.
Box 533
S-751 21 Uppsala
Sweden
Phone: 46-18-18 38 74
e-mail: WESSMAN@TSL.UU.SE
FAX: 46-18-18 38 33

T. Wise
University of Wisconsin-Madison
Department of Physics
1150 University Avenue
Madison, WI 53706
Phone: (608) 263-2438
e-mail: WISE@WISCNUC
FAX: (608) 262-3598

K. Zapfe
MEA
DESY
Notkestrasse 85
D-2000 Hamburg 52
Germany
Phone: 040/89 98-3743
e-mail: 13633::ZAP
FAX: 040/89 98-3094

A. N. Zelenski
Institute for Nuclear Research
60th October Anniversary
Prospect, 7a
Moscow, 117312
Russia
Phone: (604) 222-1047
e-mail: ZELENSKI@TRIUMFCL
FAX: (604) 222-1074

Y. Zhou
Department of Physics
University of Wisconsin
1150 University Avenue
Madison, WI 53706
Phone: (608) 263-2263
e-mail: YONG@WISCNUC
FAX: (608) 262-3598

# WORKSHOP ON POLARIZED ION SOURCES AND POLARIZED GAS TARGETS
## May 23–27, 1993, Madison, Wisconsin

### Monday, 24 May, Wisconsin Center

08$^{55}$  **Welcome**                                          W. Haeberli

**ATOMIC BEAM TARGETS**
Chairperson: A. D. Krisch

09$^{00}$  Tests of a High-Density Polarized            K. Zapfe, DESY
          Hydrogen Gas Target Storage Cell
          and Spin Filter for Stored Ion Beams
09$^{40}$  Polarized H or D Target for Experiments      A. D. Roberts, UW-Madison
          in the Indiana Cooler
10$^{00}$  Experiments with Polarized Deuterium         D. Toporkov, Novosibirsk
          Target at VEEP-3 Storage Ring: Status
          and Perspective
10$^{30}$  Coffee/Posters A
11$^{00}$  Description of a Monte Carlo Method          I. D. Boyd, Cornell
          for Simulation of Nonequilibrium Gas
          Dynamics
11$^{40}$  Optimization of the FILTEX-HERMES            F. Stock,
          Atomic Beam Source                           MPI, Heidelberg

**POLARIZED H,D TARGETS**
Chairperson: Y. Wakuta

13$^{30}$  Status of the Ultra-Cold Polarized Jet       R. S. Raymond, U. Michigan
          for Neptun-UNK
14$^{00}$  A He Film Coated Quasi-Parabolic Mirror      V. Luppov, U. Michigan
          to Focus a Beam of Ultra-Cold Spin-
          Polarized Atomic Hydrogen
14$^{20}$  Evaporation of Solid Polarized HD            A. Honig, Syracuse U.
14$^{50}$  Coffee/Posters A
15$^{20}$  Polarized Internal                           J. van den Brand, UW-Madison
          Targets for Electronuclear Experiments
16$^{00}$  On-Line Efficient Proton and Deuteron        T. B. Clegg, U. N. Carolina
          Beam Polarimetry
16$^{30}$  ROUND TABLE                                  E. Steffens
          Atomic Beam and Low Temperature              MPI, Heidelberg
          Targets

| | | |
|---|---|---|
| $10^{10}$ | A Dense Polarized $^3$He Target Under Electron Beam Conditions | W. Heil, Gutenberg Univ. |
| $10^{40}$ | Coffee/Posters B | |
| $11^{10}$ | Polarized $^3$He Internal Gas Targets | R. G. Milner, MIT |
| $11^{40}$ | Design of a Polarized $^3$He Target for the CEBAF Large Acceptance Spectrometer | W. Hersman, U. New Hampshire |

**TARGET–MACHINE INTERACTION**
Chairperson: D. Toporkov

| | | |
|---|---|---|
| $13^{30}$ | ROUND TABLE Optical Pumping of $^3$He | W. Heil, Gutenberg U. |
| $14^{30}$ | Studies of Storage Cells at NIKHEF | C. de Jager, NIKHEF |
| $15^{00}$ | Optimum Luminosity of Polarized Gas Targets in Storage Rings | B. von Przewoski, IUCF |
| $15^{30}$ | Coffee/Posters B | |
| $16^{00}$ | Interactions Between a Polarized $^3$He Target, the IUCF Cooler, and an Experiment | J. Sowinski, IUCF |
| $16^{30}$ | Beam-Induced Target Depolarization | E. R. Kinney, U. Colorado |
| $17^{00}$ | ROUND TABLE Machine–Target Interaction | H. O. Meyer, IUCF |

Thursday, 27 May, Wisconsin Center

**OPTICAL PUMPING I**
Chairperson: P. Schmor

| | | |
|---|---|---|
| $09^{00}$ | Laser-Driven Source of Spin Polarized Atomic Hydrogen and Deuterium | B. M. Poelker, ANL |
| $09^{30}$ | Measurement of $P_{zz}$ of the Laser-Driven Polarized Deuterium Target | C. E. Jones, ANL |
| $09^{50}$ | Limitations of Optically Pumped Spin-Exchange Polarized Targets | T. Walker, UW-Madison |
| $10^{20}$ | Target for Producing Polarized $^{21}$Na by Optical Pumping | P. A. Voytas, UW-Madison |
| $10^{40}$ | Coffee/Posters B | |
| $11^{10}$ | Development of the Optically Pumped Polarized Ion Source at KEK | Y. Mori, KEK, Japan |

**OPTICAL PUMPING II**
Chairperson: R. Holt

| | | |
|---|---|---|
| $13^{30}$ | Performance of the LAMPF Optically Pumped Polarized Ion Source | D. Tupa, LANL |

$14^{00}$  Polarization Diagnostics and Optical       D. R. Swenson, LANL
Pumping Studies for OPPIS at LAMPF
$14^{20}$  Status of the TRIUMF Optically              C. D. P. Levy, TRIUMF
Pumped Polarized H $^-$ Ion Source
$15^{00}$  Coffee/Posters B
$15^{30}$  Production of Polarized Heavy               M. Tanaka, Kobe College
Ions for Intermediate-Energy Nuclear
Physics
$16^{00}$  Spin-Exchange Polarization                  A. Zelenski, TRIUMF/Moscow
in H–Rb
$16^{30}$  ROUND TABLE                                 Y. Mori, KEK, Japan
Optically Pumped Polarized Ion Sources
and Targets

## POSTER PAPERS

Monday and Tuesday, Posters A

| | | |
|---|---|---|
| M. A. Ross, U. Wisconsin | Test of a Polarized H Gas Target by $pp$ Spin Correlation at 10 MeV | P-A1 |
| W. A. Kaufman, U. Mich. | Beam Transport of Low Temperature Atomic Hydrogen | P-A2 |
| T. B. Clegg, U. North Carolina | Operating Experience with an Atomic Beam Source Equipped with an ECR Ionizer | P-A3 |
| G. Clausnitzer, Giessen | The Giessen Atomic Beam Source | P-A4 |
| H. Paetz gen. Schieck, Köln | Cesium Beam for the COSY-Jülich Polarized Ion Source | P-A5 |
| J. S. Price, U. Mich. | Measurement of Polarization for Polarized Gas Targets | P-A7 |

Wednesday and Thursday, Posters B

| | | |
|---|---|---|
| A. Zelenski, TRIUMF/Moscow | Proposal for a High Current Pulsed Optically Pumped Polarized H $^-$ Ion Source | P-B1 |
| Y. Mori and M. Kinsho, KEK and GUAS | Polarized Negative Deuterium Ions with Dual Optical Pumping | P-B2 |
| L. W. Anderson, UW-Madison | Optical Pumping of Spin-Exchange Targets | P-B3 |
| C. Martin, UW-Madison | Optical Pumping of K in a High Magnetic Field by Linearly Polarized Light | P-B4 |
| O. Häusser, TRIUMF | Spin Physics with a Polarized $^3$He Target | P-B6 |
| A. Zelenski, TRIUMF/Moscow | Intensity Modulation of the Beam from OPPIS | P-B7 |

# PREFACE

The Workshop on Polarized Ion Sources and Polarized Gas Targets was held in Madison, Wisconsin from 23 May to 27 May 1993. The workshop is the most recent in a series of workshops held at about three-year intervals. Previous workshops have been held at Ann Arbor, Michigan, U.S.A.; Vancouver, British Columbia, Canada; Montana, Switzerland; and Tsukuba, Japan. This workshop was attended by about 100 scientists. There were 40 talks and 13 poster presentations. The subjects addressed in the workshop included atomic beam polarized H and D targets, ultra-cold spin-polarized H targets, targets formed by evaporation of spin-polarized HD, optically pumped H and D targets, optically pumped polarized $^3$He targets, optically pumped $^{21}$Na targets, atomic beam polarized ion sources, optically pumped charge exchange polarized ion sources, and target–machine interactions. In addition to the oral and poster presentations there were five round table discussions on various topics that focused attention on the current status of target and ion source technology and on the direction of future research.

In addition to the technical sessions the workshop participants enjoyed a picnic, a banquet, and sightseeing on the shores of Lake Mendota, around the University of Wisconsin campus and the city of Madison.

Credit for the success of the conference goes to many people. Linda Kennedy, Jean Michael, Marion Schmidt, and Julie Neitzel were indispensable in the preparation of the workshop and helping participants and speakers cope with problems during the meeting. The help of Julie Neitzel in finding housing for everyone was essential to the comfort of our visitors. In addition, she made a major contribution to the preparation of these Proceedings. The organizing committee Y. Mori, KEK; M. Kondo, Osaka U.; R. D. McKeown, Caltech; J. Cameron, IUCF; R. J. Holt, Argonne; T. B. Clegg, U. of No. Carolina; A. D. Krisch, U. of Michigan; P. A. Schmelzbach, PSI; P. W. Schmor, TRIUMF; A. Zelenski, Inst. for Nuclear Research, Moscow; M. Leduc, Saclay; W. Kubischta, CERN; E. Steffens, MPI helped develop the basic themes of the workshop and suggested topics and speakers. Professors W. Haeberli and L. W. Anderson prepared the program.

The conference was supported by grants from the NSF (PHY-9220687) and from the International Committee on High Energy Spin Physics.

# I. ATOMIC BEAM AND LOW TEMPERATURE H AND D TARGETS

# TEST of a HIGH-DENSITY POLARIZED HYDROGEN GAS TARGET STORAGE CELL and SPIN FILTER for STORED ION BEAMS

Kirsten Zapfe

*Deutsches Elektronen-Synchrotron DESY, Notkestraße 85, D-22603 Hamburg 52, Germany*
*(for the HERMES-FILTEX target group[*])*

A high-density storage cell target of polarized atomic hydrogen was installed in the Heidelberg test storage ring. A target thickness of 1.0 $\times 10^{14}$ atoms/cm$^2$ (2 hyperfine states) was achieved cooling the cell to 100 K. The target polarization was $P_T = 0.84$ and $P_T = 0.45$ when state 1 or states 1+2 were selected, respectively, or 90 % of the maximum possible value. With this target a feasibility test of a new method to polarize beams of strongly interacting charged particles circulating in a storage ring has been performed using protons of 23 MeV. After passing through the target some $10^{10}$ times a polarization build-up is clearly demonstrated in the present experiment.

## Introduction

Discussions in the last years[1] showed considerable interest in studies of spin physics using internal polarized hydrogen or deuterium gas targets in electron and proton rings. So far, experiments with polarized hydrogen gas targets are limited to one experiment with a polarized deuterium target at the electron storage ring VEPP-3 in Novosibirsk[2]. Proposed experiments include, for instance, the study of deep inelastic scattering of high energy electrons by polarized hydrogen und deuterium to study the spin structure functions of the nucleon (HERMES experiment at HERA)[3] as well as antiproton spin physics (e.g. FILTEX experiment to polarize a circulating beam of antiprotons)[4].

For these kinds of experiments a target thickness in the order of $10^{14}$ $\vec{H}/cm^2$ is required. A jet of polarized atoms, e.g. from an atomic beam source, provides at best a target thickness of the order $10^{12}$ $\vec{H}/cm^2$ [2]. The available target thickness of jet targets can be improved by more than two orders of magnitude by use of a storage cell[6], consisting of a long narrow tube through which the circulating beam passes and into which polarized atoms are injected.

## The Polarized Hydrogen Gas Target

The atomic beam source[7] and the target chamber[8] constructed at Heidelberg by the HERMES-FILTEX target collaboration were described in previous workshops. The atomic beam source was recently improved and an intensity of $I = 8.1 \times 10^{16}$ atoms/s in two substates is obtained routinely[9]. The cell (see Fig. 4 of ref 8) used for the measurements reported here consists of a T-shaped aluminum tube of 0.2 mm wall thickness, 11 mm in diameter and 250 mm in length[10]. The feed tube is 100 mm long with a diameter of 10 mm. The depolarization by wall bounces is minimized using a wall coating of 120 FEP Teflon[11].

The storage cell target together with the atomic beam source has been tested in 1991[12] by passing a 35 MeV beam of $\alpha$-particles from the Heidelberg MP tandem accelerator through the cell and measuring the left-right asymmetry of recoil protons near $\theta_{lab} = 21°$. The target polarization was found to be $0.42 \pm 0.04$ compared to the maximum possible value for hyperfines states 1+2 in a weak magnetic field (5 G) of $P_{Tmax} = 0.5$. The measurements were difficult because of a large background from scattering of beam halo by the cell wall.

## Measurements of Target Thickness and Polarization

The storage cell target was installed in the Heidelberg test storage ring (TSR). Already the first tests of the target with a stored beam in the TSR showed excellent energy spectra of the recoil protons, with a background of less than 1 %. Background measurement were made by replacing the hydrogen gas in the target cell by an amount of oxygen sufficient to cause the same amount of beam loss by scattering.

The target polarization und the target thickness were measured using an electron-cooled beam of 27 MeV $\alpha$-particles of intensity up to 0.3 mA. The target polarization was reversed by reversing the 5 G guide field over the target. During the tests, the target cell was cooled down to 100 K in order to increase the target density.

### Target Polarization

The target polarization was extracted from the count rate asymmetries with spin up and down for each counter. Fig. 1 shows a spectrum of the recoil protons at a laboratory angle of $\theta_{lab} = 21°$ for target spin up and down. The target polarization could be measured to an accuracy of 1 % in a few minutes. The value of the p-$\alpha$ analyzing power in the angular range used here is large and well known[13].

Measurements of the polarization as a function of the cell wall temperature are shown in Fig. 2. The open and closed circles refer to measurements with the RF-transition between the sixtupoles of the atomic beam source turned off (2 substates) and on (1 substate), respectively. The target polarization is constant for wall temperatures between 80 K und 300 K. Below 60 K a significant decrease in polarization is observed. The observed average values for T $\geq$ 80 K are 0.45 with RF-transition off, and 0.82 with RF-transition on. Tracking calculations for the sixtupole system and the efficiency of the RF-transition (95 %) gave ideal values of 0.49 and 0.88, respectively. In both cases the measured polarization amount to more than 90 % of the ideal value, leaving little room for polarization losses by wall depolarization. It should be mentioned, that the results may still undergo small changes in the final data analysis.

Another important result was, that after several weeks of running with alpha beam currents between 0.05 and 0.3 mA and proton beams of up to 1 mA no aging effects of the cell coating and therefore no deterioration of the target polarization have been observed.

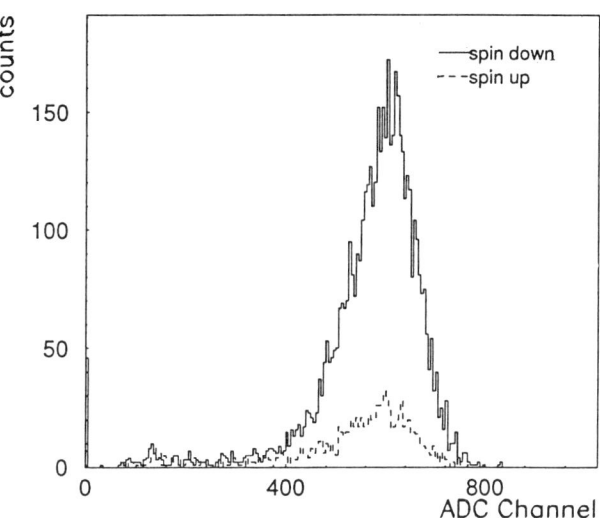

FIG. 1. Pulse height spectrum produced by recoil protons at $\theta_{lab} = 21°$ for target spin up
(- - -) and down (——) in the bombardment of the polarized hydrogen target by 27 MeV $\alpha$-
particles circulating in the storage ring. The spectra shown were obtained by selecting a single
spin state. The results indicate a target polarization of $P_T = 0.826 \pm 0.004$ (statistical error).
The run time was 6 min.

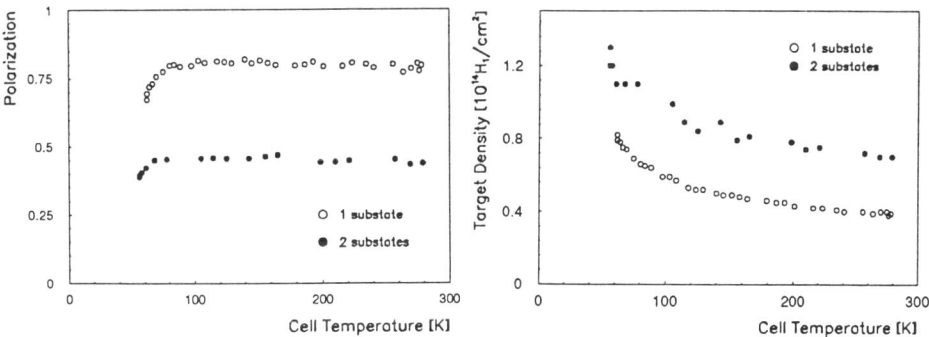

FIG. 2. Measured target polarization (left) and target thickness (right) as a function of the
cell wall temperature. Open and closed circles refer to measurements with the RF-transition
in the atomic beam source turned off and on, respectively.

## Target Thickness

The target thickness was calculated from the spin-averaged count rates and
the known cross section, detector geometry and beam current. As an independent method the deceleration rate of the beam with electron cooling switched
off has been used to determine the target thickness. The right hand side of Fig.

2 shows the decrease in target thickness from rejection of state 2 atoms, as well as the expected increase in target thickness with decreasing cell wall temperature ($n \sim 1/\sqrt{T}$). At the optimum working temperature of 100 K a thickness of $n = (1.00 \pm 0.01 \pm 0.02) \times 10^{14}$ atoms/cm$^2$ has been measured with two substates of hydrogen, and a thickness of $n = (0.60 \pm 0.01 \pm 0.02) \times 10^{14}$ atoms/cm$^2$ with one substate. This is in agreement with the values calculated from the atomic beam flux and the cell geometry.

### Test of a Spin Filter

The polarized hydrogen target described above has been used for a first feasibility test of a spin filter to polarize strongly interacting particles[14]. The method is of particular interest for the production of polarized antiprotons and can be described as spin-selective attenuation of the particles circulating in a storage ring. The particles pass through the target for a sufficiently long time that a fraction of the particles is lost by nuclear scattering in the target. The stored beam can be considered to consist of a fraction of particles with spin up and a fraction with spin down. The total strong interaction cross section of the beam with the target can be expressed as

$$\sigma_T = \sigma_0 \pm \sigma_1 P_T \, , \tag{1}$$

where the positive and negative sign applies, respectively, to the fraction of the beam whose spin is parallel ($\uparrow\uparrow$) and anti-parallel ($\uparrow\downarrow$) to the spin of the target. Here, $\sigma_0$ is the spin independent part and $\sigma_1$ is the spin dependent part of the cross section, and $P_T$ the polarization of the target in the vertical direction. If we neglect mechanisms other than interaction with the target, the intensity of the spin-up and spin-down particles in the stored beam each decrease exponentially but with different time constants. The resulting polarization build-up of the beam as a function of filter time $t$ can be expressed in the absence of depolarization as [4]:

$$P(t) = tanh\frac{t}{\tau_1} \, . \tag{2}$$

The time constant $\tau_1$, which characterizes the rate of polarization build-up, is $\tau_1 = 1/\sigma_1 P_T n f$ where $n$ is the target thickness in atoms/cm$^2$ and $f$ is the revolution frequency of the particles in the ring.

The experiment was carried out as follows. An (unpolarized) 23 MeV proton beam of up to 1 mA was stored in the ring, using multiturn stacking, while reducing the beam phase space by electron cooling. Subsequently the beam was left to circulate for periods between 30 and 90 min. After 90 min, the beam intensity had decreased to 5 % of the initial value, so that longer filter times could not be explored. The polarization of the remaining beam was subsequently detected by making use of the large difference in the pp elastic scattering cross section between parallel and antiparallel spins. For this purpose the direction of the 5 G guide field, which determines the direction of the target polarization (up or down), was reversed periodically and the pp elastic count rates for the two target orientations were compared.

Scattered protons were detected in scintillation counter telescopes located at $\Theta_{lab} = 33.3°$ above and below the plane of the storage ring. Therefore the normal to the scattering plane (y) is in the horizontal direction. For target and beam polarization in the vertical direction (x), the cross section is

$$\sigma(\Theta) = \sigma_0(\Theta)(1 \pm A_{xx}(\Theta)P_B P_T) \, , \tag{3}$$

where the positive sign applies when beam and target are polarized in the same direction, the negative sign when they are in opposite directions. The spin correlation coefficient $A_{xx} = -0.93$ was calculated from the pp phase shifts [15].

We denote the sum of the number of counts registered in the two detectors for target spin up and target spin down by $N_\uparrow$ and $N_\downarrow$, respectively, and define the count rate asymmetry e as:

$$\epsilon = \frac{N_\uparrow - N_\downarrow}{N_\uparrow + N_\downarrow} \, . \tag{4}$$

The component of beam polarization in the up direction is obtained from $\epsilon$ as:

$$P_B = \frac{\epsilon}{A_{xx}P_T} \, , \tag{5}$$

where $P_T$ is the magnitude of the target polarization during the beam polarization measurement. The asymmetry $\epsilon$ measured after filtering the stored beam for different lengths of time, is shown in Fig. 3. Each point shown is the weighted mean of a number of measurements with approximately the same filtering time. The solid dots refer to filtering with target polarization up, while the open circles refer to target polarization down during filtering. Measurements were also made for zero filtering time. As expected, the rate of polarization build-up is the same for both directions of the target spin during filtering. Instrumental effects on beam or detectors associated with reversal of the guide field could in principle cause $\epsilon \neq 0$ for filter time zero. The data indicate that such effects are small.

The straight lines in Fig. 3 show the best fit to the data, with a rate of polarization build-up of

$$\Delta P_B/\Delta t = \pm(1.24 \pm 0.06) \times 10^{-2} \text{ hr}^{-1} \, , \tag{6}$$

which implies $\tau_1 = 80$ hrs. As expected the polarization which the beam acquires is in the same direction as the polarization of the target.

It is interesting to compare the observed rate of polarization build-up (Eq. (6)) with the rate expected from Eq. (2). Since $P_B \ll 1$, the rate of build-up is $1/\tau_1 = \sigma_1 P_T n f = 2.4 \times 10^{-2} \text{ hr}^{-1}$ ($\sigma_1 = 122$ mb for 23 MeV protons,

$f = 1.177$ MHz) which is about twice the observed value.

A possible explanation for the discrepancy is that the circulating beam may have a finite polarization lifetime $\tau_p$. The storage ring was operating very near a depolarizing resonance ($Q_v = 3.81$ resp. $Q_v^{res} = 3.84$). If we assume the value of $\tau_1$ given by Eq. (7) the best fit to the measurements is obtained with a polarization life time $\tau_p = 81 \pm 7$ min (dashed curves in Fig. 3). The quality of the data at present is not sufficient to tell whether the polarization build-up is slower than expected or whether the beam depolarizes during the build-up.

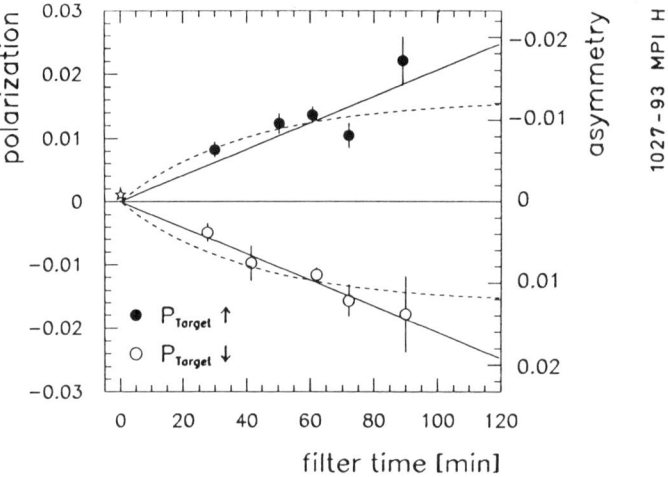

FIG. 3. Asymmetry (right-hand scale) and polarization (left-hand scale) measured after filtering the beam in the storage ring for different times t. The solid lines are based on an assumed rate of polarization build-up of $1.24 \times 10^{-2}$ hr$^{-1}$, which corresponds to $\tau_1 = 80$ hrs. The dashed lines are based on the expected build-up rate ($\tau_1 = 42$ hrs) and an assumed polarization life time of $\tau_p = 81$ min.

## Conclusions

A storage cell target for polarized hydrogen has been installed in the Heidelberg test storage ring. At 100 K a target thickness of $n = 1 \times 10^{14}$ atoms/cm$^2$ (state 1+2) has been measured. The observed target thickness is about 30 times larger than the recently reported value for the Phase II deuterium target in use at the VEPP-3 ring in Novosibirsk[14].

It is important to note, however, that in applications, where a strong magnetic field is applied over the target like the HERMES experiment, high polarization and high target thickness can be obtained at the same time by use of two spin states: states 1+4 yielding $P_T = +1$, states 2+3 yielding $P_T = -1$. In the present tests, the low beam energy prevented the use of a strong magnetic field.

Due to the very short time of about 10 min for one run, several aspects like the dependence of the target parameters on the cell wall temperature, or the long-term stability, could be studied for the first time with high precision. The results show that a polarized storage cell in a storage ring is a reliable and highly efficient tool for future experiments.

A spin filter to polarize a stored beam of strongly interacting particles has been successfully tested using stored protons. Additional difficulties arise for applications to antiprotons, where the spin dependent part of the total cross section is not known but is expected to be less than for pp scattering. On the other hand, the present results indicate that with a luminosity of more than $10^{28}$ cm$^{-2}$s$^{-1}$ after filtering and the capability of rapid spin reversal, small effects can be measured with good precision. Therefore, interesting experiments on the spin dependence should be possible, even if the achievable polarization of the stored antiproton beam were only a few percent.

## References

1. e.g. Proc. 9th Int. Symp. in High Energy Spin Physics, Bonn 1990, Vol. I: Conference Report, eds. K.H. Althoff and E. Meyer (Springer, Heidelberg, 1991).
2. R. Gilman et al.; Phys. Rev. Lett **65** (1990) 1733.
3. HERMES-Kollaboration; Proposal DESY PRC-90/01 (1990).
4. H. Döbbeling et al.; Proposal CERN/PSSC/85/80 (1985), and Addendum (1986).
6. W. Haeberli; Proc. Workshop on Nucl. Phys. with Stored, Cooled Beams, Indiana 1984, eds. P. Schwandt and H. O. Meyer, AIP Conf. Proc. **128** (1985) p. 251.
7. W. Korsch; Proc. 9th Int. Symp. in High Energy Spin Physics, Bonn 1990, Vol. II: Workshops, eds. E. Meyer, E. Steffens and W. Thiel (Springer, Heidelberg, 1991) p. 168.
8. K. Zapfe; p. 222 of ref. 8.
9. F. Stock; *Optimization of the HERMES-FILTEX Atomic Beam source*, proceedings of this conference.
10. K. Zapfe; Proc. Workshop on Polarized gas Targets for Storage Rings, Heidelberg 1991, eds. H.-G. Gaul, E. Steffens and K. Zapfe, MPI für Kernphysik, Heidelberg (1992) p. 135.
11. W. Haeberli; p. 194 of ref. 8.
12. M. Düren et al.; Nucl. Instr. and Meth. **A322** (1992) 13.
13. P. Schwandt, T. B. Clegg and W. Haeberli; Nucl. Phys. **A163** (1971) 432.
14. P. L. Csonka, Nucl. Instr. and Meth. **63**, 247 (1968).
15. R. A. Arndt, J. S. Hyslop III, and I. D. Roper, Phys. Rev. **D35**, 128 (1987); and program SAID (scattering analysis interactive dial-up).
16. D.M. Nikolenko; p. 13 of ref. 10. (1992) 32.

\* B. Braun, W. Brückner, M. Düren, D. Fick, H.-G. Gaul, G. Graw, M. Grieser, W. Haeberli, M.T. Lin, Ch. Montag, Z. Moroz, B. Povh, M. Rall, K. Rith, F. Rathmann, P. Schiemanz, E. Steffens, J. Stenger, F. Stock and J. Tonhäuser.

# POLARIZED H OR D TARGET FOR EXPERIMENTS ON THE INDIANA COOLER: DEVELOPMENT OF THE ATOMIC BEAM SOURCE

A. D. Roberts, T. Wise, W. Haeberli
*University of Wisconsin, Madison, WI 53704*

An atomic beam source (ABS) which will provide an internal polarized H or D target at the IUCF Cooler ring has been developed at the University of Wisconsin. The aim of the development was not only to achieve a large atomic beam intensity, but to produce a beam that can be injected with high transport efficiency into a long, narrow tube. The source was designed to inject the atomic beam into a storage cell through a 10mm diameter, 13cm long cylindrical tube placed 26cm from the last ABS sixpole magnet.

Figure 1 shows a simplified scale drawing of the source. The vacuum chambers are constructed of stainless steel, and the chamber seals are metal gaskets, in order to provide the required clean vacuum conditions. Each of the five vacuum chambers shown is pumped either by pairs of turbo pumps or by a cryo pump.

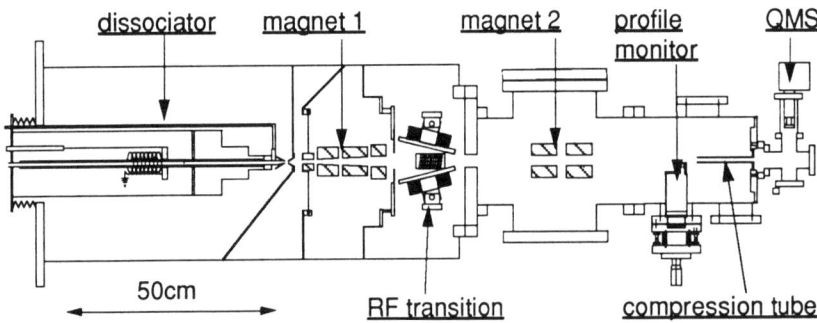

FIG. 1. Simplified scale drawing of the ABS. Vacuum pumps are not shown.

Hydrogen (or deuterium) is dissociated in a RF dissociator which is coupled to a cooled nozzle, shown in Fig. 2. The aluminum nozzle is cooled to a nominal 84K with liquid nitrogen. Atoms in the $m_j=+\frac{1}{2}$ spin states are focused by two sets of sixpole magnets. These permanent magnets are constructed of Nd-B-Fe sintered material encased in stainless steel cans. A description of the construction of similar magnets can be found elsewhere[1,2]. The poletip fields are 1.36±0.04T for the small diameter (10.5 to 22mm), 12 segment magnets, and 1.49±0.02T for the larger diameter (25mm) 24 segment magnets. Between the two sets of magnets is placed a RF transition unit (medium-field transition). A drift length of 26cm was left after the last sixpole magnet to allow space for future installation of additional RF transitions, required for producing large tensor or vector polarized D beams[3]. The beam diagnostic equipment on the right-hand side of Fig. 1 consists of a beam profile monitor and a calibrated compression tube used as the absolute beam intensity monitor.

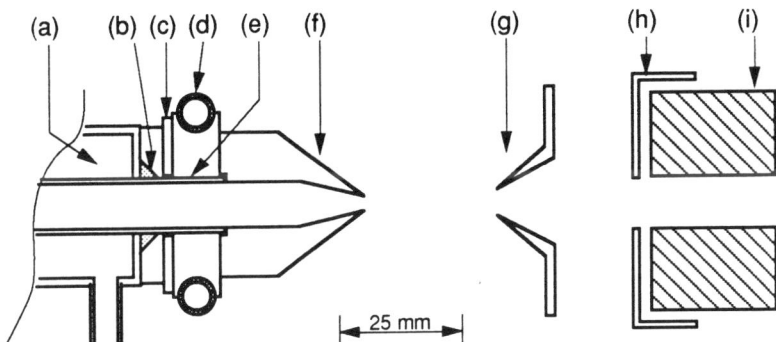

Fig.2. Nozzle skimmer detail shown to scale: a) water cooling, b) indium vacuum seal, c) thermal insulation (Teflon), d) liquid nitrogen cooling, e) copper to Pyrex indium solder joint, f) aluminum nozzle, g) skimmer, h) collimator, i) magnet element.

In recent years, the development of atomic beam sources by trial and error has been replace by a more analytic approach, where optimum design is sought from studies of the operating properties of the individual parts. Our study is based on an expression by Korsch[4], which expresses the atomic beam flux $Q_H$ entering the compression tube at the exit end of the source as:

$$Q_H = \frac{1}{2}(2\alpha Q_{H_2})ft(1-A). \qquad (1)$$

The symbols have the following meaning:
$Q_H$ = atomic beam intensity into the specified target (# atoms/s).
$\alpha$ = degree of dissociation of beam emerging from the nozzle.
$Q_{H2}$ = flow rate of molecular gas into dissociator.
$f$ = fraction of all atoms emerging from the nozzle which enter the aperture of first magnet element.
$t$ = transmission efficiency of magnet system.
$A$ = fraction of H atoms in the beam lost by scattering .

Optimization of the source parameters was based on beam intensity calculations using either measured or calculated values for each of the factors in Eq. (1). Some results of these studies are outlined below.

The degree of dissociation $\alpha$ was measured as a function of gas flow, nozzle temperature, etc., with a quadrupole mass spectrometer mounted on the beam axis. No change in $\alpha$ was found with a constant gas throughput of 1mbar $\ell$/s over a nozzle temperature range from 85K to 145K. Lower temperatures were not studied because of poor thermal linkage between the nozzle and cold head. With an 85K nozzle $\alpha$ increases by 35% for all measured flows up to 2mbar $\ell$/s with the addition of ~0.1% $O_2$ to the $H_2$ gas. The dependence of $\alpha$ on gas throughput is well described by the linear relation:

$$\alpha = 0.93 - (0.10)\, Q_{H2} \qquad (2)$$

where $Q_{H2}$ is given in [mbar $\ell$/s].

Beam attenuation $A$ was studied with molecular rather than atomic beam in order to avoid uncertainties from changes in the degree of dissociation. The beam intensity was measured with a 1cm diameter compression tube located 54cm from the nozzle. Some results of these tests are summarized in Fig. 3 which plots the molecular beam intensity as a function of gas throughput at various nozzle temperatures. A calculation of the expected forward peak intensity from purely effusive flow (cosθ distribution) without scattering is shown as a dotted line. Additional measurements were made with dummy magnets installed to simulate the scattering within real magnets. Attenuation within typical narrow bore magnets accounts for ~20% of the total scattering loss.

The transmission factor $t$ was calculated numerically. A gradient search routine was used to help find local maxima in the transport efficiency. For each magnet geometry the magnetic field gradients along the beam path were calculated according to the Halbach equations for permanent magnets[5]. The parametrization of the atomic velocity distribution as a function of nozzle temperature and gas throughput was determined from time-of-flight measurements on the Heidelberg source[6].

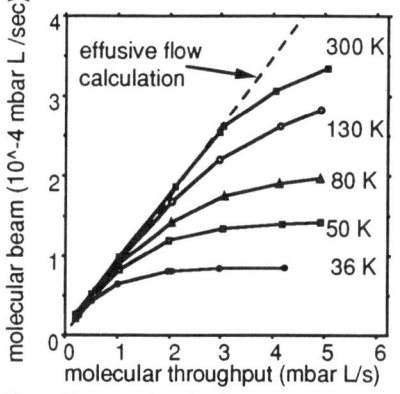

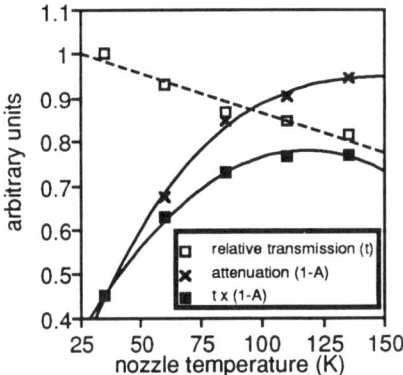

Fig.3. Free molecular beam intensity

Fig.4.  Calculated magnet systems optimized for different nozzle temperatures.

The scattering data (Fig. 3) show the substantial influence of beam temperature on forward beam intensity. Thus the well known advantage of increased acceptance of the spin-separation magnet at low  beam temperatures is obtained at the expense of much increased beam attenuation. Figure 4 shows how the interplay of transport efficiency $t$ and scattering (1-$A$) result in an optimum operating beam temperature. For each nozzle temperature, a magnet system was designed for maximum calculated transmission efficiency within the limitations of realistic available magnets and spacing restrictions for RF transitions. The product of $t$ and (1-$A$) shows a broad peak at a nozzle temperature of about 100K.

The success of the systematic approach to ABS design is measured not only by the atomic beam intensity achieved, but also by the extent to which the predicted beam output conforms to the measured performance. Measurements of atomic beam flux $Q_H$ were made with the 1cm i.d., 13cm long compression tube located 74cm from the nozzle and 26cm from the last magnet element to coincide with a proposed storage cell feed pipe geometry. In this way usable flux into a realistic storage cell target could be measured directly. Results of the intensity measurements are shown in Fig. 5. The maximum beam intensity observed so far is $6.7 \times 10^{16}$ atoms s$^{-1}$ into the compression tube. Results reproduce well from day to day even after prolonged exposure of the apparatus to air.

Calculations of predicted beam intensity into the compression tube using the measured values for $\alpha$, $Q_{H2}$, $f$ and $(1-A)$ as well as the calculated transmission $t$ agree with the measured data to within 10%.

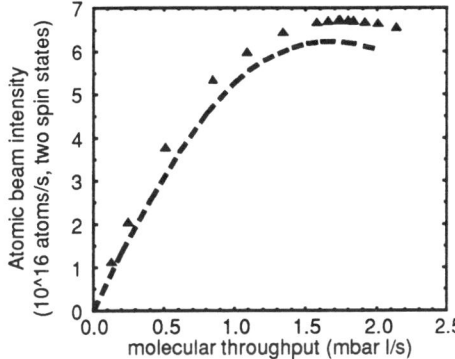

Fig. 5. Atomic beam intensity into the 1cm i.d., 13cm log compression tube 26cm from the last magnet. The dashed line the calculated intensity.

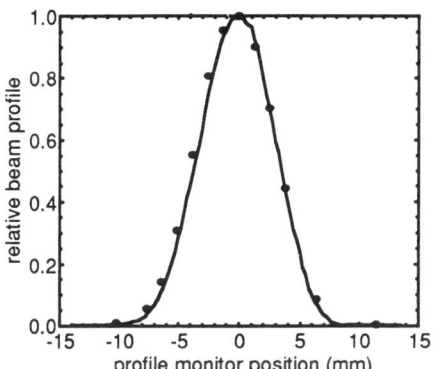

Fig. 6. Beam profile 17cm from the last magnet. The calculation (solid line) includes the spread from the 4.75mm compression tube diameter.

The beam intensity profile was measured with a 4.75mm i.d. moveable compression tube placed 17cm downstream of the last magnet element. Results are shown in Fig. 6 along with a profile calculated from the same transport code used to design the magnets. The calculations include the artificial spreading effect of the compression tube diameter. The trajectory calculation shows that the beam converges to a FWHM of 4.5mm at the 1cm tube entrance with 78% of the beam passing through the tube.

### References

1. Y. Wakuta, H. Hasuyama, S. Ikenaga and K. Yamada, Nucl. Instrum. Methods A260 (1987) 543.
2. P. Schiemenz, A. Ross, and G. Graw, Nucl. Instrum. Methods A305 (1991) 15.
3. A. D. Roberts, P. Elmer, M. A. Ross, T. Wise, W. Haeberli, Nucl. Instrum. Methods **A322** (1992) 6.
4. W. Korsch, Proc. 9th Int. Symp. in High Energy Spin Physics, Bonn 1990, Vol. II: Workshops, eds. W. Meyer, E. Steffens and W. Theil (Springer, Heidelberg, 1991) p. 168.
5. K. Halbach, Nucl. Instrum. Methods 169 (1980) 1.
6. F. Stock, Max Planck Institut für Kernphysik, Heidelberg, personal communication.

# Polarization Measurements of a Storage Cell Target

M. A. Ross, A. D. Roberts, T. Wise, W. Haeberli
*University of Wisconsin*
W. A. DeZarn, J. Doskow, H. O. Meyer, R. E. Pollock, B. v. Przewoski,
T. Rinckel, F. Sperisen
*Indiana University Cyclotron Facility*
P. V. Pancella
*Western Michigan University*

## Abstract

A storage cell has been constructed for use as an internal target at the IUCF electron-cooled storage ring (Cooler). We report on nuclear polarization measurements of hydrogen, produced by an atomic beam source (ABS), in this storage cell. The results indicate a target polarization in excess of 0.70 for atoms in a single spin state.

## Target

A thin-walled storage cell target has been constructed for use in several proposed Cooler experiments. The cell has a 1 cm x 1 cm aperture to allow for passage of the circulating beam and is 25 cm in length. The cell wall is made from 5 μm teflon foil, which allows for detection of low energy scattered particles. Polarized hydrogen (or deuterium) atoms enter the central region of the storage cell through an aluminum feed tube that is 1 cm dia. and 13 cm long. In order to ensure long polarization lifetimes of atoms in the cell, the feed tube has been coated with teflon [1].

## Measurements

In preparation for experiments on the Cooler, polarization measurements using the storage cell target described above were performed at the UW Tandem accelerator using *pp* elastic scattering. At energies less than 8 MeV and scattering angles near $\theta_{cm}=90°$, the spin correlation parameters $A_{yy}$ and $A_{xx}$ are both -0.99 ± 0.01 and the analyzing power $A_y \approx 0$ [2]. Therefore, with vertical beam and target polarization, the differential cross section to a very good approximation becomes

$$\sigma = \sigma_0(1-p_bp_t) \qquad (1)$$

where $p_b$ is the beam polarization and $p_t$ is the target polarization. The target polarization for spin up (↑) and spin down (↓) can easily be determined from the yields in the various polarization states

$$p_t^{\uparrow} = \frac{1}{p_b} \frac{Y_{\downarrow\uparrow} - Y_{\uparrow\uparrow}}{Y_{\downarrow\uparrow} + Y_{\uparrow\uparrow}} \quad \text{and} \quad p_t^{\downarrow} = \frac{1}{p_b} \frac{Y_{\downarrow\downarrow} - Y_{\uparrow\downarrow}}{Y_{\downarrow\downarrow} + Y_{\uparrow\downarrow}}. \tag{2}$$

For the yields in eq. 2, the first arrow indicates the beam polarization and the second arrow is the target polarization. The beam polarization is measured separately in a polarimeter using $p + {}^4He$ elastic scattering [3]. Typically, the value is $p_b = 0.88 \pm 0.01$.

Fig. 1 shows the ABS, storage cell, and target chamber mounted at the high energy end of the tandem. Polarized 7.6 MeV protons travel through the storage cell into the beam polarimeter. Polarized hydrogen atoms from the ABS enter the storage cell through a feed tube mounted at an angle of 60° with respect to the tandem beam line. Silicon strip detectors located 5 cm

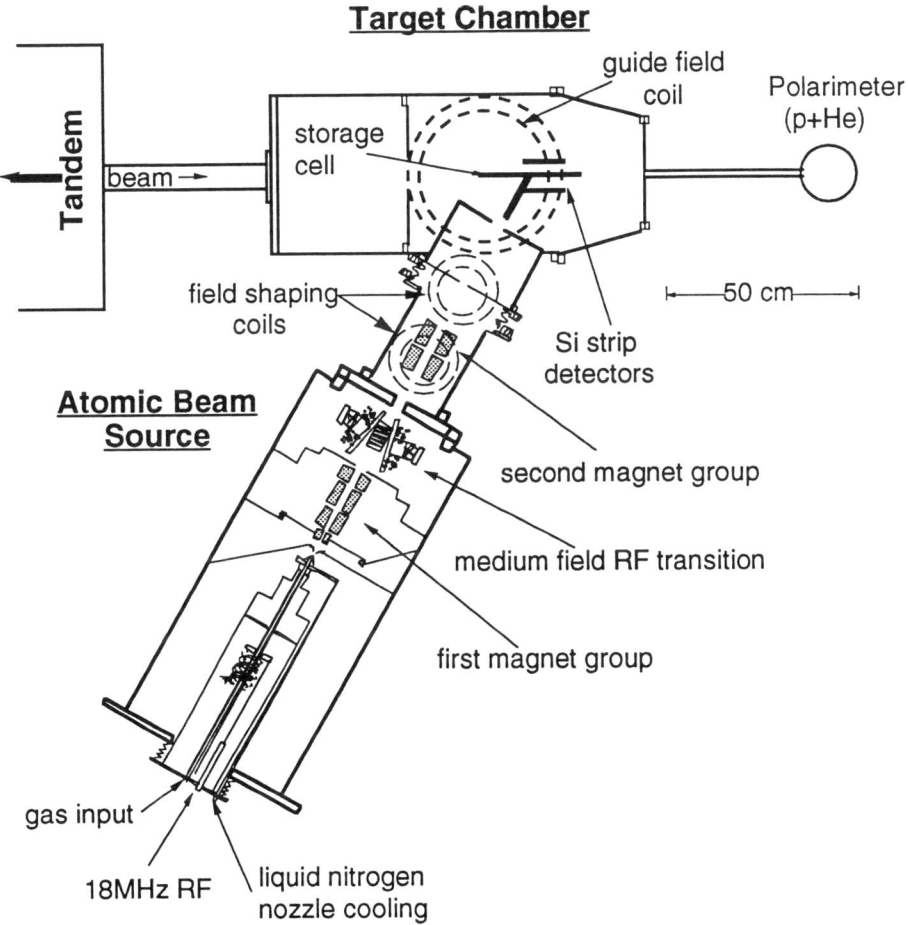

**Fig.1.** Layout of the Atomic Beam Source at the UW Tandem accelerator.

from the center of the storage cell measure the energy and position of the scattered and recoil protons.  Each detector has an active area of 4 cm by 6 cm, is 1 mm thick, and has 28 strips.  A total of four detectors are used in these measurements.  The azimuthal position of the detectors is chosen to avoid interference of the scattered protons with the feed tube assembly.  As seen by the beam, the detectors are located in the upper-left and lower-right quadrants.  Since the density of target atoms along the storage cell decreases linearly from the central region of the cell, the location of the detectors along the beam direction is chosen to give the maximum $pp$ coincident rate.

An event is defined by a coincidence between the upper-left and lower-right detectors.  For each event the energy and position signal of each detector that fired is recorded.  The beam polarization is reversed every second; the target polarization is reversed after every run (approximately once an hour).  Fig. 2 shows a plot of the energy in the upper-left detector $E_L$ vs. the energy in the lower-right detector $E_R$ for each beam spin state.  The $pp$ elastic events are clearly evident along the diagonal.  The background is typically 4% of the total number of events and is determined from measurements with an empty cell.

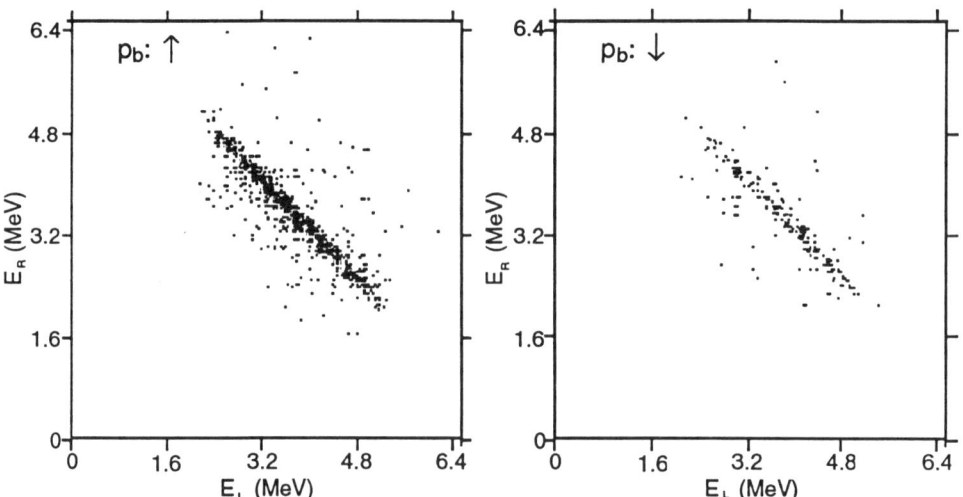

**Fig. 2.** Energy in the upper-left detectors ($E_L$) vs. energy in the lower-right detectors ($E_r$).  PP elastic event are along the diagonal.  The direction of the target polarization is ↓.

## Results

The target polarization was first measured with the medium field RF transition (see fig. 1) turned off.  In this situation, atoms in both $m_j=+1/2$ spin states are focused by the ABS magnets and enter the storage cell. The guide field coils shown in fig. 1 define the target polarization direction and produce a magnetic field of 1.5 mT. Under these conditions, the polarization of the atoms in the storage cell, assuming no polarization losses in the cell, is expected to be 0.485. The measured value is 0.436±0.022.

The target thickness, determined from the rates and the known *pp* cross section, was $2 \times 10^{13}$ atoms/cm$^2$. A new cell with a beam opening of 8 mm by 8 mm is currently under construction. With this new cell and improved vacuum conditions in the ABS, a target thickness of $7.2 \times 10^{13}$ atoms/cm$^2$ in two spin states is expected.

When the medium field RF is turned on, a transition is made from the $m_j = +1/2$, $m_i = -1/2$ atoms to the $m_j = -1/2$, $m_i = -1/2$ atoms. The second magnet group (see fig.1) then defocuses the $m_j = -1/2$ atoms. This reduces the flux of atoms into the feed tube by about a factor of two, but increases the polarization of the atoms in the cell. Assuming no losses in the cell the polarization is

$$p = \frac{1 - \varepsilon a - (1 - \varepsilon)R}{1 + \varepsilon + (1 - \varepsilon)R} \qquad (3)$$

where *a* is the correction for the 1.5 mT field over the cell, $(1-\varepsilon)$ is the efficiency of the medium field unit, and *R* is the fraction of atoms in the cell with $m_j = -1/2$ that are not sufficiently defocused by the second magnet group. For a magnetic field of 1.5 mT, $a = 0.03$. The efficiency of the medium field transition has been measured previously to be 0.96 [4]. The value of *R* is 0.052, determined from transport calculations of the magnet assembly. The polarization with the medium field transition on is therefore expected to be 0.871. The measured value is 0.722±0.021. Table 1 summarizes all polarization values.

### Table 1. Polarization Results

| Medium Field | $p_t(\uparrow)$ | $p_t(\downarrow)$ | $|p_t|$(ave) | $p_{calc}$ |
|---|---|---|---|---|
| OFF | 0.415±0.032 | -0.457±0.031 | 0.436±0.022 | 0.485 |
| ON | 0.701±0.029 | -0.743±0.030 | 0.722±0.021 | 0.871 |

The results indicate that when the medium field transition is on we obtain a polarization that is 83% of the expected maximum. The reason for the 17% loss is not known at the present time. Inefficiency in the medium field unit or polarization losses in the cell are two possibilities. However, the polarization is large enough for experiments at the IUCF Cooler to begin later this year.

**References**
1. J.S. Price, Ph.D. Thesis, University of Wisconsin, 1993.
2. R. A. Arndt, L. D. Roper, R. L. Workman, M. W. McNaughton, Phys. Rev. **D45** (1992) 3995, and the SAID database mentioned therein.
3. P. Schwandt, T. B. Clegg, W. Haeberli, Nucl. Phys. **A163** (1971) 432.
4. A. D. Roberts, P. Elmer, M. A. Ross, T. Wise, W. Haeberli, Nucl. Instr. and Meth. **A322** (1992) 6.

# MEASUREMENT OF POLARIZATION FOR POLARIZED GAS TARGETS

J. S. Price*

Department of Physics, University of Wisconsin-Madison, Madison, WI 53706

## ABSTRACT

A method for measuring the $\vec{D}^\circ$ target polarization in a strong magnetic field ($B = 4B_c, B_c = 11.7$mT) is described. Measurements of the tensor polarization of a $\vec{D}^\circ$ gas target with different cell wall materials are presented. The polarization measurements were made on cells of two different geometries in order to assess the effect of wall collisions on the polarization of the target. The temperature of the cells was varied over the range $20$ K $<$ T $<$ $300$ K. These strong field test results are compared to previous measurements of the polarization of a $\vec{H}^\circ$ gas target in cells of similar geometry and wall material in a weak magnetic field (B=0.5 mT).

## INTRODUCTION

The usefulness of polarized gas targets in storage rings depends on maintaining high target polarization as well as target thickness. The use of long narrow storage cells as internal targets in storage rings, as proposed by Haeberli [1], offers the advantage of a large increase in target thickness compared to polarized jet targets. The choice of cell wall material is important since target atoms may undergo hundreds of wall collisions on the average before leaving the cell.

## TEST APPARATUS

The polarimeter described here is based on the extraction of ions from within the target cell, followed by acceleration, mass analysis and determination of the nuclear polarization of the extracted ions. A more complete description of the apparatus is found in Ref. [2]. A diagram of the polarimeter is shown in Fig. 1.

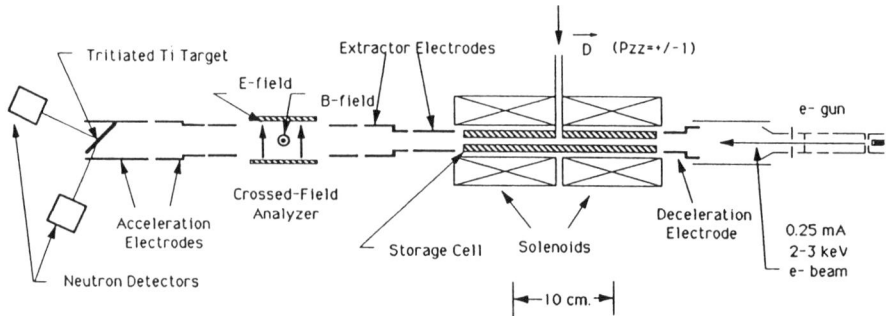

Fig. 1 Diagram of polarimeter used in strong field tests.

---

* Present address: Randall Lab of Physics, University of Michigan, Ann Arbor, MI 48109

A commercial electron gun was used to inject electrons along the axis of the storage cell. The electrons were decelerated to 80 eV before entering the cell since this energy corresponds roughly to the maximum in the ionization cross section. A solenoid supplies a longitudinal magnetic field of 470 G over the storage cell. This field provides a radial confinement of the ions formed in the cell, as well as providing confinement of the low energy electron beam.

The extraction electrodes accelerate the ions to 2.5 keV. The mass separator consists of a crossed E-B-field analyzer of 5 cm length. The magnetic field is fixed and provided by permanent magnets, and the electric field is adjusted to pass particles of different mass. After passing through the mass analyzer, the ions are accelerated to 70 keV and are incident on a tritiated titanium target. The deuteron alignment $p_{zz}$ is deduced from the anisotropy of the neutrons from the $^3H(d,n)^4He$ reaction. The two neutron detectors in Fig. 1 are positioned to be along and at right angles to the deuteron alignment direction. The 21°offset of the detectors with respect to the beam axis reflects the spin precession in the B-field of the mass analyzer.

One interesting aspect of the polarimeter is that the polarization of the extracted ions reflects the proper weighted mean of the target polarization in the cell. Diagnostic tests show that the efficiency of extraction of ions from the two halves of the cell is the same [2].

## TEST RESULTS

The target polarization $p_{zz}$was measured separately for $D^+$ and $D_2^+$ ions as a function of temperature for cells of two different geometries corresponding to two different average ages of the particles in wall collisions. The nuclei of the molecular ions are essentially unpolarized. The measured $p_{zz}$for the $D_2^+$ ions was $0.058 \pm 0.027$. The maximum possible tensor polarization in the absence of any wall depolarization, $p_{zz}=0.887$, is shown by the horizontal line in the figures below.

Fig. 2 a shows measured $p_{zz}(D^+)$ vs T. Between room temperature and 100 K the value of $p_{zz}$of the atoms is about 95 % of the maximum possible value. The actual target polarization (shown in Fig. 2 b, $p_{zz}$(total target)) is not diminished by the presence of a small fraction of $D_2$. There seems to be little wall collision dependence on the depolarization for 3170 Teflon above 100 K. Target polarization decreases rapidly for cell temperatures below 100 K and the effect of wall collisions becomes evident.

Recombination to form molecules and depolarization of the atoms at the cell wall appear to be independent for uncoated copper. A copper cell (average age 80 collisions) was tested (see Fig. 3 a). The atoms' polarization, $p_{zz}(D^+)$, is nearly 90% of the expected maximum for T > 100K but the total target polarization is below 30% because the atoms which recombine to form molecules dilute the polarization.

The advantage of the present method is that this dilution factor is determined experimentally. A measurement of the current on the tritiated foil from the $D^+$ ions compared to the $D_2^+$ ions determines the coefficient of recombination as a function of temperature, $\alpha_r(T)$. Fig. 3 b is comparison of $\alpha_r(T)$ for 3170 Teflon, which effectively inhibits recombination of D atoms, and uncoated

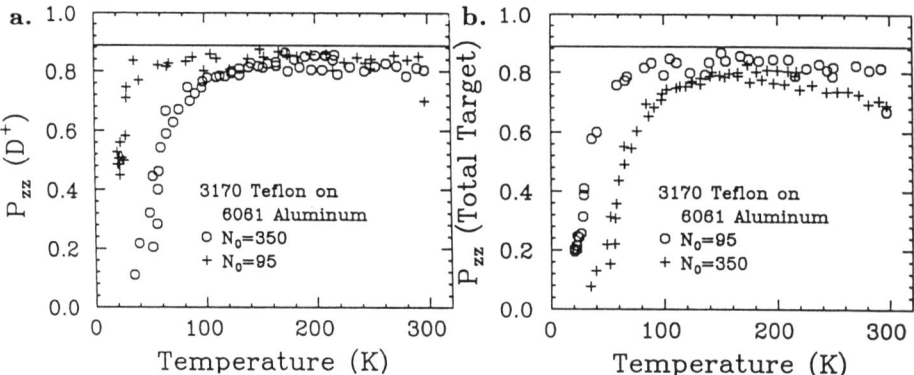

Fig. 2 a. Polarization of the atoms ($p_{zz}(D^+)$) vs cell temperature (T) b. Total target polarization ($p_{zz}$(total target)) vs T for 3170 Teflon coated aluminum cells, $N_o$=95 and 350.

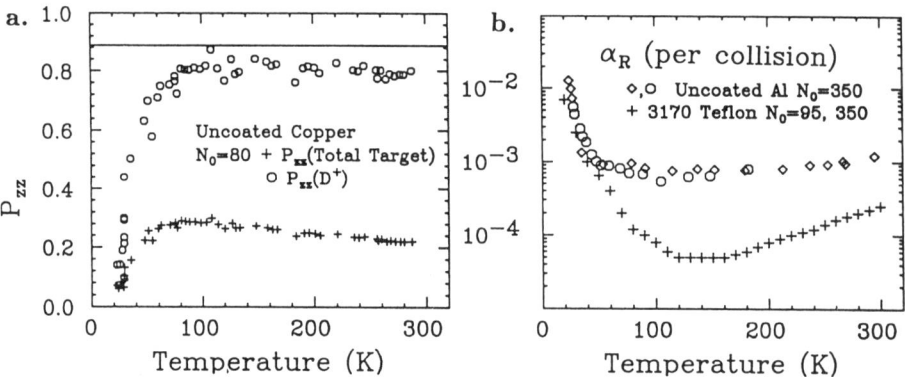

Fig. 3 a. $P_{zz}(D^+)$ and $P_{zz}$(total target ) vs T for an uncoated copper cell, $N_o$ = 80. b. Recombination coefficient, $\alpha_r$, vs T for 3170 Teflon and uncoated aluminum. Uncoated copper's $\alpha_r$ is greater than $10^{-2}$ independent of T.

aluminum which has much larger recombination. Clearly, large target polarization can be maintained in spite of the increased number of wall collisions by protecting the wall with suitable wall coatings similar to polarized $\vec{H}^\circ$ gas targets in a weak magnetic field [3].

A comparison of weak field $\vec{H}^\circ$ and strong field $\vec{D}^\circ$ polarization measurements reveals that the application of a magnetic field of 4 $B_c$ is enough to inhibit depolarization at cell wall surfaces significantly (see Fig. 4 a). Uncoated aluminum $N_o$=380 cells show severe depolarization in a weak field, but an $N_o$=350 cell performs well in a 4 $B_c$ magnetic field. Fig. 4 b shows that a similar improvement in target polarization can be seen for 3170 Teflon coated cells, although the increase from weak field to strong field (80% to 90% ) is not nearly so pronounced.

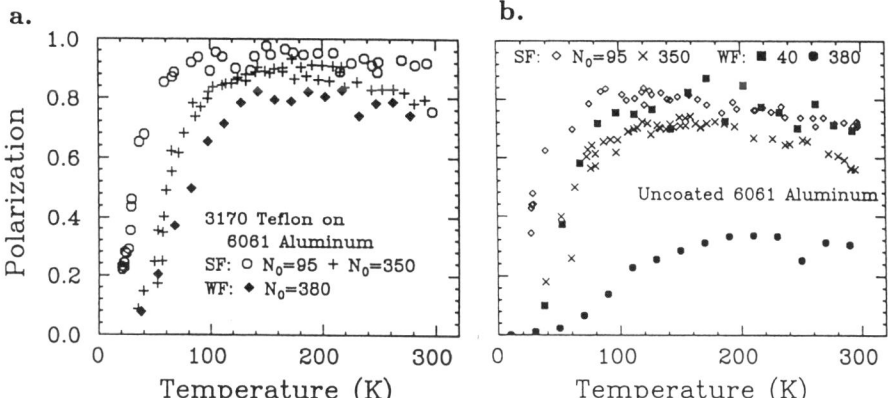

Fig. 4 Comparison of weak and strong field results. a. Weak field polarization (fraction of calculated maximum) and strong field $p_{zz}$(total target) vs T for 3170 Teflon coated aluminum cells, $N_o$=95, 350. b. Uncoated aluminum cells were tested.

Although the magnitude of the improvement seems to correlate with recombination, the data are not sufficient to support the conclusion that recombination alone is responsible for the improvement in depolarization at $B=4B_c$ as compared to low field. No measurements of recombination coefficient were made at low magnetic field.

## CONCLUSIONS

A $\vec{D}^o$ tensor polarimeter method was used to measure the depolarization at cell walls as well as the recombination of D atoms on different materials. Results show that increasing the magnetic field inhibits depolarization although the exact mechanism responsible for this improvement is not clear. Although atoms can survive many hundreds of collisions on surfaces such as Teflon and copper, recombination on a surface produces molecules of the target gas which retain little polarization and thus reduce the target polarization by dilution.

The present work was supported by the U.S. Department of Energy.

[1] W. Haeberli, Nuclear Physics with Stored, Cooled Beams, P. Schwandt and H. O. Meyer eds., AIP Conference Proceedings No. 128 (1985), p. 251.

[2] J.S. Price and W. Haeberli "Polarization Measurement for Polarized Gas Targets," Nucl. Instr. and Meth. in Phys. Res. **A326**, 416 (1993).

[3] T. Wise, A. Converse, J.S. Price, Proc. Eighth International Symposium High Energy Spin Physics, J. K. Heller, ed. AIP Conference Proceedings No. 187, p. 1565; W. Haeberli, Proc. Ninth International Symposium High Energy Spin Physics, W. Meyer, E. Steffens, W. Thiel eds. (Springer-Verlag, Berlin, 1991), vol. 2, p. 194.

# OPTIMIZATION OF THE FILTEX/HERMES ATOMIC BEAM SOURCE

F. Stock

Universität Erlangen and
MPI–Kernphysik, Postfach 10 39 80, D–69029 Heidelberg, Germany

## ABSTRACT

The FILTEX Atomic Beam Source (ABS) for polarized hydrogen and its optimization are described. After the optimization the output flow rate into a 100 mm long and 10 mm wide cylindrical compression tube is $8.1 \cdot 10^{16} \vec{H}_1$ in two substates. The output flow rate can be determined with a precision of 5 %, it showed to be constant within 2 % in a longterm measurement over 16 h. At the FILTEX test experiment, the target density in the storage cell fed by the ABS was constant over a period of four months. The modifications in progress for the additional production of a vector and tensor polarized deuterium beam for HERMES are outlined.

## INTRODUCTION

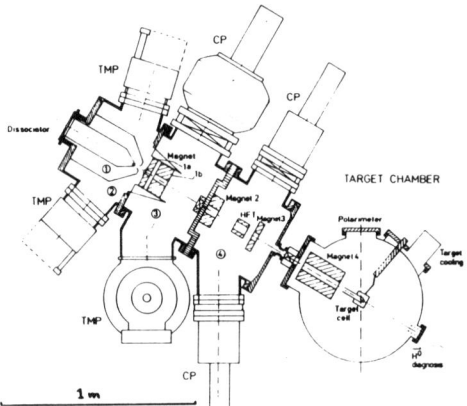

Figure 1: The FILTEX ABS and target chamber.

The FILTEX Atomic Beam Source (ABS) is a high intensity source for nuclear spin polarized hydrogen based on Stern–Gerlach spin separation of the electron spin in an inhomogenious magnetic field in combination with a high frequency transition (HFT). It was designed to feed a storage cell. Internal target experiments like FILTEX [1] or HERMES [2] employ storage cells to achieve the required target density which is about $10^{14}/cm^2$. At the FILTEX test experiment the achieved target density and polarisation was sufficient to proof the feasability of spin filtering [3].

The HERMES experiment at the HERA electron ring will measure the spindependent structure functions of the nucleons via deep inelastic scattering of longitudinally polarized electrons from transversally or longitudinally polarized nucleons ($\vec{H}, \vec{D}$ or $^3\vec{H}e$). It requires a target density of $10^{14}/cm^2$ hydrogen or deuterium atoms of maximum polarisation in a strong guiding field. That means an output flow rate of the ABS of about $8 \cdot 10^{16}$ Atoms/s is required.

The FILTEX ABS [4] consists of a powerful four stage differential pumping system, an RF dissociator with cooled Al nozzle, two skimmers and a system of 5 permanent sextupole magnets. The HFT is placed between magnet 2 and magnet 3. Fig.1 shows a schematical drawing of the ABS and the target chamber as it was set up for the FILTEX test experiment in summer 1992.

## MEASUREMENT OF THE OUTPUT FLOW RATE

In order to give precise predictions of the target density in the storage cell the output flow rate is measured with a compression tube having an injection tube of the same geometry as the feed tube of the storage cell (FILTEX: 10 mm diam., 100 mm length).

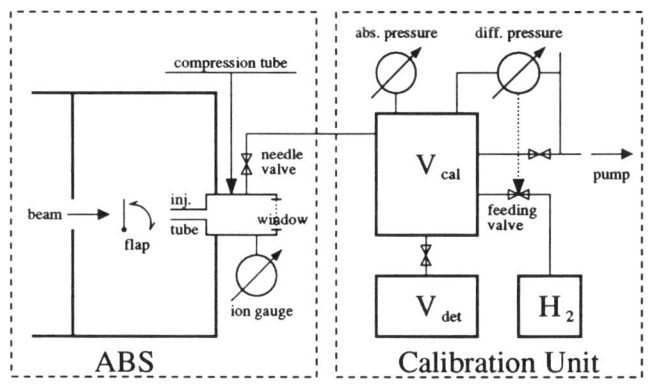

Figure 2: The compression tube and the calibration unit. The pressures in the particular parts are $p_{ABS} \approx 10^{-7}$ mbar, $p_{comp} \approx 10^{-4}$ mbar and $p_{cal} \approx 10$ mbar.

Fig.2 shows a scetch of the compression tube and the calibration unit. The output flow rate of the ABS is measured by reproducing $p_{comp}$ by means of a calibrated flow. A precise calibration yields the flow rate as function of the volume $V_{cal}$ and the pressure $p_{cal}$:

$$Q = \frac{V_{cal}}{\tau} \cdot p_{cal}, \tag{1}$$

while $\tau$ is the time constant of the pressure drop in $V_{cal}$. The accuracy of this measurement is 5 %.

## OPTIMIZATION OF THE BEAM FORMING SYSTEM

The beam forming system consists of the nozzle and two skimmers called skimmer and collimator (see fig.3). The nozzle is made of 99.5 % Al.

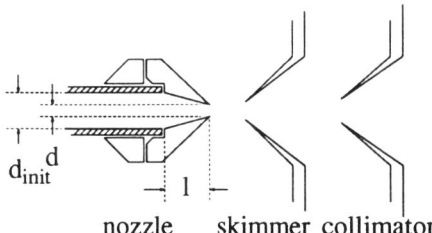

Figure 3: The beam forming system of the ABS consisting of nozzle, skimmer and collimator.

While the initial diameter $d_{init}$ is fixed by the diameter of the discharge tube, both length and final diameter were varied. Table I shows the optimum output flow rate for different nozzle diameters. The measurements were done with half the old magnet system and the injection tube placed instead of magnet 3 (see fig.1). The highest flow rate is obtained with the smallest diameter. Variation of the nozzle length did not show large differences in the output flow rate.

Table I: Output flow rate for different nozzle diameters.

| d/l [mm] | 2/17 | 4/17 | 5/17 |
|---|---|---|---|
| $Q(\vec{H_1}$, 2 subst.) $[10^{16}/s]$ | 5.4 | 4.1 | $\approx 3.5$ |

The influence of the distance nozzle to skimmer was investigated. Fig.4 shows the flow rate into the compression tube as a function of the dissociator throughput for different nozzle–skimmer distances. There is an optimum distance, which can be easily adjusted. Varying the collimator width by means of an iris aperture instead of the collimator proofed that there is an optimum aperture.

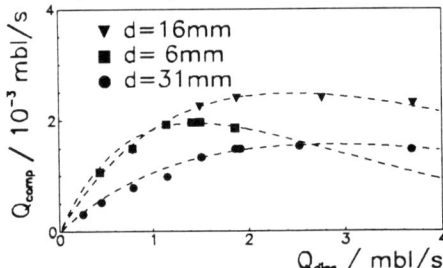

Figure 4: The output flow rate as function of the dissociator through-put for different nozzle–skimmer distances d.

Since the $H_2$ throughput through the dissociator is $1 - 3$ mbl/s the question was investigated wether the pumping system is powerful enough to handle this large gas load. For that purpose, the output flow rate was measured at a dissociator throughput of 1.5 mbl/s as function of the residual gas pressure of each pumping stage. The pressure was increased above the working pressure by an external $H_2$ inlet. The measured data were extrapolated to zero pressure which means infinite pumping speed. Fig.5 displays the $H_1$ flux into the compression tube as function of the pressure in each pumping stage. The numbers give the percentage of the flow rate at the working point in comparison to the flow rate at infinite pumping speed. The combined result of all stages shows that the output flow rate at the working point is 80.6 % of the value for infinite pumping speed. Later improvements gave 82 %. This means that the output flow rate can be improved only by about 20 % even with infinite pumping speed. This proofs that the pumping system is very powerful and well balanced.

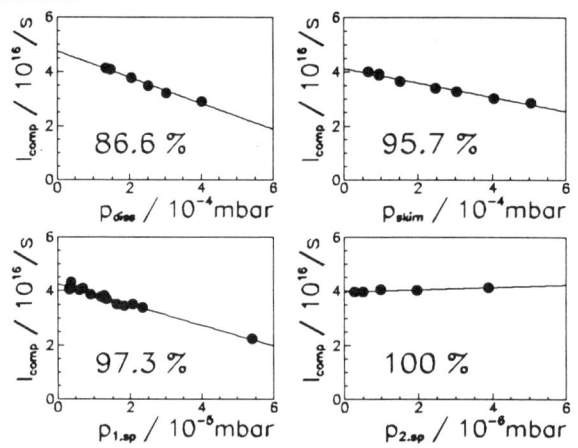

Figure 5: The attenuation by residual gas. The ouput flow rate as function of the pressure in a particular pumping stage is shown. The pressure is varied by introducing additional $H_2$ by means of a needle valve.

The nozzle is cooled in order to increase the angular acceptance of the magnet system. For its large cooling power a liquid nitrogen cooling line was installed. The lowest achievable nozzle temperature with discharge switched on is 80 K. Fig.6 shows the temperature dependence of the output flow rate measured with the whole magnet system after completion of the optimization. The measurement was done with increasing nozzle temperature after switching off the nozzle cooling. The maximum value for the flow rate is $Q = 8.1$ $H_1$/s in two substates into the compression tube at the lowest nozzle temperature of $T_n = 80$ $K$. The output flow rate strongly decreases with increasing nozzle temperature. Drops in the output flow rate are visible at 150 K and 240 K. The reason for this may be an increasing background pressure due to degassing oxygen and alcohol respectively from the nozzle. The alcohol may come from a small leak at an O–ring sealing between the cooling cycle for the discharge tube and the dissociator chamber.

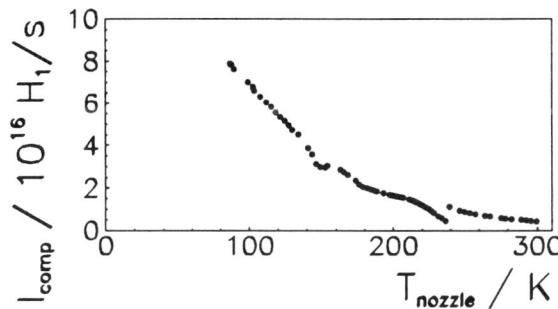

Figure 6: The output flow rate as a function of the nozzle temperature.

This result is the best ever measured at an ABS with the given precision and meets well the HERMES design figure. Since liquid nitrogen cooling is limitted to 78 K the effective use of a closed cycle He–refrigerator is planned. For $T_n = 60\ K$ an increase of the flow rate by about 40 % is expected.

## DESIGN OF AN IMPROVED MAGNET SYSTEM

For the design of the FILTEX magnet system, a track tracing code using Monte Carlo techniques has been employed. The track tracing code takes into account a parametrization of both the atomic beam temperature and the drift velocity of the particles as a function of the nozzle temperature. The parametrization was extracted of several time of flight measurements of the atomic beam, fitted with a modified Maxwell velocity distribution. It is for beam temperature and drift velocity

$$T_{beam} = 0.403\ T_n - 22.57 \tag{2}$$
$$v_{drift} = 2.880\ T_n + 1312. \tag{3}$$

Allowing tapered magnets predicted for the transmission factor $\Omega \cdot T$ an improvement from $6.8 \cdot 10^{-3}$ for the old system to $11.8 \cdot 10^{-3}$ for the new system. This is in agreement with the compression tube measurements. Table II shows some parameters of the new FILTEX magnet system. The magnets are permanent magnets and made of 12 or 24 segments [5].

Table II: Parameters of the new magnet system of the FILTEX–ABS

| No. | # Segments | Outer Diam. [mm] | Inner Diam. [mm] | Length [mm] | Tip Field [T] |
|---|---|---|---|---|---|
| 1 a | 12 | 80 | 5.4 – 7.0 | 34 | 1.419–1.401 |
| 1 b | 12 | 80 | 8.0 – 10.0 | 83 | 1.388–1.355 |
| 2 | 24 | 80 | 12.5 | 60 | 1.497 |
| 3 | 24 | 80 | 12.5 | 30 | 1.497 |
| 4 | 24 | 80 | 12.5 | 200 | 1.497 |

## PERFORMANCE DURING THE FILTEX TEST EXPERIMENT

A longterm compression tube measurement showed a stability of the output flow rate of 2 % over 16 h at the maximum flow rate of $8.1 \cdot 10^{16}\ H_1/s$. During the internal target experiment the output flow rate can be monitored looking at the count rate of events which is correlated to the target density. Having the target density and the conductance of the storage cell the output flow rate of the ABS can be calculated. Online evaluations showed no drop in the output flow rate over a period of four months. The longest running periods without switching off the discharge were two weeks. The ABS proofed to be very reliable and very stable in operation.

## MODIFICATIONS FOR HERMES

For HERMES, the ABS is being prepared to produce both hydrogen and deuterium beams of maximum polarization. A strong guiding field in the target region implies the use of two substates for any particular polarization. A change in the sign of the polarization is done by selection of different substates by means of sextupole magnets and rf–transitions. Table III shows a complete scheme of the transitions for hydrogen and deuterium.

Table III: System of rf–transitions. WFT – weak field transition, MFT – medium field transition, SFT – strong field transition.

| Substates | Hydrogen (subst. 1–4) | | Deuterium (subst. 1–6) | | | |
|---|---|---|---|---|---|---|
| | Vectorpolarization | | Vectorpolarization | | Tensorpolarization | |
| After 1st Sextupoles | 1 + 2 | 1 + 2 | 1 + 2 + 3 | 1 + 2 + 3 | 1 + 2 + 3 | 1 + 2 + 3 |
| 1st transition | off | off | MF (3–4) | MF (3–4) | MF (1–4) | MF (1–4) |
| Subst. | 1 + 2 | 1 + 2 | 1 + 2 + 4 | 1 + 2 + 4 | 2 + 3 + 4 | 2 + 3 + 4 |
| After 2nd Sextupoles | 1 + 2 | 1 + 2 | 1 + 2 | 1 + 2 | 2 + 3 | 2 + 3 |
| 2nd transition | SF (2–4) | off | SF (2–6) | off | SF (2–6) | SF (3–5) |
| Subst. | 1 + 4 | 1 + 2 | 1 + 6 | 1 + 2 | 3 + 6 | 2 + 5 |
| 3rd transition | off | WF (1–3) | off | WF | off | off |
| Subst. | 1 + 4 | 2 + 3 | 1 + 6 | 3 + 4 | 3 + 6 | 2 + 5 |
| $P_z$ | +1 | −1 | +1 | −1 | 0 | 0 |
| $P_{zz}$ | − | − | +1 | +1 | +1 | −2 |

Table IV: Different types of rf–transitions for the ABS.

| Transition | Selection Rule | Frequency (MHz) | Orientation RF–field vs. $B_0$ | Resonator |
|---|---|---|---|---|
| WFT (weak field trans.) | $\Delta F = 0$ $m_F \leftrightarrow -m_F$ | 10 | $\perp$ | coil |
| MFT (medium field trans.) | $\Delta F = 0$ $\Delta m_F = \pm 1$ | 50 (H), 20 (D) | $\perp$ | coil |
| SFT (strong field trans.) | $\Delta F = 1$ $\Delta m_F = 0$ | 1430 (H), 335 (D) | $\parallel$ | tuned line |

The transitions work on the principle of adiabatic passage by Abragam and Winter. Table IV gives an overview over the different types of rf–transitions. A complete set of rf–transitions is designed and under construction. Two SFT–cavities are already successfully tested for the required resonance frequency. The design of a modified chamber for focussing magnets and rf–transitions is completed. This chamber is optimized to the geometrical constraints of the HERMES experiment. Completion of the modified HERMES target source is anticipated for spring 1994.

## REFERENCES

[1]   H. Döbbeling et al., Proposal CERN/PSCC/85/80 (1985)

[2]   HERMES Proposal, DESY–PRC 90/01.

[3]   K. Zapfe: contribution to this workshop.

[4]   W. Korsch: Proc. Workshop on High Energy Spin Physics, Bonn 1990; Springer Verlag, Berlin Heidelberg 1991, p. 168.

[5]   P. Schiemenz et al.: Nucl. Instr. Meth. **A305**,15 (1991).

# EXPERIMENTS WITH POLARIZED DEUTERIUM TARGET AT VEPP – 3 STORAGE RING : STATUS AND PERSPECTIVE.

*V.V. FROLOV, S.I. MISHNEV, D.M. NIKOLENKO, S.G. POPOV,*
*I.A. RACHEK, A.V. SUKHANOV, D.K. TOPORKOV,*
*E.P. TSENTALOVICH and B.B. WOJTSEKHOWSKI*[1]
*Budker Institute for Nuclear Physics, Novosibirsk, Russia*

*K.P. COULTER, R. GILMAN,*[2] *R.J. HOLT, E.R. KINNEY,*[3] *R.S. KOWALCZYK,*
*B.M. POELKER, D.H. POTTERVELD, L. YOUNG and A. ZGHICHE*[4]
*Argonne National Laboratory, Argonne, Illinois, USA*

*V.V. NELUBIN and V.V. VIKHROV*
*St.Petersburg Institute for Nuclear Physics, Gatchina, Russia*

*V.N. STIBUNOV and A.V. OSIPOV*
*Tomsk Polytechnic Institute, Tomsk, Russia*

*C.W. de JAGER, G. RETZLAFF*[5], *J. THEUNISSEN and H. de VRIES*
*NIKHEF, Amsterdam, The Netherlands*

## Abstract

A status of the internal polarized deuterium target experiment at VEPP-3 storage ring at 2 GeV electron energy is presented. A value of $T_{20}$ at q = $3.7 fm^{-1}$ was obtained by using an active storage cell in conjunction with an atomic beam source. Substantial progress is expected by introducing the high-density laser-driven target.

## Introduction

The deuteron is the simplest nucleus to test fundamental physical questions. Polarization experiments allow access to new observables through asymmetry measurements which are expected to be sensitive to effects of deuteron wave function, meson exchange currents, relativistic corrections, quarks degree of freedom, etc.

At BINP, Novosibirsk, experiments with internal tensor-polarized deuteron target began in 1984. New data on spin observables were obtained both for elastic [ 1,2 ] and inelastic [ 3,4 ] electron scattering. The first stage of the experiment was successfully completed in 1988. At present there are data taking at second phase of the experiment when an active storage cell for polarized atoms is used [ 5 ]. The third phase is expected to start next year.

### Internal tensor-polarized deterium target

At present the conventional atomic beam source is used [ 6 ]. It produces a flux of polarized atoms about $1.5 \cdot 10^{16} atoms/sec$ with tensor polarization $P_{zz} = \pm 0.95 \pm 0.05$. Approximately one third of atoms directed into the active storage cell which could be opened to provide loading of electron beam in the ring and then be closed to provide a high density of deuterium target. The cross section of the closed cell is $10 \times 20$ mm and its length 570 mm. Four synchronous linear-motion drives are used to close and open the cell.

### Detector

The detector designed for asymmetry measurements consists two identical arms placed in the vertical plane. There is a vertex detector to pinpoint event vertices. Scattered electrons are detected in the forward direction ( $\theta_e = 25°\pm 5°$ , $\varphi_e = \pm 30°$ and $180° \pm 30°$ ) by wire chambers, trigger scintillators and a layered CsI / NaI calorimeter with total thickness of 16 radiation lengths ( CsI : $6 + 6 + 6$ cm, NaI : $5 + 11$ cm ). This detector also detects protons from the knockout, disintegration and pion production reactions. At large angles ($\theta_d = 65° \pm 5°$ , $\varphi_d = 0° \pm 30°$ and $180° \pm 30°$ ) there is another set of wire chambers followed by layers of plastic scintillators ( $1 + 1 + 12 + 12.5 + 2$ cm ) for deuteron and proton detection. There are thick bars ( 20 cm ) of scintillators directed at $\theta_n = 135°$ and used as neutron detectors. It is planned to collect elastic scattering events in the momentum transfer range q = 3 - 5 $fm^{-1}$ , photodisintegration , quasi-free scattering and pion production processes. The event trigger and data acquisition system was designed to provide simultaneous detection and readout of all these processes.

### Background

The storage cell technique provides a substantial increase of the luminosity of an internal target experiment, but new problems appear. The most serious of them are (i) depolarization processes, which were investigated in earlier measurements [ 7 ] ; and (ii) new sources of background.
The highest background process for experiments with a small-aperture storage cell, is that produced by electrons which strikes the cell elements. These events occur close to the interacting point and simulate real events. Electron, executing betatron oscillation which hit the vacuum chamber elements are nearest to the beam. In the case of the active cell, the cell wall is hit. An electron penetrates inside and produces an electromagnetic shower. Lower energy electrons and positrons from the shower have a probability to scatter at large enough angle to reach the detector.
To suppress this background an active tungsten collimator was installed. It consists of 2 - cm thick blocks in the radial and vertical directions. This collimator is located about half a period of betatron oscillations upstream from the cell and takes advantage of massive elements in the ring to effectively shield the particle detectors from the electromagnetic showers produced at the collimator. The amount of suppressed background with the collimator is in a good agreement with the expected results ( factor 3 - 10 for various detectors ).

### Data - taking runs and recent results of data analysis

Two data runs of data taking were performed. During the first ( January-June 1992 ), the total charge that crossed the target was 150 kC and the direction of the holding magnetic field was fixed. In the second run ( November 1992-April 1993 ) the total charge was 170 kC and the direction of the magnetic field was switched every 2 - 3 hours during the injection of electrons into the storage ring. In both measurements the sign of $P_{zz}$ of the atomic jet injected into the cell was reversed every 200 seconds and its magnitude was continuously monitored.

In the earlier run the data was recorded only for elastic scattering and the proton knock-out reaction with a proton detected in the large-angle detector arm. This was due to the high background rate during that run. After the installation of the collimator and the upgrade of the data acquisition system, it was possible to record data in other channels. Presently, only the analysis of elastic scattering data of the first run is complete. The results are presented in the Fig. 1. The momentum transfer range was divided into two parts. The asymmetry obtained for the lowest value of q ( $3 \pm .25 \ fm^{-1}$ ) was used to determine the average tensor polarization along the target by normalizing the data to the theoretical value of $T_{20}$ given by Paris potential as well as to the value measured in earlier experiment [ 2 ]. Thus, it was found that the target polarization was $0.67 \pm 0.19$. For the second point the obtained result for $T_{20}$ seems to disagree with the theoretical predictions, however, the statistical accuracy is not sufficient to make a strong claim.

The target thickness was extracted from the counting rate of the elastic scattering events and known geometry of the experiment. For the part of the cell which is viewed by the detector, the target thickness was determined to be $(2.0 \pm .5) \cdot 10^{12} atoms/cm^2$, in a good agreement with the calculated value.

### Perspective

A novel source of polarized deuterium atoms which is based on the spin-exchange optical pumping has been developed at Argonne National Laboratory to be used in the third phase of experiment [ 8,9 ]. The new source delivers $1 \times 10^{18} atoms/sec$ with a polarization of 50 %. Most parts of the source have been already shipped to Novosibirsk. The new storage cell will be passive with a 23×12 mm opening and 400 mm length and should provide the target thickness of $1 \times 10^{14} atoms/cm^2$ at flux $2 \times 10^{17} atoms/sec$. Special quadrupole lenses will be installed in the straight section of the VEPP-3 ring to provide the possibility to fill VEPP-3 with such a small aperture.

The new polarized deuterium source and new magnetic elements are planned to be installed at the VEPP-3 straight section during the summer shut-down of the ring in 1994. The top view of the designed straight section is shown at Fig. 2.

The detectors and the data acquisition system will be the same as at the present stage, just few improvements will be done.

### Conclusion

The first experience with an active storage cell for polarized atoms to increase the target thickness in an electron storage ring proved the effectiveness of this method. An active collimator installed in the ring has allowed to reduce the background by a factor 3 - 10. New data for $T_{20}$ in electron-deuteron elastic scattering has been obtained, although with low statistical accuracy. Substantial increase of luminosity of the experiment is expected in the next phase with the use of more intense laser driven spin-exchange source of polarized atoms.

(1) Present address: RPI, Troy, NY, USA
(2) Present address: RU, Piscataway, NJ, USA
(3) Present address: UC, Boulder, CO, USA
(4) Present address: CRN-F67037, Strasbourg, Cedex
(5) Present address: SAL, Saskatoon, Canada

# References

[1] V.F. Dmitriev et al., Phys. Lett. **157B** (1985) 143.

[2] R. Gilman et al., Phys. Rev. Lett. **65** (1990) 1733.

[3] M.V. Mostovoy et al., Phys. Lett. **188B** (1987) 181.

[4] S.I. Mishnev et al., Phys. Lett. **302B** (1993) 23.

[5] K.P. Coulter et al., Nucl. Instr & Meth., to be published.

[6] A.V.Evstigneev, S.G.Popov and D.K.Toporkov, Nucl. Instr& Meth **A238** (1985) 12.

[7] R.Gilman et al., Nucl. Instr & Meth. **A327** (1993) 277.

[8] K.P. Coulter et al., Phys. Rev. Lett. **68**, (1992) 174.

[9] B.M. Poelker et al., A Laser Driven Source of Polarized Atomic Hydrogen and Deuterium, these proceedings.

[10] R. Schiavilla and D. Riska, Phys. Rev. **C43** (1991) 437.
    R. Dymarz and F.C. Khanna, Nucl. Phys. **A507** (1990) 560.
    I. Pauschenwein et al., Nucl. Phys. **A508** (1990) 253c.
    E. Nyman and D. Riska, Nucl. Phys. **A468** (1987) 473.
    C. Carlson, Nucl. Phys. **A508** (1990) 481c.
    V. Burov and V. Dostavalov, Z. Phys. **A326** (1987) 245.

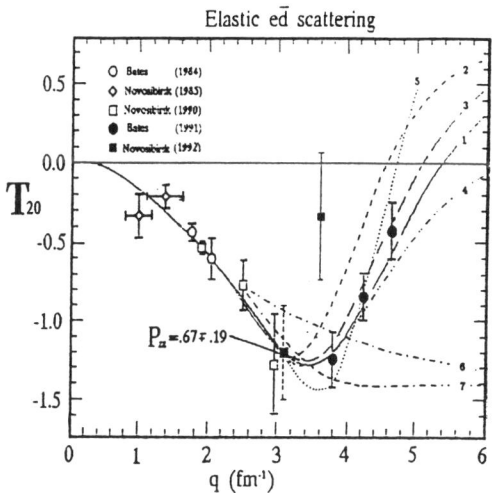

FIG. 1. Experimental results and theoretical predictions [ 10 ] for the tensor analyzing power $T_{20}$.

1 - IA $AV_{14}$
2 - IA+MEC $AV_{14}$
3 - coupling channel
4 - new Bonn+MEC
5 - Skyrme
6 - PQCD
7 - six-quark

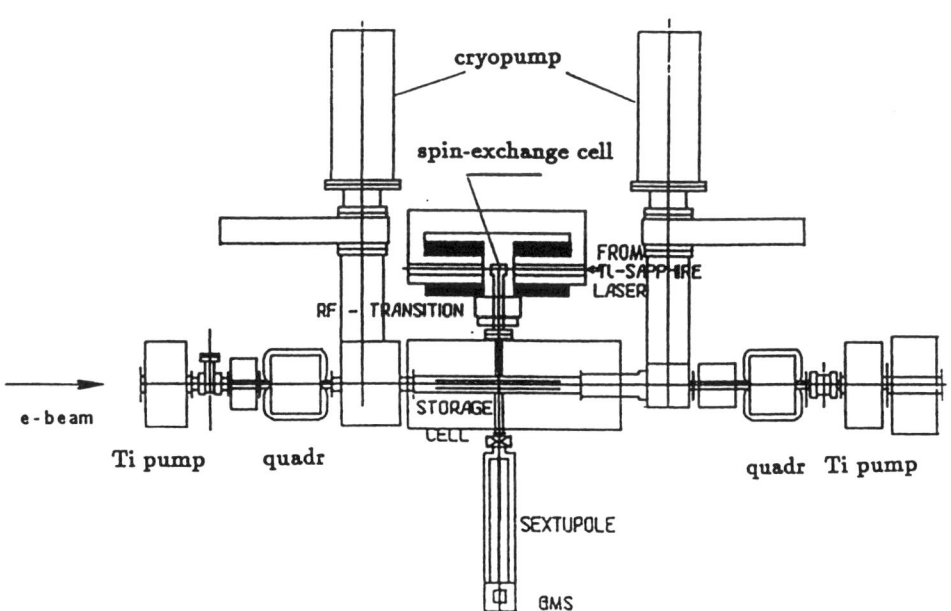

FIG. 2. Top view of the proposed straight section of VEPP-3 with spin-exchange laser driven source.

# STATUS OF THE ULTRA-COLD POLARIZED JET
# FOR NEPTUN AND NEPTUN-A AT UNK

R. S. Raymond

Randall Laboratory, University of Michigan Ann Arbor, MI 48109-1120 USA

## ABSTRACT

After tests of the prototype ultra-cold polarized jet, the jet assembly for use in the experiment NEPTUN-A at UNK is being designed, built, and tested at the University of Michigan. Planned improvements include a more powerful refrigerator, a 12 T high-gradient solenoid, and a superconducting focusing sextupole. In this talk the design and present status are described.

The experiments NEPTUN[1] and NEPTUN-A[2] at UNK, IHEP require a polarized hydrogen gas target. Expected beam size and limits on background gas density set by the experiments preclude the use of the storage cell technique to increase target thickness. We therefore decided to concentrate efforts on an ultra cold free jet.

In this type of atomic beam source atoms are cooled from about 25 K to about 0.3 K in a strong magnetic field gradient. For the present tests, for example, cooling took place at about 6 T in the gradient at one end of a solenoid with a maximum central field of 8 T. Once cooled, the thermal energy of the atoms, kT, is much lower than their magnetic energy, $\mu_e B$, so the field gradient separates the atoms according to electron spin direction. Atoms in the two upper hyperfine states are accelerated out of the field to form a beam, with an energy corresponding to a temperature of several kelvins.

For tests of this concept we used a prototype device[3]. The refrigerator in this device had a cooling power of 25 mW at 350 mK at the mixing chamber, inside of the solenoid. The beam transport and analysis systems consisted of a concave mirror, a focusing sextupole magnet and a compression tube. Drift lengths were similar to those required for the working target. The highest measured flow into the 0.5 cm$^2$ aperture of the compression tube was 3.7 x $10^{15}$/s, corresponding to a density of atoms of 3 x $10^{11}$/cm$^3$.

We know that the flow achievable in the present system is limited by, among other things, the refrigerator cooling power, the efficiency of cooling of the atoms, and transport efficiency of the beam. In order to increase focused beam density, cryogenic efficiency, and reliability of operation a new jet, shown in the Figure and intended for use at UNK, is being designed and built. Among the features of the new device are the following:

— a fully vertical design to match required experimental access.

— full shielding of liquid helium-temperature components by liquid nitrogen-cooled shields.

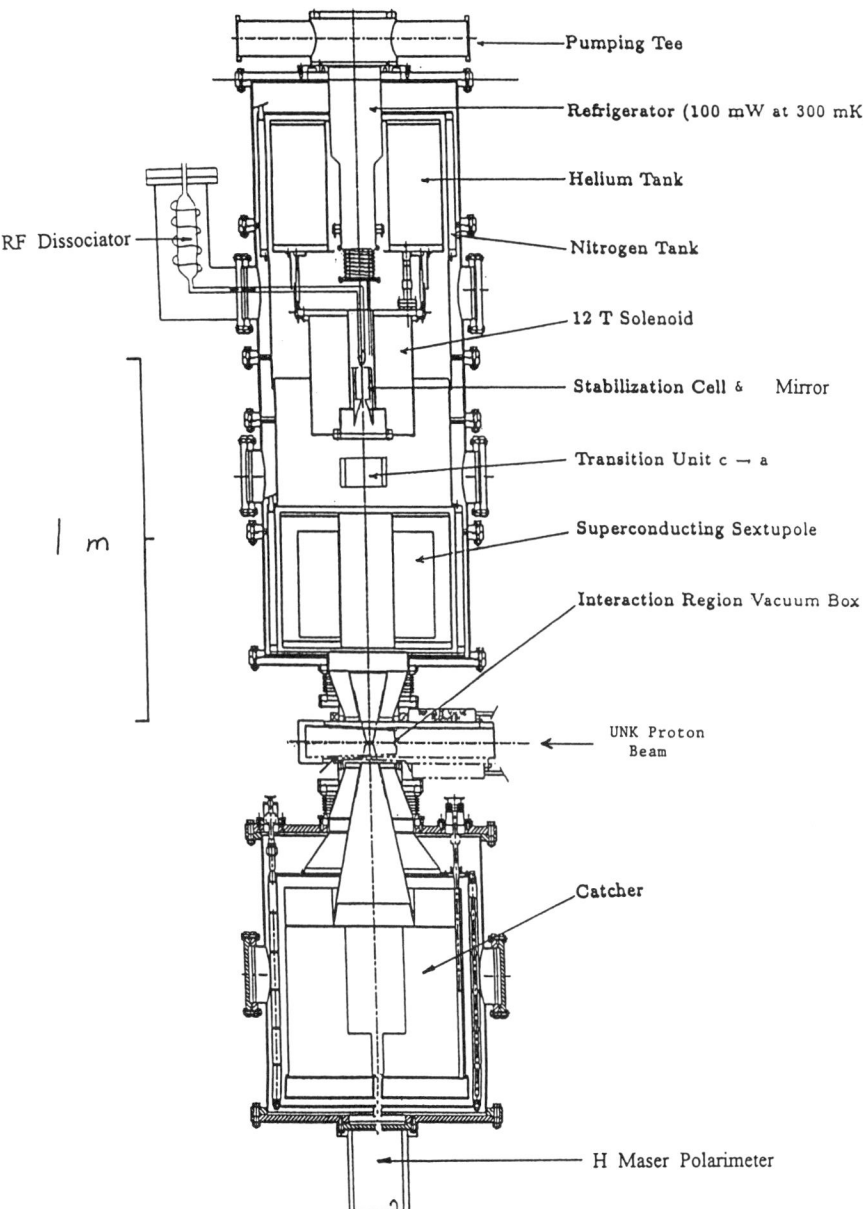

Figure 1. The layout of the polarized hydrogen jet for the experiments NEPTUN and NEPTUN-A at UNK, Protvino, Russia. Scale is approximate.

- a dilution refrigerator with a cooling power of 100 mW at 300 mK, circulating 30 millimole/s of $^3$He.

- a refrigerator mixing chamber designed to provide the maximum possible cooling of the exit aperture and mirror. The mirror shape itself is being designed using more complete calculations than were used for the test mirror.

- a separation solenoid with a maximum central field of 12 T, compared to 8 T used in the tests. This solenoid also has a bucking coil at the beam-output end, resulting in a fast fall off of the axial field. This allows the possibility of installing a 2 to 4 rf transition unit. The design of this unit is being studied.

- a focusing sextupole magnet, 20 cm long and having an 11 cm diameter bore, with superconducting coils on iron pole pieces. The maximum pole-tip field will be about 0.8 T. Use of a cold magnet allows cryopumping along most of the atomic beam path.

- a catcher, consisting of a large cryocondensation pump operating at 3 K. There will also be baffles and cryopanels above and below the UNK proton beam. With this arrangement we expect to achieve a background gas pressure in the interaction region of 3 $x$ $10^{-9}$ torr.

- a maser polarimeter below the catcher to monitor atomic state populations and thus electron and proton polarizations. Proton polarization will also be measured by measuring the elastic scattering asymmetry in the Coulomb-nuclear interference region.

With these improvements we expect to reach atomic densities of over 5 $10^{12}$/cm$^3$ in the area where the high energy beam crosses the atomic beam. We expect to have the atomic beam 2cm wide along the high energy beam direction, giving a target thickness of over $10^{13}$/cm$^2$.

Most of the vacuum chambers and the upper jet helium and nitrogen reservoirs have been built an tested. These were mounted and used last fall for a successful test of the 12 T solenoid. Parts of the refrigerator are being designed and manufactured now, with assembly to begin this summer and cryogenic testing expected to begin late this fall. The conceptual design of th sextupole is complete. Discussions with one possible vendor of the coils are continuing. The catcher design is complete and parts are being machined. The maser polarimeter is now undergoing final tests at MIT. Depending on how quickly tests of refrigerator can be completed, we hope to begin atomic beam tests early in 1994.

## ACKNOWLEDGEMENTS

This work is supported by the U. S. Department of Energy

REFERENCES

1  V. L. Solovianov, Proc. of the Workshop on the Experimental Program at UNK (Protvino, Russia 1987), p. 191.

2  A. D. Krisch, Proc. of the Workshop on Physics at UNK (Protvino, Russia 1989), p. 152.

3  T. Roser et al., Nucl. Instrum. Methods **A301**, 42 (1991).

# BEAM TRANSPORT OF LOW TEMPERATURE ATOMIC HYDROGEN

W. A. Kaufman

Randall Laboratory of Physics, University of Michigan, Ann Arbor, MI 48109

## ABSTRACT

Analytic calculations and particle tracking simulations[1] are presented for a polarized atomic hydrogen beam produced by extraction from an ultra-cold (T=300 mK) helium film coated cell in a large solenoidal magnetic field (12 T). Initial focusing of states 1 and 2 by the solenoidal field and subsequent focusing by a sextupole are examined within the constraints imposed by the requirements of the polarized jet for the experiments NEPTUN and NEPTUN-A at UNK.

## INTRODUCTION

The transport of a low temperature (T=300 mK) electron-spin polarized hydrogen beam formed in the gradient of a large solenoidal field (12 T) and subsequently focused by a sextupole is modeled analytically. The analytic model serves as a check on the more realistic particle tracking simulation presented.

The layout of the transport system (Fig. 1) is chosen to satisfy space requirements for a weak field transition unit and 3 K cryopumps along the beam path.

## THE ANALYTIC MODEL

The modeled system is a beam of state 1 and 2 atoms effusing from an aperture placed at 10 cm as measured from the magnet center. Here, the solenoid field (gradient) is 6 T (1 T/cm). The entrance face of the 20 cm long sextupole is placed at 35 cm. The sextupole field is adjusted to deliver the beam to the target region at 90 cm. For simplicity, the aperture is taken to be a monochromatic point source whose atoms' velocity (9360 cm/s) corresponds to the average velocity in a beam effusing from a 300 mK volume of gas.

The treatment of the solenoid field is radically simplified by setting the axial gradient to a constant. The solution to the atoms' equation of motion in the solenoid field[1] allows the radius of the trajectory to be expressed in terms of $z$:

$$\rho(z) = 2(z_2 - z_0)\frac{\dot{\rho}_0}{\dot{z}_m} \sin\left[\frac{1}{\dot{z}_m}\left(\sqrt{\dot{z}_0^2 + \dot{z}_m^2(\frac{z - z_0}{z_2 - z_0})} - \dot{z}_0\right)\right], \tag{1}$$

where the point source at $z_0$ is located in a field $B_0$, and $z_2$ is where the linear field falls to zero. The axial velocity that might be gained in the solenoid gradient is

$$\dot{z}_m = \sqrt{\frac{2\mu_B B_0}{m}}. \tag{2}$$

If the radial force is neglected entirely, the trajectories become parabolic.

After the solenoid field, the trajectories are straight lines that traverse a drift space, $z_2 < z < z_3$, until they enter the sextupole at $z_3$. A sharp edge approximation is assumed, so that no longitudinal components of the sextupole field are considered. The magnitude of this field depends on $\rho$ only,

$$B = B_6 \frac{\rho^2}{\rho_6^2}, \tag{3}$$

so that only a radial force acts on the atoms. $B_6$ and $\rho_6$ denote the sextupole pole-tip field and radius, respectively.

The state 1 equations of motion yield the well known result

$$\rho(z) = \rho_i \cos \kappa \left(\frac{z - z_i}{\dot{z}_i}\right) + \frac{\dot{\rho}_i}{\kappa} \sin \kappa \left(\frac{z - z_i}{\dot{z}_i}\right),$$

$$\kappa = 2\mu_B B_6 / m\rho_6^2, \tag{4}$$

where the initial values $\rho_i$, $\dot{\rho}_i$, $z_i$, and $\dot{z}_i$ refer to the trajectory parameters at the sextupole entrance. The motion of state 2 atoms in a sextupole field is not as simple; the dependence of the effective magnetic moment on the field cannot be ignored. State 2 equations of motion in a sextupole field are

$$\ddot{\rho} = -\kappa^4 \rho^3 / \sqrt{(\kappa\rho)^4 + (a/m)^2}, \tag{5a}$$

$$\ddot{z} = 0. \tag{5b}$$

Eqn. (5a) may be integrated once to yield

$$\dot{\rho}^2 = \dot{\rho}_i^2 + \sqrt{(\kappa\rho_i)^4 + (a/m)^2} - \sqrt{(\kappa\rho)^4 + (a/m)^2}. \tag{6}$$

It is simplest to solve Eqn. 6 numerically. Use of Eqns. 1,4 and the solution of Eqn. 6, the fact that the trajectories are straight lines in the drift spaces, and matching at the boundaries between the various regions, allows a full description of the trajectories emanating from a point source at $z_0$ and traversing the arrangement pictured in Fig. 1. A sextupole field $B_6(\rho = 5 \text{ cm}) = .473$ T focuses the paraxial state 1 trajectories to the target at $z = 90$ cm.

Fig. 2 displays state 1 trajectories emanating from a point source for several different initial angles. Two characteristics of Fig. 2 are noteworthy: the solenoid field acts to compress the trajectories by a large amount, while the sextupole focuses them, with a certain amount of resulting radial aberration at the paraxial focus. Fig. 3 displays state 1 and 2 trajectories with initial angles between 0° and 30°, in 5° increments. It was assumed that from the source to the sextupole they were identical, since an appreciable difference in the trajectories only occurs where the field is small ($B \leq 506$ G) and the gradients large. At angles greater than $\sim 30°$ the trajectories nearly coincide in the sextupole, whereas for smaller initial angles state 2 is more weakly focused than state 1. Some state 2 trajectories ($\theta_0 \leq 5°$) do not return to cross the magnetic axis. Provided that only the smaller angles are accepted, the sextupole acts as a nuclear spin filter, with state 1 atoms well focused in a relatively diffuse background of state 2 atoms.

## PARTICLE TRACKING SIMULATION

Populations of atomic trajectories in the solenoid-sextupole system were computed with a tracking simulation that numerically integrates the equations of motion using the Adams-Bashforth-Moulton "predictor-corrector" method.[2] The solution computed for the initial steps, via Runge-Kutta, is used to polynomially extrapolate (predict) the solution one step advanced; the extrapolation is then corrected using derivative information at the new point. An adaptive step size algorithm was used to keep the error within specified bounds. A substantial part of the tracking routine was based on a program written by Ellilä, Niinikoski, and Penttilä.[3]

A realistic solenoid field and its derivatives were produced numerically on a predetermined mesh,[4] and used as input to the tracking routine. Local values of the field and its derivatives were provided by spline interpolation. The sextupole field was generated analytically within the tracking program, and a sharp edge approximation used. The sextupole field was assumed to be entirely shielded from dilution by the solenoid field.

The initial data for the trajectories were generated by taking a random uniform deviate and, where feasible, applying the transformation method[2] to produce a random deviate with the desired distribution function. Otherwise, the rejection method[2] was employed. Specifically, the initial positions lay randomly and uniformly within a $.25 \times .75$ cm$^2$ rectangular aperture at $z = 10$ cm. The initial directions and speeds were chosen according to an effusive distribution of velocities at $T = 300$ mK. The resulting trajectories (5000 per simulation) were abandoned if they struck the effective 10 cm sextupole bore. A sextupole field $B_6(\rho = 5$ cm$)=.413$ T maximized the state 1 tranmission to the $0.5 \times 2.0$ cm$^2$ target region located at $z = 90$ cm.

Figs. 4 and 5 display hits in the target plane for simulations of state 1 and state 2 transport. The transmission of state 1 (2) was .21 (.14) to the target region. In a weak guide field the resulting nuclear polarization is .60.

The transmission is expected to improve with the inclusion of a helium film coated mirror in the transport system.[5]

This research was supported by the U. S. Dept. of Energy.

## REFERENCES

1. W. A. Kaufman, to be published in Nucl. Instr. Meth. A.
2. W.H.Press, B.P.Flannery, S.A.Teukolsky, W.T.Vetterling, *Numerical Recipes*, (Cambridge University Press, 1989).
3. M.Ellilä, T.O.Niinikoski, and S.Penttilä, Nucl. Instr. Meth. B14 (1986) 571.
4. See for example M.T.Menzel and H.K.Stokes, User's Guide to the Poisson/Superfish Group of Codes, Los Alamos Natl. Lab. publication LA-UR-87-115 (1987).
5. See V. G. Luppov, these proceedings.

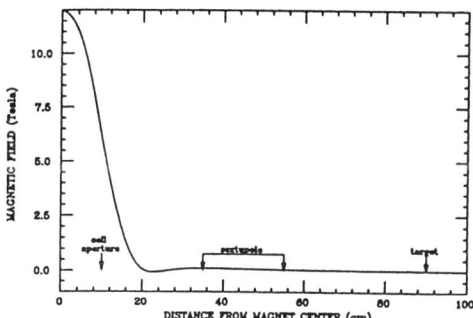

Fig. 1. The layout of the solenoid-sextupole system.

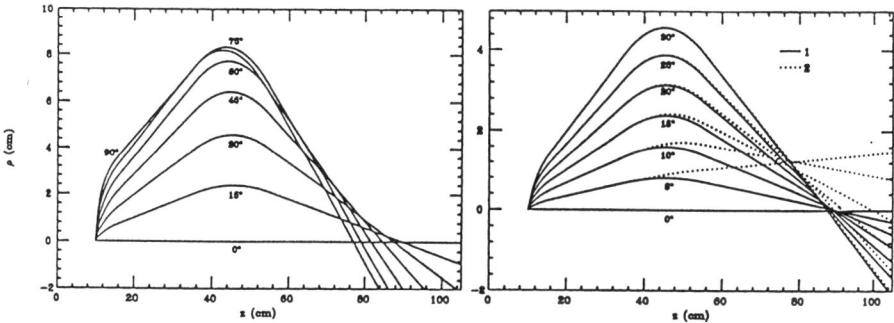

Fig. 2. State 1 trajectories' dependence on initial angle in 15° increments.

Fig. 3. State 2 (dotted curves) and state 1 (solid curves) trajectories' dependence on initial angle in 5° increments.

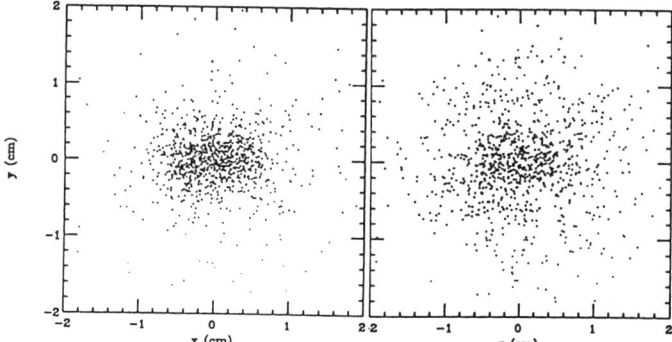

Fig. 4. Simulation of state 1 transmission to target.

Fig. 5. Simulation of state 2 transmission to target.

# A HELIUM FILM COATED QUASI-PARABOLIC MIRROR TO FOCUS A BEAM OF ULTRA-COLD SPIN POLARIZED ATOMIC HYDROGEN

V.G. Luppov*, W.A. Kaufman, K.M. Hill, R.S. Raymond, A.D. Krisch

Randall Laboratory, University of Michigan, Ann Arbor, MI 48109-1120 USA

## ABSTRACT

A 350 mK helium-4-coated mirror was used to increase the intensity of an ultra-cold electron-spin-polarized atomic hydrogen beam. The mirror uses the observed specular reflection of atomic hydrogen from a superfluid-helium-covered surface[1]. A quasi-parabolic polished copper mirror was installed with its focus at the 5 mm diameter exit aperture of an atomic hydrogen stabilization cell in the gradient of an 8 T solenoid field. The four-coned mirror shape, which was designed specifically for operation in the gradient, increased the beam intensity focused by a sextupole magnet into a compression tube detector by a factor of about 7.5.

Using the Michigan prototype jet[2] we investigate a relatively new ultra-cold method which uses a temperature about 0.3 K and a magnetic field above 5 T to produce a high density electron-spin-polarized atomic hydrogen beam. Depolarization and recombination into molecular hydrogen are strongly suppressed because the average thermal energy is much too small to flip the electron spin. The method, called "no microwave" extraction, uses the sharp gradient of a strong magnetic field to separate the cold hydrogen atoms of different electron spin states.

The Michigan prototype jet using the "no-microwave" extraction is shown in Fig. 1.

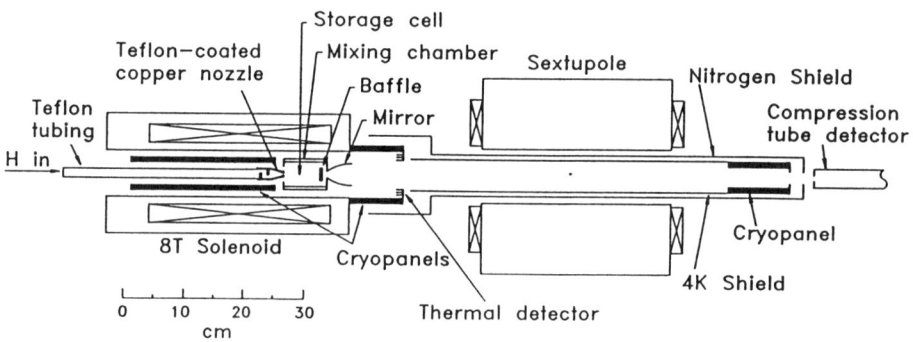

FIG. 1. Schematic diagram of the Michigan prototype ultra-cold spin-polarized atomic hydrogen jet.

The atomic hydrogen was produced in a room temperature rf dissociator and guided to an ultra-cold stabilization cell through a teflon tube with a teflon coated copper nozzle held at about 20 K. The double walls of the cell formed the horizontal mixing chamber of the dilution refrigerator; its cooling power was about 25 mW at 350 mK. The cell's entrance and exit apertures were respectively located at 95% and 65% of the central field of the 8 T superconducting solenoid. The cell typically operated at a temperature of 350 mK, and was completely covered with a superfluid $^4$He film to suppress the surface recombination of hydrogen atoms.

After the hydrogen atoms were sufficiently thermalized by collisions with the cell surfaces, the magnetic field gradient physically separated the atoms according to their two different electron-spin states. The atoms in the two lowest hyperfine states (high field seekers) were attracted toward the high field region. Most of these atoms eventually escaped from the cell through the 50 mm$^2$ annular gap around the entrance nozzle. These atoms then recombined on bare surfaces; the resulting molecular hydrogen was pumped away by cryopanels and other cold surfaces. The atoms in the two higher energy hyperfine states (low field seekers) were repelled toward the low field regions, where they collected and then effused from the 5 mm diameter exit aperture. After emerging from the exit aperture, the electron-spin-polarized atoms were magnetically accelerated by the remaining field gradient.

We measured the extracted atomic hydrogen beam flux, using a compression tube (CT) detector mounted downstream of both the cell and the sextupole magnet. A tracking simulation for our prototype jet with a 7.3 T solenoid field indicates that typically 1% of the beam effusing from our ultra-cold cell can be focused into the CT by our sextupole. Increasing either the sextupole aperture or the solenoid field would improve the beam transport, but requires a serious modification of the apparatus.

The quantum reflection of cold hydrogen atoms from a helium-film-covered surface below 0.5 K was demonstrated by Berkhout et al.[1] They measured about 80% specular reflectivity for normal incidence on a hemispherical optical quality concave quartz mirror coated with a 100 mK saturated $^4$He film.

We designed a "parabolic" mirror to use the specular reflection as an "atomic optics" focussing technique in our ultra-cold spin-polarized atomic hydrogen jet target. Assuming specular reflection and a point source, a parabolic mirror should form a parallel beam of monochromatic atomic hydrogen. Such a mirror could significantly increase the beam available for focussing by a sextupole magnet. Used quasi parabolic mirror has a special shape taken into account that the magnet field gradient accelerates atoms and bent them trajectory before reflection. The copper four-coned approximation mirror was installed with its focus at the 5 mm diameter exit aperture of the atomic hydrogen stabilization cell (see Fig. 2).

We first made baseline measurements with no mirror. The maximum signal was observed at the largest central solenoid field of 7.3 T (see Fig. 3); this gave the largest gradient which increased both the electron spin separation inside the cell and the solenoid fucusing outside.

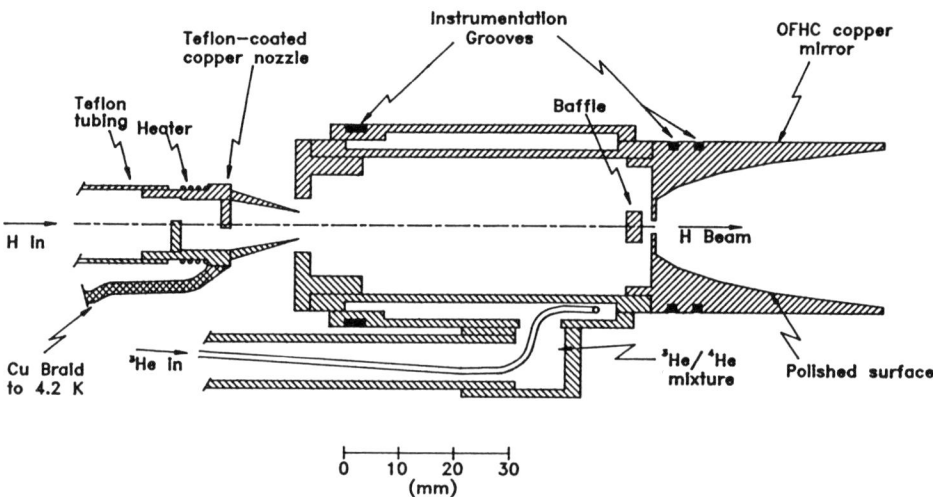

FIG. 2. Schematic diagram of the stabilization cell and mirror. The Teflon-coated copper nozzle is also shown.

The measured CT signal for the mirror polished with a 1,000 nm abrasive versus the solenoid magnetic field are plotted in Fig. 3. At a field of 7.5 T the four-coned mirror's CT signal was $2.4 \cdot 10^{15}$ H s$^{-1}$, this was about 4.8 times larger than the extrapolation to 7.5 T of the signal with no mirror.

For formation of a low divergence beam using the specular reflection, the mirror surface should be at least no rougher than the thickness of the helium film on the surface, which is typically about 10 nm. Therefore, we repolished the mirror surface with a 50 nm, alumina suspension; this significantly reduced the surface roughness. With this repolished surface the measured peak CT intensity at 8 T increased to $3.7 \cdot 10^{15}$ H s$^{-1}$. Comparison with the no mirror baseline extrapolation to 8 T indicates that the repolished mirror increased the measured beam intensity by a factor of about 7.5.

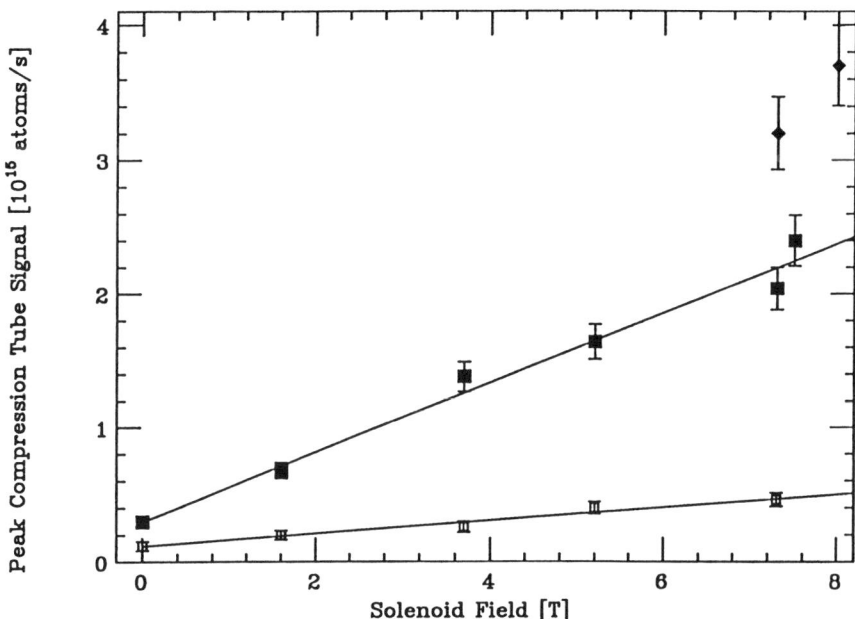

FIG. 3. The peak compression tube signal at each optimal sextupole current is plotted versus the solenoid field. The open squares are with no mirror; the black squares are the four-coned mirror with a rougher surface; the black diamonds are with the repolished four-coned mirror. The curves are linear least square fits to the data.

In summary, this work demonstrates the possibility of forming an ultra-cold atomic hydrogen beam by using a polished mirror coated with a superfluid $^4$He film. We plan to design and test a field gradient mirror for our new 12 T solenoid.

ACKNOWLEDGEMENTS

This work was supported by a research grant from the U.S. Department of Energy.

REFERENCES

* Permanent address: Joint Institute for Nuclear Research, Dubna, Russia.

1. J. J. Berkhout et al., Phys. Rev. Lett. **63**, 1689 (1989).

2. T. Roser et al., Nucl. Instrum. Methods **A301**, 42 (1991).

# SIMULATION OF GAS DYNAMICS IN ATOMIC BEAMS USING A MONTE CARLO PARTICLE METHOD

Iain D. Boyd
Cornell University, Ithaca, NY 14853

Werner Kubischta
CERN, Geneva, Switzerland

## ABSTRACT

Computational results obtained using a Monte Carlo particle method are presented for flows typical of the nozzle-skimmer configurations employed to generate polarized beams of atomic hydrogen. The effects of flow rate and nozzle wall temperature on the terminal state of the atomic beam are investigated. The Monte Carlo method is used to compute velocity distributions. Comparison with experimental data shows qualitatively good agreement. Requirements for improvement of the present numerical results are discussed briefly

## INTRODUCTION

Modern experimental facilities for generating polarized beams of atomic hydrogen operate under the combined conditions of low-densities and small geometric length scales[1,2]. These conditions produce flow in the transition regime between the continuum and free-molecular limits where the reduced number of collisions may create nonequilibrium energy distributions. The degree of nonequilibrium in such flows is characterized by the Knudsen number, Kn: the ratio of the mean free path to a characteristic length. At large values of Kn (>0.01) nonequilibrium phenomena are significant. This is certainly true of the atomic hydrogen beams discussed in Refs. 1 and 2. Flows with high Knudsen numbers are also of interest to the aerospace community for high-altitude hypersonic flows (e.g. the Space Shuttle), and for low-thrust satellite propulsion. A very robust and successful numerical technique has been developed to simulate the nonequilibrium gas dynamics that occurs under these nonequilibrium flow conditions. This is the direct simulation Monte Carlo method (DSMC)[3]. It is the purpose of this paper to describe application of the DSMC method to the problem of atomic hydrogen beam formation under conditions similar to those investigated experimentally in Refs. 1 and 2.

The DSMC technique simulates the large number of bodies in the real gas (e.g. $10^{20}$ molecules) by a much smaller collection in the computer (e.g. 1 million model particles). Each simulated particle possesses individual properties including 3 velocities (u, v, w), up to 3 spatial coordinates (x, y, z), and internal energy (rotation, vibration). The particles move through a computational grid in physical space. Time is discretized into small steps over which collisions are computed among the particles occupying each computational cell. When 2 particles interact, elastic collision mechanics are simulated in which the velocity vectors of the 2 particles are re-calculated under the assumption of momentum and energy conservation. In addition, models for simulating inelastic behavior such as rotational and vibrational excitation, and chemical reactions, have been included in the DSMC formulation[4]. The particles also collide with solid surfaces. This is usually accomplished using the simple specular and diffuse reflection models. The numerical algorithms for the DSMC method are under constant development and reflect new opportunities in computer hardware. A successful implementation on a vector machine has been reported[5] and parallel architectures are currently under investigation.

## COMPUTATIONS

In the present study, a 2-d axisymmetric nozzle-skimmer configuration is modeled using the vectorized DSMC code described in Ref. 5. A schematic diagram of the flow geometry is given in Fig. 1. Particles are introduced into the simulation along the inlet boundary AB. Particles traveling out of the computational domain across any of the boundaries AB, CD, DE, or EA are removed. It is assumed at the nozzle inlet that the hydrogen is fully dissociated with a temperature of 340 K. Intermolecular collisions are simulated using a temperature-dependent form for the cross-section[6]:

$$\sigma = \sigma_\infty (1 + T^*/T), \tag{1}$$

where $\sigma_\infty$=4.52x10$^{-20}$ m$^2$ and $T^*$=31 K. This is a Sutherland-law potential with an equivalent energy-dependent form for DSMC[7]:

$$\sigma = \sigma_\infty (1 + 6kT^*/E_c), \tag{2}$$

where $k$ is Boltzmann's constant and $E_c$ is the energy of the collision. The interaction of particles with both the nozzle and skimmer walls is simulated using the diffuse model with complete thermal accommodation to the temperature of each surface. It is assumed that the beam expands into a perfect vacuum.

Parametric studies for the flow of atomic hydrogen are performed by variation of the temperature of the nozzle wall, $T_w$, and the nozzle flow rate, $\dot{q}$. Room temperature (300 K) and a very low temperature (40 K) are the values investigated for the nozzle wall temperature, representative of the ranges used in Refs. 1 and 2. The skimmer surface temperature is always taken to be 300 K. Four different atomic flow rates in the nozzle inlet are investigated: 0.25, 0.50, 1.00, and 1.50x10$^{20}$ atoms/s. To achieve a given flow-rate in the simulation, a small stream velocity and corresponding density are chosen as input parameters. Generally, this procedure produces flow rates lower than desired due to the escape of particles backwards across the inlet plane. The density is therefore increased until the flow-rate given by the balance of the atoms entering and leaving across the inlet plane corresponds to the required value.

The simulations typically employ 5,250 computational cells and 250,000 particles. To obtain flow field solutions requires about 15 minutes on a Cray Y/MP supercomputer. More detailed results such as velocity distribution functions require much longer simulations to be performed in order to reduce statistical noise in the computed data.

## RESULTS AND DISCUSSION

A contour plot of number density for $\dot{q}$=0.50x10$^{20}$ atoms/s and $T_w$= 300 K is shown in Fig. 2. Note that the computational domain is extended to distances of 15 cm from the nozzle exit, and 6 cm radially from the axis. In the expansion of the gas through the skimmer and beyond, the decay in density on the axis is observed to obey the familiar inverse square relationship with distance.

One interesting aspect of the computations is the effect of the nozzle wall temperature. At $T_w$=300 K, the gas and surface are at similar temperatures. However, simulating the nozzle as a low temperature atomic accommodator significantly alters the flow field. The very cold temperature acts as an energy sink through which the velocity of the gas is severely retarded. Simultaneously, the density in the nozzle is increased dramatically due to the slower moving gas.

Thus, for the same flow rate condition, the character of the gas at the exit of the nozzle is quite different for the two values of $T_w$ investigated. Specifically, at $T_w=40$ K, the Knudsen numbers are about 5 times smaller than when the nozzle wall is at room temperature. The local Knudsen numbers obtained at the nozzle exit for each of the conditions investigated are shown in Fig. 3 as a function of flow rate. These results show that, at the same flow rate, nonequilibrium collisional phenomena are significantly affected by the nozzle wall temperature.

The forward intensity of the beam close to the axis at a distance of about 10 cm from the nozzle exit is shown in Fig. 4 as a function of flow rate for the two wall temperatures studied. In each case, there is close agreement and the increase of intensity with flow rate is linear. The curve for purely effusive flow is also shown in Fig. 4 and is about 50% of the DSMC results. Note that the quantitative results of the present study may be affected by including the influence of the finite back-pressure of the vacuum chambers.

The DSMC computation provides very detailed information on the gas at the microscopic level. As each simulated particle possesses an individual vector of velocity components, it is possible to extract distribution functions directly from the simulation. For each computation performed, the distribution of the axial velocity component is obtained near to the flow axis at a distance of about 10 cm from the nozzle exit. The distribution computed in this manner for $\dot{q}=0.50\text{x}10^{20}$ atoms/s and $T_w=300$ K is shown in Fig. 5. The velocity is normalized by $\zeta$, the most probable thermal speed at the nozzle wall temperature. Included in Fig. 5 is the velocity distribution predicted by simple effusive theory for a gas expanding from an orifice at the nozzle wall temperature. Note that the peak of the effusive distribution occurs at $v/\zeta=1$. For the case considered in Fig. 5, the flow rate is relatively low and the nozzle wall temperature is close to the inlet gas temperature. Under these conditions, the distribution predicted by DSMC is in close agreement with that predicted by the simple theory.

The velocity distribution obtained at $\dot{q}=1.5\text{x}10^{20}$ atoms/s and $T_w=40$ K is shown in Fig. 6. Here it is observed that the DSMC distribution is shifted to higher velocities and is thinner than the effusive result: both of these trends are in qualitative agreement with phenomena reported experimentally in Refs. 1 and 2. The departure of the distributions from the effusive model may be characterized in terms of the most probable velocity ($v_{pp}$) and the width of the distribution (FWHM), $\Delta v$. The present numerical results for these quantities are compared with the experimental data from Ref. 1 and separate data taken at CERN (Ref. 2) † in Figs. 7 and 8 for the 2 different nozzle wall temperatures. In Figs. 7a and 7b are shown the most probable velocities found in the computations and experiments divided by the effusive result. Generally, the DSMC results are in qualitative agreement with the experimental data: the most probable velocity increases with flow rate. Quantitative differences may be attributed in part to differences in geometry and species. In Figs. 8 the widths of the distributions are compared. Again, DSMC gives the correct trends: as the flow rate is increased, the width of the distribution decreases by a small amount. Note that the experimental data of Ref. 1 shows the opposite trend at $T_w=40$ K.

The changes in the peak velocity and width of the distributions are caused by two effects. First, the gas is not fully accommodated to the temperature of the nozzle wall. There are particles close to the flow axis which never collide with the cool nozzle, and these are characterized by the much higher inlet temperature

---

† Data from Ref. 1 are for molecular beams and slightly smaller nozzle and skimmer; Ref. 2 used a rectangular nozzle and skimmer.

of 340 K. Second, collisional effects close to the nozzle exit lead to expansion cooling that is the critical feature of supersonic beams.

The present DSMC computations are encouraging in that detailed nonequilibrium phenomena observed experimentally are predicted. There are three major areas of concern in these first calculated results. There is some uncertainty over the precise form of the collision cross-sections for atomic hydrogen, particularly at low energies. The temperature-dependent nature of these flow conditions is clear from Figs. 8. The velocity distribution functions computed by the DSMC technique will be quite sensitive to the exact form of the cross-sections employed. Secondly, the role of the model employed for the interaction of the gas with the solid surfaces of the nozzle and skimmer walls is clearly of great significance. Indeed, for these flows, this aspect of the simulations is of equal importance to the intermolecular collisions. The sensitivity of the computed results to the use of more sophisticated wall models should be investigated. Finally, the influence of the background gas in the vacuum chambers must be included in the numerical analysis. This is an extremely important aspect of these types of flows. The finite pumping capacity of the experiments means that the data reported in Ref. 1 was particularly affected at the higher flow rates. Including the effects of background pressure in vacuum chamber experiments has been performed in a previous DSMC study of jet flows from small helium rockets[8]. It is therefore a goal for future work to include such effects in the atomic beam simulations.

## REFERENCES

1. D. Singy, P.A. Schmelzbach, W. Gruebler and W.Z. Zhang, Nuc. Instr. A278 349 (1989).

2. E. Bosisio, Tesi di Laurea, Univ. di Milano, 1991 (unpublished).

3. G.A. Bird, *Molecular Gas Dynamics*, (Clarendon, Oxford, 1976).

4. I.D. Boyd, Phys. Fluids A 4 178 (1992).

5. I.D. Boyd, J. Comp. Phys. 71 217 (1991).

6. J.M. Dickson, Progress Nucl. Techn. Inst. 1 105 (1965).

7. I. Kuscer, Physica A 158 784 (1989).

8. I.D. Boyd, Y.R. Jafry, and J. Vanden Beukel, J. Spacecraft (in press).

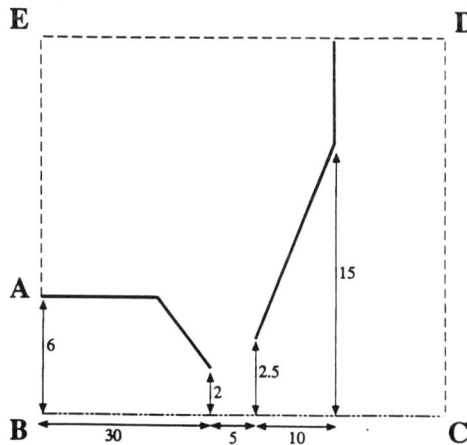

Fig. 1. Nozzle-skimmer geometry: vertical lengths are radii (dimensions in mm).

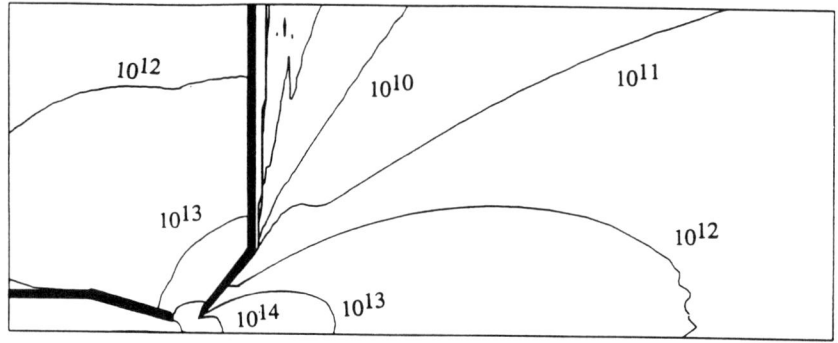

Fig. 2. Number density contours (atoms/cm$^3$):   $\dot{q} = 0.50$ x $10^{20}$ atoms/s,  Tw=300 K.

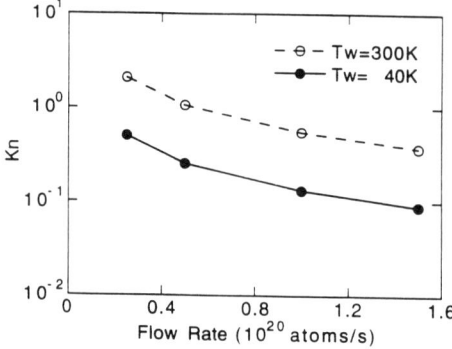

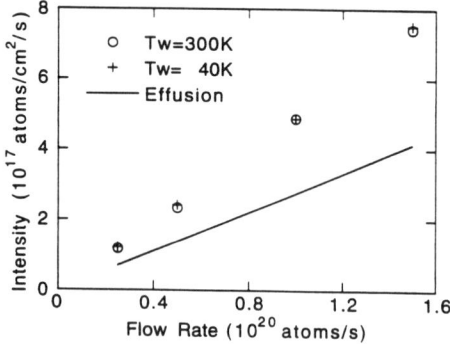

Fig. 3. Variation with flow rate of the Knudsen number at the nozzle exit.

Fig. 4. Variation with flow rate of forward intensity at a distance of 10cm from the nozzle.

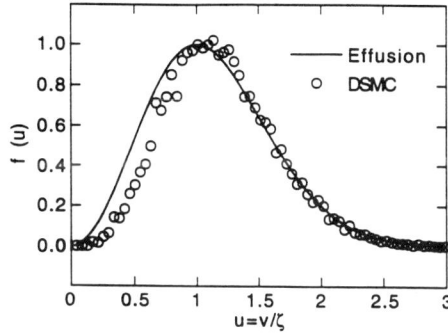

Fig. 5. Velocity distribution on the axis for $\dot{q}$=0.5x$10^{20}$ atoms/s and Tw=300 K.

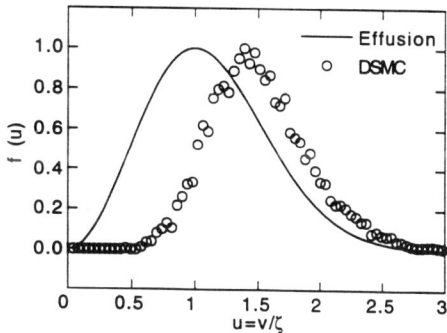

Fig. 6. Velocity distribution on the axis for $\dot{q}$=1.5x$10^{20}$ atoms/s and Tw=40 K.

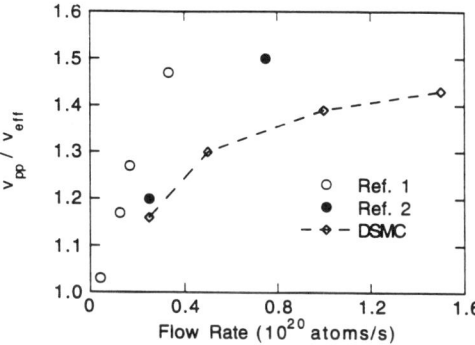

Fig. 7a. Variation of most probable velocity with flow rate at Tw=40 K (see footnote on experiments).

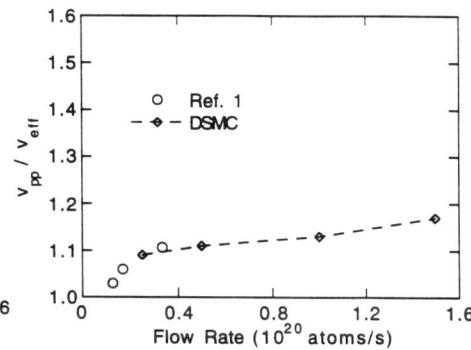

Fig. 7b. Variation of most probable velocity with flow rate at Tw=300 K.

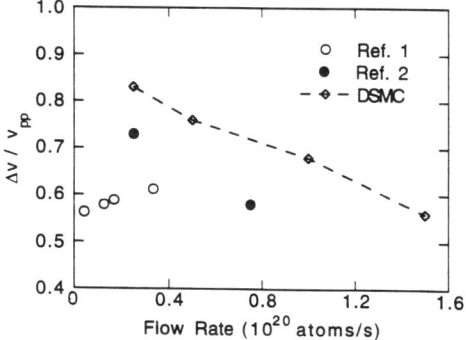

Fig. 8a. Variation of velocity distribution width with flow rate at Tw=40 K (see footnote on experiments).

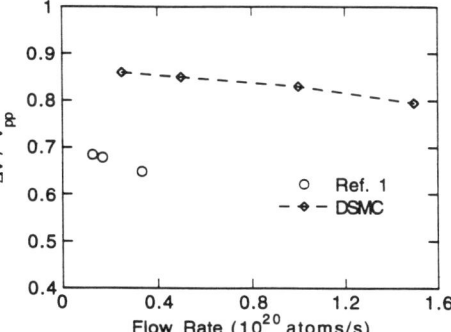

Fig. 8b. Variation of velocity distribution width with flow rate at Tw=300 K.

# ACHIEVEMENT OF TRANSPORTABLE POLARIZED D, IN SOLID HD, WITH A ONE DAY PASSIVELY MAINTAINED POLARIZATION

A. Honig, N. Alexander[*], Q. Fan, X. Wei and Y. Y. Yu[**]
Physics Dept., Syracuse University, Syracuse, NY 13244

## ABSTRACT

At a previous workshop[1], we discussed evaporating solid HD with spin-polarized deuterons to produce a high density polarized deuteron gaseous internal target. Since then, we have achieved in solid HD 38% polarized D, whose spin-lattice relaxation time at 1.5K in a field of 0.1T is of the order of a day. Optimization of the procedure with the present apparatus should result in 60% D polarization, and longer polarization holding times. The polarized sample of approximately 0.2 cm$^3$ volume used here is extractable from the dilution refrigerator with a cold-transfer apparatus which maintains the sample at or below 5K, insuring retention of the high polarization. It is subsequently insertable into a variety of systems, and employable as a polarized solid, liquid or gas. We are exploring the possibility of extending the polarization maintenance time to about a month (with a matched 1 month preparation time), of polarizing metastably H as well as D, and of producing much larger samples, of the order of 100 cm$^3$.

## INTRODUCTION

Spin-polarized solid HD has many advantages as a polarized target. It is at present the simplest molecule available for polarization, has the largest percentage of free protons among current competitors, allows independent polarization direction reversal of H and D which can result in an effectively polarized neutron target, and can be made very large since the polarization is passively maintained. Although temperatures near 10mK and magnetic fields near 15T are necessary for polarization production, the resultant high polarization can be maintained for long times at temperatures near 4K and fields of only about 0.1 T, due to the long relaxation times (polarization maintenance times). A relaxation switch[2] which permits polarization in an acceptable time at mK temperatures and subsequently "turns off" enabling the polarization to endure at 1 - 4K and modest 0.1T fields was proposed in 1967. Before that and in the years since, many spin-relaxation studies were undertaken encompassing the parameter space of T, B, and the J = 1 concentrations of hydrogen (ortho-$H_2$) and deuterium (para-$D_2$) impurities, $c_1^H$ and $c_1^D$ respectively, which serve as relaxation agents for the bulk of the H and D in the HD molecules which are essentially all in the molecular rotational J = 0 state in our operational T < 5K region. These studies enable us to optimize the polarized targets, and recent development of means of cold-transferring[3] (at 4K) targets out of dilution refrigerators and into any other desired apparatus makes the system deployable. At the time of the 1988 Eighth International Symposium on High Energy Spin Physics, sufficient work had been done under the strong impetus of

* Present address: General Atomics, San Diego, CA 92186
** Present address: Intel Corp., Chandler, AR 85226.

utilizing polarized D in solid HD as a nuclear fusion fuel, that we explored the idea of gasifying it to produce a high density polarized internal beam target[1]. Liquifaction of solid polarized $D_2$ has also been demonstrated[4]. In this report, we emphasize the present status of metastably polarized D in HD, and comment on efforts underway to produce a much larger and more highly polarized (both H and D) target, in connection with the BNL Laser-Electron Gamma Source (LEGS)[5] and the GRenoble-Anneau-Accelerateur-Laser (GRAAL)[6] programs. A demonstrated large volume, long duration, highly polarized H and D solid HD sample, easily transportable to various apparatuses, may have a strong impact on polarized internal beam work, on solid polarized targets and on polarized-fuel fusion in magnetic-confinement fusion machines, in addition to the already successful mating to an inertial-confinement-fusion (ICF) chamber[7].

## CURRENT STATUS OF POLARIZED TARGET DEVELOPMENT

A comprehensive description of the polarization method has recently been presented[8]. Here we give an account of the production of an actual polarized sample[9].

The starting HD solid is condensed from doubly distilled, very pure (of $H_2$ and $D_2$) HD which is intentionally doped to a $c_1^H$ of about $5 \times 10^{-4}$, to insure a relaxation rate not much lower than $1 \ d^{-1}$ at the polarization conditions near 10mK and 12 - 13 Tesla. The corresponding $T_1^H$ at 1.3K and 0.1T, $T_1^H[1.3;0.1]$, is about 20 seconds, $c_1^H$ is about $3.5 \times 10^{-4}$ and $c_1^D$ is about $2.5 \times 10^{-4}$. These dependences were inferred from relaxation measurements on incrementally doped samples[10]. The $c_1$'s can be reduced by J = 1 to 0 conversion at liquid helium temperatures, which proceeds exponentially in solid HD with temperature-independent rates of $0.157 \ d^{-1}$ and $0.055 \ d^{-1}$, respectively for $H_2$ and $D_2$. They can also be increased by back-conversion at room temperature. At 1.2K and 0.1T for H, and 1.2K and 0.6T for D (same resonance frequency), convenient reference points for a target, and within a $[c_1^H, c_1^D]$ domain applicable to this experiment, we found

$T_1^H[1.2;0.1]$ approximately proportional to $[(c_1^H)^{-2.1} (c_1^D)^{-1.6}]$,

and $T_1^D[1.2;0.6]$ proportional to $[(c_1^H)^{-1.0} (c_1^D)^{-3.9}]$, which combined with the exponential decay of the $c_1$'s, gives us $T_1^H[1.2;0.1]$ increasing as (exp 0.418 t) and $T_1^D[1.2;0.6]$ increasing as (exp 0.372 t), with t in days. After 15 days at liquid helium temperatures, $T_1^H[1.2;0.1]$ and $T_1^D[1.2;0.6]$ are expected to increase by factors of 530 and 270 respectively. $T_1^H$ may only be a few hours at the end of this experiment, but $T_1^D$ exceeds a day.

Fig. 1 shows the time line for the polarization production. After loading the sample, the temperature is lowered to 9mK and the field is raised to 12T, foregoing in this early effort raising the field to its maximum 13T value by pumping the helium bath to 2K. After 4.5 days under the polarizing conditions, the field is lowered to 0.6T and then to 0.1T, where the D and H NMR signals are measured, giving the acquired polarizations. It is evident we began with too long a $T_1^H[1.2;0.1]$ value (48 s), since the H polarization reached only 55% of its equilibrium value. For D, the acquired polarization was 87% of its equilibrium value. The field is next lowered, and a forbidden-

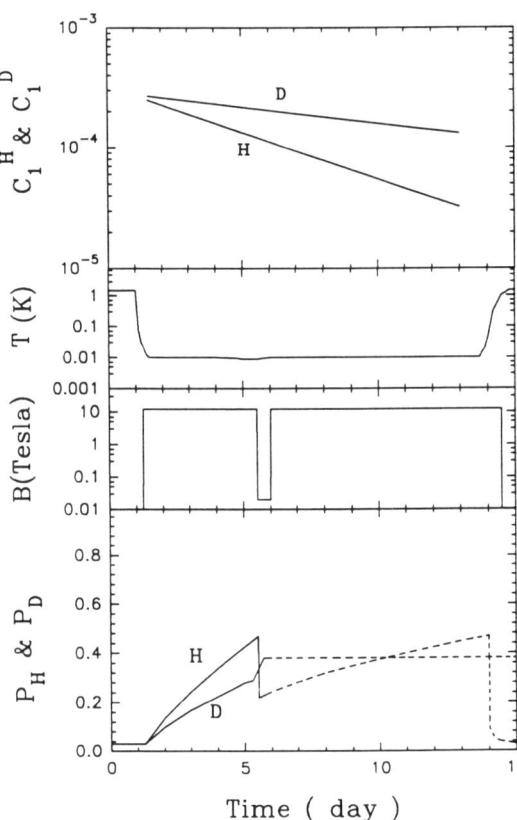

Fig. 1. Time line of polarization
process.

Time ( day )

transition adiabatic rapid passage[11] at 0.0329T effects a polarization transfer from the H to the D, reducing the former and enhancing the latter. NMR of H and D are quickly measured, and we return to the high fields, where relaxation times are long. The H partially repolarizes and the D retains most of its (above equilibrium) polarization. The dotted lines show qualitatively what is expected during the aging process of the next 9 days. $T_1^D$ and $T_1^H$ were actually measured at 1.2K at the end of the run, but in the region of the dotted lines, complicated field and temperature maneuvers (not shown) took place to survey additional relaxation times. The success of the experiment is evidence that the proton relaxation time does not increase exponentially with decreasing temperature as was conjectured in earlier experiments[12], and that the D relaxation time does not rise as fast at that of H with decreasing T and increasing B. The 38% $P^D$ is a fair value, but starting with a higher $c_1^H$ and waiting a few extra days would have resulted in more than a 30% increase in $P^H$, and a 4% increase in the static D polarization build-up. That, and using the full 13T field, brings the $P^D$ to near 60%. Additional gain in $P^D$ is expected from lowering the temperature to an already achieved 8 mK, and increasing the efficiency of the forbidden-transition adiabatic rapid passage from its present value of 50% of the theoretical maximum.

The removal of the sample is carried out with a cold-transfer apparatus which has recently been described[3] in connection with small ICF target shells. For larger samples, gas bags of several mm diameter are available which can be permeated with hydrogens and then cryocondensed, or sample can be loaded into target holders, aerogels, or polymer foams. A quickly evaporated polarized solid $^3$He target may be amenable to a protocol similar to that used with solid HD. The size of removable polarized samples is limited by refrigerator geometry, but also by the J = 1 to J = 0 conversion heat, mostly from the hydrogen impurities. This amounts to 1.3 µW/mole at $c_1^H$ = 5 X $10^{-4}$. Our refrigerator can remove 2 µW at 10 mK. Removing the conversion heat from the interior of the sample must also be done, with thin conducting wires or possibly with $^3$He mixed with the HD.

It is apparent that longer production times lead to longer final relaxation times, although there is uncertainty as to the validity of the rate of $T_1$ increase with decreasing $c_1$'s at the lower $c_1$ values, and uncertainty regarding whether another relaxation mechanism might impose an upper limit on $T_1$'s. If the target is used in a process where the polarization is not rapidly destroyed by the beam, such as the LEGS[5] or GRAAL[6] experiments, then 1 month preparation for one month usage is a feasible strategy. We are conducting experiments in which the relaxation times are monitored over a long period without the sample being removed from the helium bath, so as to avoid back-conversion problems. In addition, we plan to acquire a new dilution refrigerator with 10 $\mu$W heat-removal capability at 8 mK and a 5 cm bore access accommodating a sample of about 5 moles, together with a 17T magnet. H polarizations to 95% and D polarizations above 70% should be attainable. In addition to providing a desirable target for the LEGS experiments, such a sample may be useful as a long term supply of internal polarized target, by chipping off and using small volumes, as suggested in Ref. 1.

This work has been supported by the U. S. Dept. of Energy under Grant DE-FG03-92SF19201.

## REFERENCES

1. A. Honig, Workshop for Polarized Gas Targets for Storage Rings, Intl. Conf. on High Energy Spin Physics, Minneapolis, 1988. AIP Conf. Proc. No. 187, 1554 (1989).
2. A. Honig, Phys. Rev. Lett. 19, 1009 (1967).
3. N. Alexander, J. Barden, Q. Fan and A. Honig, Rev. Sci. Instrum. 62, 2729 (1991).
4. Y. Y. Yu, X. Wei and A. Honig, Bull. Am. Phys. Soc. 36, 1354 (1991)
5. The LEGS collaboration consists principally of Brookhaven Nat'l. Lab. (A. M. Sandorfi et al) and Univ. of S. Carolina (S. Whisnant et al). Highly polarized and monochromatic gamma radiation is produced by backscattering of laser radiation from a high energy electron beam.
6. The GRAAL and LEGS proposed project with our polarized HD solids is an international collaboration, guided by J. P. Didelez, of Orsay.
7. A cold-target (unpolarized) has been transported from Syracuse and mated in target position to the OMEGA ICF fusion chamber at Laboratory for Laser Energetics (LLE) at the Univ. of Rochester. Polarized shots unfortunately will not be possible there for several years, since the machine is undergoing an upgrade.
8. A. Honig, N. Alexander and S. Yucel, Muon Catalyzed Fusion Workshop, Sanibel Is.,FL,May 1988. AIP Conf. Proc. No. 181,199(1989)
9. For a preliminary report, see N. Alexander, Y. Y. Yu, Q. Fan, X. Wei and A. Honig, Bull. Am. Phys. Soc. 35, 925 (1990).
10. H. Mano. PhD thesis, Syracuse University, 1978. Unpublished.
11. A. Honig and H. Mano, Phys. Rev. 14, 1858 (1976).
12. J. A. Brown, PhD thesis, State University of New York at Stony Brook, 1977. Unpublished. [Univ. Microfilm, Ann Arbor, MI 48106].

# POLARIZED INTERNAL TARGETS FOR
# ELECTRONUCLEAR EXPERIMENTS

J.F.J. van den Brand
University of Wisconsin-Madison, Madison, WI 53706

## ABSTRACT

Polarized internal gas targets represent a unique opportunity for the measurement of spin observables in electro-nuclear physics. Two measurements will be discussed. Firstly, spin observables have been measured in elastic and quasi-free scattering of 45, 200, 300, and 415 MeV polarized protons from a polarized $^3$He internal gas target at the Indiana University Cyclotron Facility Cooler Ring. The data obtained constitute the first measurement of spin correlation parameters using a storage ring with polarized beam and polarized internal gas target. Secondly, a quasi-free (e,e'p) experiment using tensor polarized deuterium will be discussed. Here, the goal is the measurement of the S- and D-state parts of the proton spectral function by scattering 700 MeV electrons off polarized deuterium. A storage cell is fed by deuterium atoms from an atomic beam source. Large acceptance detectors have been used in both experiments. The internal-target technique has broad applicability in nuclear and particle physics.

## INTRODUCTION

Electron scattering is a powerful tool for a precise study of hadronic structure. The interaction takes place through the exchange of a virtual photon. The leptonic tensor can be treated exactly using QED and the virtual-photon hadron interaction is generally described in the impulse approximation. The mean free path of the photon is long compared to that of hadronic probes which enables a study of the entire nuclear volume. However, the cross sections are generally small which explains why few experiments have been performed using polarized electrons and polarized targets[1,2].

One of the goals of electron scattering studies is to understand hadrons in terms of QCD where the constituents are pointlike quarks and gluons. Here, there are only two degrees of freedom: flavor and spin. By adding spin measurements to the study of hadronic structure one obtains two important advantages. Firstly, the total number of independent observables is increased, and secondly because of interference effects some of these observables show enhanced sensitivity to important small components of nucleon and nuclear structure. The significance of these advantages has been recognized by the electron scattering community and measurement of spin observables forms a principal ingredient in the physics programs at CEBAF, DESY, MAINZ, MIT-Bates, NIKHEF and SLAC.

The charge form factor of the neutron is a good example of an observable which may shed light on the importance of quarks-gluon versus nucleon-meson degrees of freedom. In the latter framework the charge distribution partly arises from a $|p\pi^- >$ wave function component while in the former picture it is due to the color-hyperfine interaction in the one-gluon exchange potential. Various measurements have been performed on this fundamental quantity but the present error bars ($\approx 50$ %) on the data exclude any definite conclusions. In the absence of a free neutron target the data have been obtained from a subtraction of deuteron and proton cross sections. In unpolarized electron scattering the neutron charge form factor (estimated at maximum to be about $G_E^n = 0.05$, compared to $G_E^p = 1$) enters squared into the cross section which makes the subtraction method extremely susceptible to systematic errors. With spin-dependent electron scattering using reactions such as $^2\vec{H}(\vec{e}, e'n)$ and $^3\vec{H}e(\vec{e}, e')$ one can measure[3]

$$W_1^{TL'} = -2\sqrt{2}F_{C0}F_{M1} \tag{1}$$

which involves the interference between two form factors. In quasi-free scattering the asymmetry will be sensitive to the product of $G_E^n$ and $G_M^n$. This constitutes a powerful measurement since one deals with the product of form factors where one is of small magnitude whereas the other is large.

Polarized internal gas targets in storage rings offer significant promise for measurement of spin observables in nuclear and particle physics[4] in that they realise the interaction between polarized beam and polarized target in the most ideal way. They comprise a source which generates a flux of chemically and isotopically pure polarized nuclei directed into a windowless conductance limiter (the storage cell), through which the circulating beam of the storage ring passes. This cell increases the dwell time of the polarized atoms near the interaction region and significantly enhances the target thickness. Since the target nuclei are present as pure polarized atomic species there is no dilution of the beam-target asymmetry. The target polarization is high ($> 50\%$), is directed with low ($\sim$ 10 gauss) holding fields, and reversible on a rapid time scale ($\sim$ seconds). Consequently, the systematic errors associated with this technique are small. In addition, the storage cell wall is thin which allows for the detection of heavily ionizing recoil particles. Further, because the luminosities associated with polarized internal targets are relatively low compared to external experiments, large acceptance detectors can be used to detect scattered particles over a broad kinematic range. Experiment[5] CE-25 at IUCF demonstrated all the major aspects of this powerful technique. Subsequently, experiment[6] 91-12 attempts to measure target analyzing powers in an electron scattering environment. Next I will discuss both experiments in some detail.

## EXPERIMENT CE-25 AT IUCF

Measurements were performed at elastic and quasi-free scattering kinematics at proton incident energies of 45 MeV (unpolarized beam), and 200, 300 and 415 MeV (polarized beam). Here the spin-dependent cross section for both beam and target spins oriented normal (n) to the scattering plane can be written as

$$\sigma = \sigma_0(1 + A_{00n0}P_b + A_{000n}P_t + A_{00nn}P_bP_t),\qquad(2)$$

where $P_b$ and $P_t$ are the polarizations of the beam and target, $A_{00nn}$ is the spin-correlation parameter, and $A_{00n0}$ and $A_{000n}$ are the beam and target analyzing powers, respectively. The experiment was carried out at the IUCF Cooler Ring.

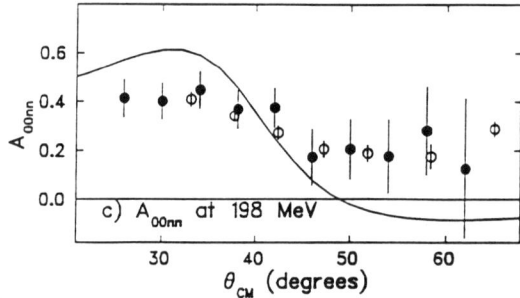

Fig. 1. The spin correlation parameter $A_{00nn}$ for p - $^3$He elastic scattering at 200 MeV. The present experiment is indicated by the full circles; the data given by the open circles correspond to Ref. 8. The curve is a prediction[9] from Ray.

To carry out the experiment a new type of polarized $^3$He internal gas target was developed[7]. The storage cell is an open-ended tube of 40 cm length and of rectangular cross-section ($13 \times 17$ mm$^2$) with sides of 1.7 $\mu$m thick aluminized mylar foil. The flow rate of $1.2 \times 10^{17}$ atoms/s resulted in the world's most dense polarized internal gas target with a thickness of $1.5 \times 10^{14}$ atoms cm$^{-2}$. The average polarization in the pumping cell was $0.45 \pm 0.02$ and was measured using detection of the 667 nm line in the $^3$He discharge. The target polarization was reversed every 300 s by reversing the circular polarization of the laser light. In addition, the orientation of the 10 gauss magnetic field was reversed approximately every hour.

The spin observable $A_{00nn}$ at 200 MeV incident proton energy is shown as a function of $\Theta_{CM}$ in Fig. 1 in comparison to data[8] obtained at TRIUMF using a spin-exchange polarized $^3$He target. The agreement between both data sets is good. The curve corresponds to a calculation[9] using the reaction model by Ray. The disagreements indicate that for these kinematics a full four-body calculation may be required.

The spin correlation parameter $A_{00nn}$ measured in experiment CE-25 constitutes the first measurement of spin observables with a polarized beam incident on a polarized internal target. A more detailed account of the experiment has been given by Sowinski elsewhere in these proceedings. For a discussion on the polarized target I refer to the contribution by Milner.

## EXPERIMENT 91-12 AT NIKHEF

Here, I report on a study of the deuteron spin structure by scattering electrons with energies up to 700 MeV off a polarized internal deuterium gas target in the Amsterdam Pulse Stretcher Ring (AmPS). Electrons and protons from quasi-free scattering will be measured in coincidence to determine the tensor target analyzing power as a function of missing momentum and missing energy. The tensor analyzing power is related to the ratio of the S- and D-state momentum distributions. The spin structure of the deuteron ground-state wave function is a central question in intermediate energy nuclear physics. Such an understanding is crucial when one intends to extract experimental information on the charge and spin structure of the neutron by employing spin-dependent lepton scattering off polarized deuterium.

Polarized deuterium nuclei from an atomic beam source are fed into an open-ended storage cell, cooled to 80 K, operated in a strong magnetic holding field ($\approx$ 300 G) for maximum tensor polarization. A flux of $1.6 \times 10^{16}$ deuterium atoms $s^{-1}$ into a 15 mm diameter feed tube has been measured[10]. The scattered electrons are detected by a CsI-calorimeter, covering a scattering angle range of 20° to 50°. The calorimeter consists of 60 CsI blocks, arranged in six layers. Each block has dimensions of $15 \times 6 \times 6$ cm$^3$. The detector covers a solid angle of 140 msr and is mounted below the scattering chamber. Only the central quarter of the acceptance will be used in order to limit leakage of the shower out of the sides of the calorimeter. Simulations studies predict an energy resolution of 5 %. Two plastic scintillators, 50 and 10 mm thick, are used to define the trigger signal. Vertex reconstruction is done by two sets of wire chambers with a wire spacing of 2 mm. The knocked out protons are detected with a range telescope, consisting of 15 layers of plastic scintillator $30 \times 50$ cm$^2$, each with a thickness of 10 mm, preceded by one layer of 2 mm, so that deuterons can be separated from protons. The instrument is preceded by two sets of wire chambers. The range telescope covers a solid angle of nearly 300 msr. It will stop protons with an energy up to 150 MeV and has an energy resolution of about 3 %.

To measure the polarization of the deuterium nuclei in the storage cell, several devices have been or are being constructed: a Breit-Rabi polarimeter, a Balmer polarimeter and a T$_{20}$ polarimeter. The measurement of the average degree of polarization over the storage cell is essential to the determination of the analyzing powers. Several test measurements were performed with a stored electron beam to obtain information on the spatial distribution of the stored beam, so that the diameter of the storage cell could be selected. For this I refer to the contribution by de Jager elsewhere in these proceedings.

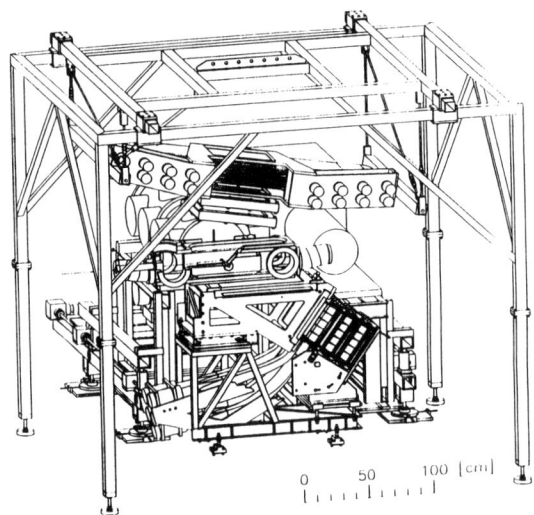

**Fig. 2.** Internal target setup at the AmPS electron scattering facility at NIKHEF.

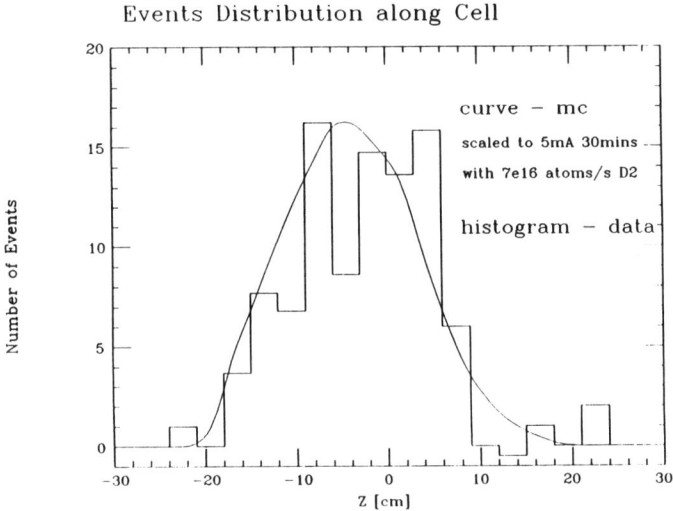

**Fig. 3.** Distribution of the (e,e′p) coincidence events as function of their vertex position. The curve is the result of a Monte Carlo calculation which takes the dependence on cross section, target density distribution, and detector acceptances into account.

   The complete set-up, including a 20 mm diameter storage cell, has been tested with a stored beam at an energy of 400 MeV, an injected intensity of 20 mA and a life time of 300 s. All detection systems and the ABS were operated under realistic beam conditions and coincident electron-proton events were detected. Fig. 3 shows the distribution of the (e,e'p) coincidence events as function of their vertex position. The shape of the distribution reflects the triangular target density distribution of the 400 mm long storage cell weighted by the scattering cross section. Negative vertex positions correspond to the upstream part of the storage cell. Here the average scattering angle is smaller which results in a higher cross section.

In summary, the first experiment with a polarized beam incident on a polarized gas target internal to a light-ion storage ring has been successfully completed. Preparation for a subsequent internal target experiment in an electron scattering environment is ongoing. We have shown that the internal-target technique leads to practically background-free measurements of undiluted asymmetries. This work has significant implications for future experiments at nuclear and high-energy physics facilities.

The author's research is supported by the National Science Foundation under Contract No. PHY-9019983.

## REFERENCES

1. C.E. Jones *et al.*, Phys. Rev. **47**, 110 (1993).
2. A.K. Thompson *et al.*, Phys. Rev. Lett. **68**, 2901 (1992).
3. T.W. Donnelly, in the book '*Modern Topics in Electron Scattering*', editors Frois and Sick; World Scientific 1992.
4. K. Lee, M.A. Miller, J.-O. Hansen, A. Smith, J.F.J. van den Brand, H.J. Bulten, R. Ent, W.W. Jacobs, C.E. Jones, W. Korsch, W. Lorenzon, D. Marchlenski, H.O. Meyer, R.G. Milner, J.S. Neal, S.F. Pate, W.K Pitts, B. von Przewoski, J. Sowinski, F. Sperisen, E. Sugarbaker, C. Tschalaer, O. Unal and Z.-L. Zhou

   *Phys. Rev. Lett.* **70**, 738 (1993)
5. J.F.J. van den Brand and J. Sowinski, Cooler ring proposal CE-25 for IUCF (1990).
6. J.F.J. van den Brand and C. de Jager, Electron Scattering proposal 91-12 for NIKHEF-K, Amsterdam (1991).
7. K. Lee, J.-O. Hansen, J.F.J. van den Brand, R.G. Milner, MIT-Bates preprint, August 1992, accepted for publ. in Nucl. Instr. and Meth.
8. O. Hausser, Proceedings of the 7th International Conference on Polarization Phenomena in Nuclear Physics, Paris, July 9 - 13 (1990); Edited by A. Boudard - Y. Terrien.
9. L. Ray *et al.*, Phys. Rev. **C37**, 1169 (1988).
10. Z.-L. Zhou, private communications (1993).

# REPORT OF THE PANEL DISCUSSION ON
# "ATOMIC BEAM AND LOW TEMPERATURE TARGETS"

E. Steffens

Max-Planck-Institut für Kernphysik

D-69029 Heidelberg

Panel members: A. Honig/Syracuse, W. Kubischta/CERN, V. Luppov/Michigan and Dubna, E. Steffens/Heidelberg (Chair), D. Toporkov/Novosibirsk, T. Wise/Wisconsin

The subject of the discussion was the production of polarized hydrogen and deuterium atoms by in principle three different methods:

1. spin-dependent focussing of a cold atomic beam in the strong field of multipole (e.g. sextupole) magnets ("Atomic Beam Source" ABS);

2. separation of the fine structure components of ultra-cold atoms (kT $<<$ $\mu$B) in the fringe field of a superconducting solenoid ("Ultracold Source");

3. evaporation of solid polarized HD.

The polarized atoms may be used in two different ways as a target for an intense electron or ion beam stored in a ring:

- A dense spot is produced by strong focussing of the atoms. Such a target is very attractive because narrow obstructions are absent.

- Atoms are injected into a storage cell and confined by its thin walls, thus giving rise to an increase in areal density along the beam axis by about two orders of magnitude.

The target proposed for the HELP experiment at LEP was introduced by W. Kubischta. Cold atoms from a dissociator are focussed by two large-bore superconducting sextupoles to a dense target spot. In order to get maximum overlap with minimum gas load, a slit nozzle ($1 \times 4mm^2$) is employed. The areal density expected is $10^{13}/cm^2$ within 4 cm in length. A crucial point is the efficient dumping of this intense gas flow. The expected background pressure in the target is $5 \cdot 10^{-7}mb$. From the length of about 15 cm seen by the detector, a background rate of about 4% of the total one is calculated. It was pointed out that this high gas load may present a problem to low energy storage rings. Only prototype studies of some source components have been done so far because approval is still lacking. The anticipated working point of the dissociator is 3 $mb\ell/s$ and 40 K. The most critical point seems to be whether the required efficiency of 90% of the beam catcher can be obtained or not.

The least understood design aspect of an ABS are intensity limitations and how to overcome them. D. Toporkov gave a possible interpretation of measurements[1] performed at the FILTEX ABS[2]. The pressure rise in each of the differential pumping stages showed a linear (2nd stage) or a saturation type (3rd and 4th stage) dependence on the nozzle throughput. These curves reflect the gas flow of the beam entering the different stages, which strongly decreases further downstream. This may be due to intrabeam scattering (IBS) caused by the broad velocity distribution within the atomic beam. Apart from attenuation, IBS might also modify the velocity distribution, which could explain the discrepancies seen between measurements and the predicted performance based on velocity measurements. No such effects are seen in the Monte Carlo calculations on gas dynamics during expansion by I.D. Boyd[3].

Depolarization caused by the walls of the storage cell was discussed by T. Wise. A clear dependence of depolarization on the strength of the guide field has been seen in the study of the Wisconsin group. Most interesting is the finding that on an uncoated Al surface about 70% of the polarization is conserved in strong field, compared with 30% in weak field. This may allow us to use uncoated cells if coatings are not stable, e.g. teflon coating in an electron ring.

The status of the work at Dubna on the ultracold source[4] was summarized by V. Luppov and compared to the Michigan results[5]. The Dubna set-up does not include any focussing yet, so it is difficult to compare their result of $10^{17}$ H-atoms/pulse (300 ms) detected by a thermal detector close to the cold cell and the Michigan $3.7 \cdot 10^{15}$ H/s dc-beam focussed into a compression tube of $0.5 cm^2$ opening area. Improvement of the present density of $3 \cdot 10^{11}/cm^3$, which is still below the present FILTEX density of $5 \cdot 10^{11}/cm^3$, is anticipated for the new Michigan Mark-II source with 12 T solenoid and a large-acceptance sextupole magnet for focussing. Densities up to $5 \cdot 10^{12} H/cm^3$ are expected, which would result in a target density of $\sim 10^{13} H/cm^2$. But it was pointed out that there are possible limitations from the dilution fridge cooling power and the build-up of solid hydrogen, and experimental confirmation is necessary.

Finally the discussion turned to the proposed target based on evaporation of solid polarized HD samples[6], which are polarized by "brute force" typically at T = 20 mK and B = 13 T. Proton polarization in excess of 50% can be achieved whereas the deuteron spins in solid HD have low polarization. High deuteron polarization may be obtained by means of adiabatic fast passage.

Polarized gas may be introduced into the storage cell by evaporation of frozen pellets, e.g. at $T \approx 7K$ with $3 \cdot 10^{16}$ mol./s or larger fluxes at higher temperature. Wall depolarization or recombination may be substantial at these low temperatures. - The present apparatus at Syracuse is able to polarize 0.2 moles in 30 days, which corresponds to an average flow rate of $5 \cdot 10^{16}$ mol./s. Much higher flows are anticipated for a new apparatus (5 moles/30 days, $10^{18}$ mol./s), which is well matched to the needs of polarized gas targets. As some of the basic steps still need to be demonstrated, a proof-of-principle experiment is urgently required.

The discussion has shown that atomic beam sources, in connection with storage cells, yield high target densities and very high polarization with maximum flexibility in the choice of different polarization states. The ultracold sources have achieved good progress, but the present results are still a factor of about 30 below the design value of $10^{13}/cm^2$. The evaporation method has still not been applied and needs experimental confirmation.

I am indebted to Kirsten Zapfe for taking notes during the discussion. I would like to thank the members of the panel for their cooperation.

## REFERENCES

1. D. Toporkov, in Proc. Workshop on Pol. Gas Targets for Storage Rings, Heidelberg 1991, H.G. Gaul, E. Steffens and K. Zapfe (Eds.), MPI Heidelberg (1991), p. 79.

2. W. Korsch, in Proc. High Energy Spin Physics, Bonn 1990, Vol. II: Workshops, W. Meyer, E. Steffens and W. Thiel (Eds.), Springer (1991), p. 168.

3. I.D. Boyd: Description of a Monte-Carlo Method for Simulation of Nonequilibrium Gas Dynamics, these proceedings.

4. Yu. Pilipenko et al., contributed paper, these proceedings.

5. V. Luppov, A He Film Coated Quasi-Parabolic Mirror ..., these proceedings.

6. A. Honig, in Proc. High Energy Spin Physics, Minneapolis 1988, K.J. Heller (Ed.), AIP Conf. Proc. No. 187 (1989), p. 1554.

# II. ATOMIC BEAM POLARIZED ION SOURCES

# ECR IONIZERS AND OPERATING EXPERIENCE AT PSI

P. A. Schmelzbach
Paul Scherrer Institute, CH-5232 Villigen-PSI, Switzerland

## ABSTRACT

The basic features of the ECR ionizer for the atomic beam polarized ion source are remembered. The present situaton is analysed in considerations based on the operating experience at PSI and on recent progress in the understanding of the ECR discharge. Possible improvements in the design of a second generation of ECR ionizers are suggested.

## INTRODUCTION

The increasing number of ECR ionizers operating at present sources or currently in the planning stage shows that this device is becoming a standard piece of equipment for modern atomic beam polarized ion sources. The performances achieved so far certainly justify this choice. However, one should realize that this technique is very young and that the optimal design has not been yet realized. While ionization efficiency and beam quality correspond to our wishes, the beam polarization partly failed to reach the standard previously set by sources equipped with a strong field electron bombardment ionizer or by colliding beam systems. There is therefore room for improvements since the problems encountered with the first generation of ECR ionizers are of technical nature and do not question the validity of the basic approach. Unfortunately, progress is made difficult by the fact that some operating parameters are almost definitively fixed by the design options. The optimization of elements important for a good beam polarization like layout of the vaccuum system or basic magnetic field configuration requires a new construction and test of major components of the device. In addition, criteria to define the best approach were missing. For these reasons the performances of the ionizers currently in operation have not significantly improved since the last report at the Tsukuba Workshop in 1990 [1].

The object of this paper is to analyse parts of the knowledge accumulated during the operation of one of the first device of this type and to propose some improvements to be implemented in a second generation of ECR ionizers. First a few general principles will be reminded.

## BASIC PRINCIPLES

The function of an ECR ionizer is identical to that of an ECR ion source, from which it has been derived. A plasma is confined in a magnetic field with a minimum B structure generated by the superposition of an axial field with a suitable mirror ratio and a multipolar (generally hexapolar) radial field. In such a structure, there exists closed surfaces with constant values of the total field. If a microwave field of a suitable frequency is injected into the system, the condition for electron cyclotron resonance is satisfied on such a surface. Upon crossing this zone the electrons become stochastically heated to a temperature up to several tens of keV. High energy of the electrons and long storage time of the ions allowing for a stepwise ionization to high charge states are mandatory for an ECR heavy ion source. Microwave frequencies above 10 GHz, microwave powers around one kW and B fields up to one T are typical figures for this kind of devices. Conversely, low charge states can be efficiently produced in lower frequency systems, typically 2.5 to 6.4 GHz. This type of source is especially characterized by a high gas efficiency and a good quality of the extracted beams.

The present designs are to a large extent the result of an empirical approach because several aspects of the function of the ECR source are not yet fully understood. For example, it was generally accepted in the past that in such a device the plasma chamber acts as a multimode cavity. In fact it has been recently demonstrated by measurements of the plasma density that the latter follows the amplitude distribution of a strongly dominant mode and that the location of the amplitude maxima with respect to the resonance surface strongly influences the performance of the source [2].

The feasibility studies [3,4] and the design of the first ECR ionizer [5] based on operating parameters of small ECR sources like Jülich pre-ISIS1[6]. In several points, however,  major departures from the basic construction are inavoidable to integrate this technique in a polarized ion source. The systems remains nevertheless so similar that the ECR ionizer can benefit from any step in the understanding and in the development of the ECR ion source.

## SPECIFIC PROBLEMS

The much fundamental differences between a ECR heavy ion source and a ECR ionizer for an atomic beam polarized ion source are as follow:
- Only the ionization of hydrogen atoms is of interest. There is no need for very hot electrons, a maximum of the energy distribution below 1 keV being optimal. Therefore microwave of low frequency (typically 2.5 Ghz, the corresponding B-field at resonance being 87.5 mT) and moderate magnetic fields are suitable for this purpose, with the advantage of low cost and lower level technology.
- The atoms to be ionized are introduced into the ionizer by the means of an atomic beam. The flow is obviously not a free parameter, the density is marginal compared to the operating condition in a ECR ion source (pressure of several $10^{-6}$ mbar) and its distribution may cause matching problems with the plasma geometry best suited to optimal conditions of the plasma discharge.

- The atomic beam is produced in an apparatus at ground potential, enters the ionizer through a large aperture, is partly ionized in a plasma at a potential of several kV, and, to avoid recombination on the wall of the plasma chamber and on the extraction system, leaves the system through very large apertures, also. The main problems associated with this geometry are: possible plasma discharge and extraction of ions towards the atomic beam apparatus, difficulty to design an optimal beam optical system at the extraction, microwave leakage in the pumping compartment (cryopumps !).

- The ionizer has to preserve polarization. While the occurence of rf transitions as the atoms are moving through the resonance zones can be ruled out, the need to reduce the effect of transverse components set limits on the hexapol field at the atomic beam location. This, again, results in matching problems with the optimum plasma geometry. A further problem of the mirror field configuration is the shielding of the rf transitions units against the stray field from the ionizer.

- The ECR sources are working at pressure of the order of $10^{-6}$ mbar or higher, generally with a large amount of buffer gas. As a result, the total extracted beam reaches several milliampères. If the polarized source is installed on the injection axis of the accelerator, i.e. without beam analysis after the source, a background beam one order of magnitude larger than the beam to be transported to the machine is not acceptable. The ECR ionizer should therefore work at a very low pressure, where experience shows that the smallest disturbancies, arising for example from parasitic discharges in the extraction system, are sufficient to extinguish the plasma discharge.

In the next sections, the PSI ionizer is taken as an example to present possible solutions to the abovementioned problems and to illustrate the influence of a particular design on the performance of the source.

## THE PSI ECR IONIZER

The PSI ECR ionizer [1,7] is a final version of the prototype developped in collaboration with KfK [5]. The mechanical layout is shown in fig. 1. The vaccum chamber surrounding the plasma region is a 3 mm thick Pyrex tube of 8.4 cm inner diameter. A copper cylinder 10 cm in diameter and 23 cm long forms the microwave cavity. Cooling is provided by air blown between cavity wall and Pyrex tube. The microwave power (20-30W, 2.46 GHz) is feeded by a conical wave guide between two pole pieces of the hexapol magnet. Stricking features are the large diameter (5 cm) of the beam extraction system, which allows for the free passage of the atomic beam, and the fact that the plasma potential is defined solely by the first electrode.

The ionizer is connected to the atomic beam apparatus by a 30 cm long, 2 cm diameter glass tube surrounded by three rf transition units. This (traditional) configuration is unfavourable for the transmission of the atomic beam, but the very long insulated section has the advantage of inhibiting the extraction of ions towards the atomic beam apparatus. In the KfK/PSI prototype where the distance to the next grounded element was shorter a small permanent dipole magnet just at the entrance of the ionizer was used with some success to suppress or, at least, to locally dump the beam extracted in this direction.

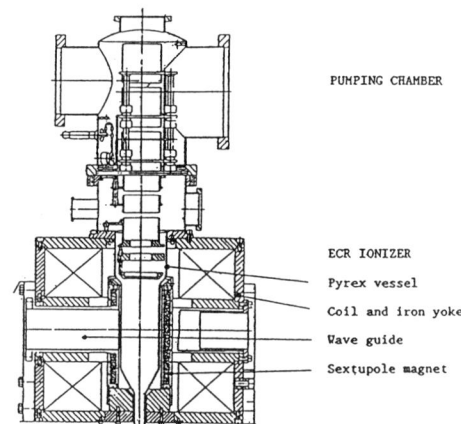

The pressure measured at the chamber just above the extraction is typically $3 * 10^{-8}$ mbar rest gas, $10^{-7}$ mbar with atomic beam on and $1.5 * 10^{-7}$ mbar with the addition of the $N_2$ buffer gas. A very stable plasma discharge is observed even at this low pressure. To achieve this result special attention was paid during the development of the ionizer to configure the extraction and focussing system in such a way that parasitic Penning discharges, electrode loading and secondary electron backstreaming are minimized.

Fig. 1. The PSI ECR ionizer with the associated pumping chamber.

A plot of the measured total magnetic field is shown in fig. 2 together with a sketch of the central part of the ionizer. The shape of the axial mirror field was optimized on the prototype ionizer by varying the distance between the solenoid coils. In the final version, the current in the entrance and exit solenoid coils can be adjusted independently. The transversal components at the location of the atomic beam are smaller than 10 mT.

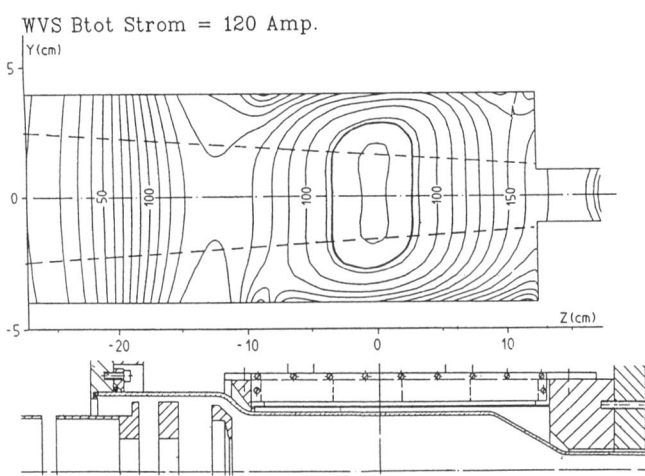

Fig. 2. Plot of the total magnetic field in the PSI ECR ionizer. The central resonance surface at 87.5 mT is indicated by the thicker curve. The dotted lines indicate the maximum possible radial extension of the atomic beam. The detail of the ionizer construction is shown at the horizontal scale.

The 3 electrode accel-decel extraction system shown is suitable for extraction of beams larger than 1 mA. For operation at low pressure, the total extracted current is lower and the best results are obtained with the second electrode closer to the plasma electrode and set at a positive potential   slightly higher than needed for a constant field between electrodes 1 and 3.

The interpretation of the influence of the magnetic field structure on the source performances  is model dependent. Since the extension to the ECR ionizer of the model of ref. 2 shortly introduced above leads to a better consistensy in the analysis of the observed behaviour it will be accepted for the following discussion.

## THE BEAM INTENSITY

The intensity expected from a ECR ionizer have been discussed by Clegg et al. in previous papers [3,4]. By comparing average values of the electron density it is found that an ECR ionizer operated at 2.5 GHz would have a 3 times higher ionization efficiency than the ANAC electron beam ionizer[8] which previously equipped the PSI polarized ion source. In fact the beam intensities measured along the axial injection line to the Philips Cyclotron are 2 to 2.5 times  higher, but the significantly smaller energy spread allows for an improved bunching factor and a better injection efficiency. The performance of the system source+cyclotron has increased by a factor 3 to 6 depending on the beam energy.

The dependence of the beam intensity on the axial mirror field was investigated in a range of settings varying up to 25 % from the optimum values. The stability of the beam is excellent at the axial mirror field for maximum intensity. This stability is conserved in a large domain if the change in the excitation of the solenoid coils results in a simultaneous shift of the position of the resonance zones on the axis. A deviation from the 6 cm axial distance between the resonance zones reduces the beam intensity dramatically and introduces violent and fast instabilities causing a very noisy  beam. Obviously the best location of the resonance zones follows the $\lambda/2$ rule of ref. 2. A further evidence that the highest ionization efficiency is concentrated on the resonance surface rather than in the enclosed volume is given by the beam intensity dependence on the hyperfine state. This modulation is distorted by an additional effect apparently due to the contribution of the lateral part of the resonance zone leading to a slightly increase of the ionization probability of the atoms defocussed by the hexapol field of the ionizer.

## THE BEAM POLARIZATION

The beam polarizations reaches 75 (80) % for protons, i.e. significantly less than the 85-90 % with the previously used ANAC electron beam ionizer. The axial field setting for maximum beam intensity does not coincide with the setting for maximum polarization. Increasing the excitation of the solenoid coil at the entrance side of the ionizer by 20 % let the proton polarization increase from 75 to 80 % at the cost of a noisy beam with an intensity reduced by more than a factor 2. A modest increase of the hexapole field (10 %) at the entrance side help to reduce the instabilities.

The polarization can be reduced by two effects which, unfortunately, are not detectable by unique signatures. With the proposed model it is now possible to understand the influence of the two contributions on the performance of the ionizer.

First, the proton polarization suffers from the to low holding field in the ionizing volume. Taking into account a dominant ionization efficiency at the resonance surface the effective field is 87.5 mT and does not depend significantly on the shape of the axial field. The maximum achievable polarization of substates 2 and 4 is then 86 %, or 93 % for a beam with substates $1+4$ or $2+3$ (providing the Stern-Gerlach state separation and the rf transitions are perfect). In this model an increase of the average magnetic field does not lead to higher limits.

The second and main contribution reducing the polarization arises from the ionization of recombined molecules. The resonance surface extends radially to a diameter at least 2 times larger than the atomic beam and causes therefore a preferential ionization of the molecular background reflected from the atomic beam dump. This effect is particularily dramatic in the case of a ionization efficiency peaked at the resonance surface.

The increased excitation of one solenoid not only shifts the axial position of the resonance zone, it also partly compress the radial extension of the resonance surface to smaller radii. The first effect diminishes the general ionization efficiency and introduce unstabilities by destroying the overlap of the resonance surface with the microwave amplitude distribution, the second one reduces the contribution from the region where only background beam is produced. This conclusion is confirmed by the observed variation of the beam intensity contributions from the atomic beam (a probe for the ionization efficiency in a radially limited domain) and from the buffer gas alone (a probe for the total contribution) as a function of the axial field settings.

## CONCLUSION

The present analysis of the performances of the PSI ECR ionizer delivers a consistent picture of the behaviour of the intensity and of the polarization of the extracted beam. The reduced polarization is essentially due to the background contribution generated in a ionization zone of to large a radius compared with the atomic beam extension. Because of the intensity and stability behaviour of the system, a cure to this problem not only requires a stronger hexapole field (at the limit of the tolerable transversal field components) to reduce the plasma diameter but also a more sophisticated cavity design in order to establish the necessary overlap between ECR surface and microwave field distribution. Ideally, a tunable geometry will be the best approach since otherwise the present difficulty to satisfy several conditions with the axial field distribution as only adjustable parameter will remain. In any case the design of the ECR ionizer could benefit from the discussion of the ECR source presented in ref. 2.

The design of the vaccuum system has not been addressed here. Improvements are possible but they are not expected to be the key to the solution of the present problems. Other parameters (microwave power, buffer gas) have not been discussed, also. Their influence on the efficiency and stability of the ECR ionizer is important but does not fundamentally modify the behaviour described here.

It is hoped that the light shed by the present analysis on some properties of the ECR ionizer will help in improving the design of future devices and make it possible to reach the long awaited breakthrough in the proton polarization before the next Workshop.

## REFERENCES

1.  P.A. Schmelzbach, Proc. Intern. Workshop on Polarized Ion Sources and Polarised Gas Targets, KEK-Tsukuba,(ed. Y. Mori, KEK Report 90-15, 1990), 329
2.  M. Delaunay, B. Jacquot and M. Pontonier, Nucl. Instr. and Meth. A305, 223 (1991)
3.  T.B. Clegg, V. König, P.A. Schmelzbach and W. Grüebler, Nucl. Instr. and Meth. A238, 195 (1985)
4.  T.B. Clegg and M.B. Schneider, Proc. Intern. Workshop on Polarized Sources and Targets, Montana, (eds. S. Jaccard and S. Mango), Helv. Physica Acta 59, 53 (1986)
5.  L. Friedrich, G. Huttel and P.A. Schmelzbach, Nucl. Instr. and Meth. A272, 906 (1988)
6.  H.G. Mathews, H. Beuscher and C. Mayer-Böricke, Physica Scripta T3, 52 (1983)
7.  P.A. Schmelzbach, Proc. 7th Conf. on Polarization Phenomena in Nuclear Physics, Paris, (ed. A. Boudard and Y. Terrien, Les Editions de Physique, Paris, Colloque C6, 1990) p. 545
8.  H.F. Glavish, Proc. Workshop on Higher Energy Polarized Proton Beams, Ann Arbor,MI,(ed. A.D. Krisch) Am. Inst. Conf. Proc. 42, 47 (1977)

# POLARIZED ION SOURCE DEVELOPMENT AT IUCF

V. Derenchuk, R. Brown, M. Wedekind
Indiana University Cyclotron Facility
Bloomington, IN 47408

## ABSTRACT

The IUCF high intensity polarized ion source (HIPIOS) has been completed and installed in the 600 kV terminal. The design was originally based on the source in operation at TUNL, which employs cold ($\sim$ 30 K) atomic beam technology and an electron cyclotron resonance ionizer. Development of the reliability and intensity of the atomic beam source will be described. An atomic beam flux of 5.0 x $10^{16}$ atoms/sec is estimated to be incident on the ECR ionizer. Preliminary testing of a multiple grid, high conductance, extraction system on the ECR ionizer has resulted in an extracted current of 500 $\mu$A attributable to the atomic beam, as measured 30 cm downstream of the ECR ionizer. After a mass analysis magnet, 245 $\mu$A are measured with a normalized emittance of .78 $\pi$-mm-mrad.

## INTRODUCTION

The design for HIPIOS was based on the source at TUNL,[1] developed by Tom Clegg and associates, which utilizes cold ($\sim$ 30 K) atomic beam technology and an electron cyclotron ionizer (ECR). HIPIOS (Fig.1) deviates from the TUNL configuration in three significant ways: the first sextupole is 50% longer to provide stronger focussing, the cesium charge exchange canal is replaced by a gridded, single-gap RF buncher with a ramp waveform, and ion beam extraction from the

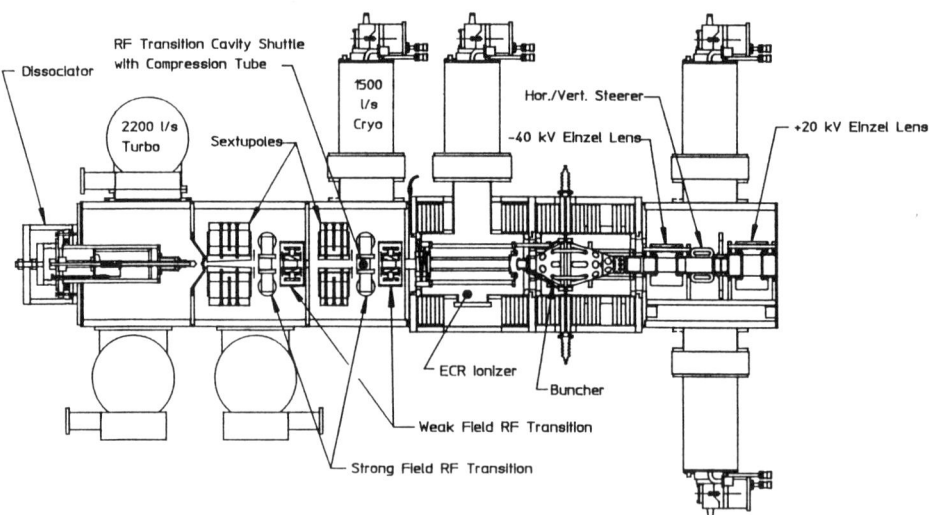

**Figure 1** General layout of the IUCF High Intensity Polarized Ion Source.

source is achieved by raising the internal structure of the ECR ionizer to 20 kV potential. The latter was dictated by the necessity of operating the source assembly at local ground potential in a 600 kV terminal and has significantly complicated the design. The HIPIOS project status and results of the initial off-line testing have been reported in the 1992 Cyclotron Conference Proceedings[2] and at the 1993 Particle Accelerator Conference.[3]

## ATOMIC BEAM DEVELOPMENT

### Apparatus

Atomic beam intensity was measured with a compression tube (Fig.2) mounted on a shuttle also designed to hold two strong field transition units. The compression tube and the two transition units can be positioned transversely from outside the vacuum system. Horizontal atomic beam profile measurements were measured using this apparatus. The entrance of the compression tube is 9.3 mm in diameter with a length of about 4 cm. The length was chosen to match the space available between the sextupole and the shuttle. With a flux of $2 \times 10^{16}$ atoms/sec and a base pressure of $2 \times 10^{-7}$ Torr, a pressure difference of $5 \times 10^{-6}$ Torr is measured with a hot cathode ion gauge. The compression tube was calibrated with a calculated flow rate through a needle valve from a fixed volume of $H_2$.

The atomic beam section is similar to the design at TUNL and has been described elsewhere.[2,5] The exit of the cold nozzle is located 20 mm from the skimmer and 46 mm from the entrance aperture to the first sextupole. The total distance from the cold nozzle to the 3 cm diameter ionizer entrance is 103.5 cm. The compression tube is located between the second sextupole and the ionizer and has been mounted in two axial locations to estimate the atomic beam divergence.

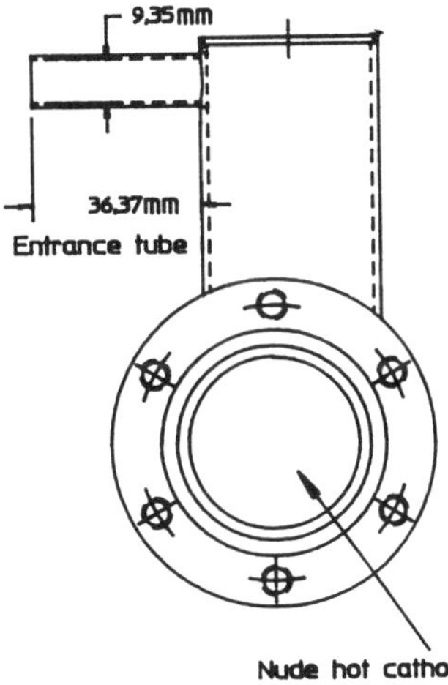

Figure 2 Top view of the compression tube mounted on the movable RF transition shuttle.

### Results

During initial testing of the ECR it was noticed that the beam intensity from the dissociator dropped by a factor of two during the first 24 hours of operation. This drop in intensity with time was correlated to the coating of the cold nozzle with $SiO_2$.[2] Several design modifications were made to the dissociator in order to address this problem: a -20°C closed loop alcohol cooling system was installed in place of the water cooling system, the dissociator tube was modified to provide cooling within 1 cm of the accommodator, and the RF coil position was adjusted for minimum operating power. $N_2$ gas to prevent recombination in the cold nozzle was mixed with the $H_2$ upstream of the dissociator, rather than being introduced directly into the accommodator.

The dissociator now runs stably without any powder formation under the following operating conditions: 25 sccm $H_2$, .030 sccm $N_2$, 33.5 K nozzle temperature, -10° C alcohol return temperature and 100 W RF power (with 23 W reflected) for periods of about one week. Other groups have increased the diameter of the discharge tube in order to address the $SiO_2$ problem.

After about 7 days of operation, the cold nozzle begins to occlude with frozen ammonia and the beam intensity drops by 30%. Warming the nozzle to room temperature then cooling it under vacuum will return the dissociator output to its original intensity. Since we find that adding $N_2$ with the $H_2$ does not result in any intensity increase, we will return the $N_2$ feed to the accommodator to decrease the formation of frozen ammonia. Also, adding .1% to .01% $O^2$ to the $H^2$ gave no improvement in the beam intensity.

With nominal atomic beam source operating parameters, an atomic beam intensity of $2.0 \times 10^{16}$ atoms/sec with a density of $3.3 \times 10^{11}$ atoms/cm$^3$ has been consistently measured in the compression tube. Atomic beam profiles have been measured at the second sextupole exit and 9 cm downstream. This data has been used to determine that the diameter of the atomic beam is totally contained within 22 mm and has a total estimated flux of $5.0 \times 10^{16}$ atoms/sec at the ionizer entrance.

## ECR IONIZER DEVELOPMENT

Operation of the ECR at ground potential while the permanent magnet sextupole was biased at 20 kV presented several problems - primarily one of radial and upstream ion extraction resulting in excessive current draw on the high voltage power supplies. Attempts to solve this problem by reverse biasing an entrance cone and screen around the permanent magnets was not entirely successful.[2] Inspired by the ECR ionizer design at PSI,[4] a quartz tube was installed as a liner for the permanent magnet sextupole to further reduce the radial extraction of beam from the ECR plasma.

A new extraction system using three molybdenum grids with 95% transmission through a diameter of 4.45 cm was also installed (Fig.3). The high conductance through the grids allows for improved axial pumping to compensate for the loss of radial pumping due to the quartz tube. The three grids are electrically isolated and attached to independent high voltage supplies to allow for testing of accel-accel and accel-decel extraction from the ECR.

Testing of the modifications has just begun, with the following preliminary results. The extraction system was operated in the accel-accel mode with voltages of 10 kV, 9 kV and 8 kV on the three grids. The 2 keV beam was transmitted through the buncher chamber, accelerated to 10 keV and transported 30 cm to a Faraday cup. With an atomic beam intensity of $2.0 \times 10^{16}$ atoms/sec in the compression tube, a beam current of 0.68 mA was measured with the sextupoles on and 0.18 mA with the sextupoles off. Downstream of a mass analysis magnet, a proton current of 295 $\mu$A was measured with the sextupoles on and a proton current of 10 $\mu$A with the sextupoles off. The normalized emittance at this location was measured to be 0.78 $\pi$-mm-mrad with 245 $\mu$A of beam.

Tuning of the ionizer solenoid currents and extraction grid potentials strongly affected the stability of the beam current. Some tunes resulted in a Penning discharge between grids which caused 100% beam modulation on the order of one kilohertz. Back streaming electrons impinging on the ECR plasma caused the beam intensity to jump several times per second.

## CONCLUSION

The HIPIOS atomic beam section has stable, constant output for a period of one week at a time without adjusting any operating parameters. The occlusion of the nozzle with $SiO_2$ has been eliminated and the total flux incident on the ionizer is estimated to be 5.0 x $10^{16}$ atoms/sec. A three grid extraction system has been designed and installed and preliminary measurements have been made. A mass analyzed proton current correlated to the atomic beam of 205 $\mu$A has been measured in a normalized horizontal phase space of 0.78 $\pi$-mm-mr. A design for a 5 keV to 10 keV Lyman $\alpha$ polarimeter was passed on to us by Anatoli Zelenski from the INR. Construction of a polarimeter will begin soon.

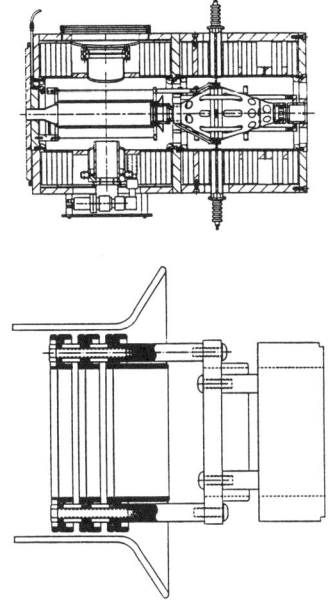

**Figure 3** ECR and Buncher assembly with quartz tube and gridded extraction system detailed.

## ACKNOWLEDGEMENTS

The authors express their appreciation to Tom Clegg, C.D.P. Levy, Andy Roberts, and Anatoli Zelenski and especially to Alexander Belov for many useful and patient conversations and help with the design of the extraction system. Also, special thanks to Howard Petri for doing the emittance measurements, Ron Kupper for building the alcohol cooling system and the entire technical staff at IUCF. The HIPIOS project is supported by NSF Grant PHY-891440.

## REFERENCES

1. T.B. Clegg et al.,"A New Atomic Beam Polarized Source for the Triangle Universities National Laboratory: Overview Operating Experience and Performance," submitted to NIM.
2. V. Derenchuk et al, Proceedings from the 13th International Conference on Cyclotrons and Their Applications, Vancouver, 1992, (World Scientific, Singapore, 1993) pp. 330-333.
3. M. Wedekind et al, "IUCF High Intensity Polarized Ion Source", presented at the 1993 Particle Accelerator Conference, Washington, D.C.
4. P. A. Schmelzbach, "ECR Ionizers and Operating Experience at PSI", this workshop.
5. M. Wedekind, et al., Conference Record, 1991 IEEE Particle Accelerator Conference, Accelerator Science and Technology (IEEE, N.Y., 1991), pp.1922-1925.

# THE GIESSEN ATOMIC BEAM SOURCE

G. Clausnitzer, R. Baumann, B. Feuerstein, H.P. Flierl,
M. Horoi, G. Keil, M. Mildner, D. Ochs, E. Pfaff, M. Preiß
Strahlenzentrum, University of Gießen, Germany

## ABSTRACT

The Gießen Lambshift Source[1] will be converted into an Atomic Beam Source utilizing the existing vacuum system and the Wien Filter. One additional chamber with two turbopumps contains the dissociator, 30 K accommodator and the first sixpole magnet. Radio frequency transitions are mounted before and after the second multipole magnet for different working modes (single or multiple hyperfinestate selection). Ionisation in an ECR discharge and double charge exchange in Cs is used for negative ion production. The atomic beam unit has been optimized successfully in a separate test stand, the actual conversion is scheduled for summer 1993.

## ATOMIC BEAM FORMATION

Hydrogen molecules are dissociated in a 27 MHz - gas discharge in a water-cooled Pyrex-tube, following the design of Jaccard/Schmelzbach[2]. Hydrogen flow rates of 30 sccm/min are used with a small admixture of $O_2$ (0,1 sccm/min) to improve the dissociation degree. The front end of the 12 mm O.D. x 1mm discharge tube extends into the vacuum chamber approximately 1 cm, a teflon shroud assures a connection to a short glass tube located in the cold copper nozzle (conical 10 mm to 3 mm diameter, 20 mm length). A small amount of $N_2$ ( < 0,1 sccm/min) is fed through the teflon for optimum accommodation at 25 K.

A two stage refrigeration system (10 W at 20 K) with a short flexible connection to the nozzle plate allows a temperature of 20 K without super-insulation after one hour cooling. The complete dissociator assembly can be moved under operating conditions; a 2200 l/s turbomolecular pump maintains a pressure of $2\text{-}3 \times 10^{-4}$ mbar.

A skimmer (diameter 4 mm) with a $90^0$ inner cone angle at a distance of approx. 10 mm from the nozzle separates the second chamber in front of the sixpole magnet, which is pumped by a 500 l/s turbo pump to $2 \times 10^{-5}$ mbar. The 10 mm diameter entrance to the magnet chamber concludes the atomic beam formation.

## SPIN SEPARATION

A schema of the installation is shown in Fig.1. The first sixpole magnet is 126 mm long with 10 mm aperture at the entrance, 25 mm in the center and straight from there out, all manufactured from soft (Armco) iron. A current of 250 amps through 15 windings per pole (5 x 5 copper with 3 mm diameter hole) result in a pole tip field of 0,9 T at the entrance and 0,7 T at the exit. A second spin separation (compression) magnet has a constant aperture (25 mm), 126 mm length and is mounted at a distance of 200 mm from the first. Atomic beam intensity and profile measurements were performed with a movable ion gauge equipped with different compression tubes. $\triangle$p-values for multipole magnets on-off were taken and the number of atoms was calculated assuming a complete recombination and thermalization in the ion gauge. Profile measurements at different distances from the first sixpole magnet revealed a distribution consistent with $v_{max}$ =1000 m/s, $v_{drift}$ =850 m/s and $T_{beam}$ =10 K. $\triangle$p-measurements behind the second (fourpole) magnet were made at a distance of 70 mm. The maximum observed $\triangle$p-value with a 5 mm I.D. compression tube was $3.10^{-5}$ mbar (corrected for hydrogen), which results in a hydrogen flux of $3\ 10^{16}$ atoms/s.cm $^2$.

Radio frequency transitions are used between the multipole magnets for a single hyperfine state selection or for the production of tensor polarized deuterons. Computer simulations show, that a four-pole magnet is advantageous in this case; compression tube measurements of the atomic beam flux support this facts.

Two radio frequency transitions are used after the second magnet for the usual two state selection mode (for hydrogen ). In this case a second sixpole magnet shows a slightly better $P^2$ x I value in the simulation compared to a fourpole for the same pole tip field; the fourpole allows a higher excitation with a (10%) higher field. Therefore the fourpole magnet was chosen for the present design.

## RADIO FREQUENCY TRANSITIONS

Three radio frequency transition units are necessary for the vectorpolarized deuteron production, mounted between the fourpole magnet and the ECR-ionizer. A compact construction was chosen, where the static B-field is in the longitudinal (beam) direction,less sensitive to stray fields from the ECR ionizer. The usually used cavities (330 MHz and 460 MHz for the deuterium transitions) are mounted "sideways", i.e. with their shortest dimension (27 mm) in beam direction. The weak field transition unit (30 MHz) has been tested successfully; the complete assembly (three units) will occupy a length smaller than 120 mm.

## IONIZATION AND CHARGE EXCHANGE

The electron-cyclotron-resonance (ECR) discharge with a 2,45 GHz micro-wave power of less than 100 watts allows an optimal ionization of thermal hydrogen. The ionizer is designed following Schmelzbach[3], using a Pyrex-tube as a discharge container. A Lisitano antenna[4] for the excitation of the discharge has the advantage of a higher density of low energy electrons compared to the usual cavity. Optimum performance is achieved, when the ECR-region is near the B - minimum.

The plasma potential is defined by the first electrode with an opening of 40mm positioned approximately at the second field maximum; a slow extraction is achieved with a second electrode (same opening) operating at a very critical potential near (above) that of the plasma electrode. A careful adjustment of this potential allows a significant reduction of the background hydrogen component and therewith an increase of the polarization. The third electrode is the entrance of the Cs-channel at a potential difference of 500 V to the plasma. All high voltage electrodes are mounted inside the solenoids, which are held at ground potential.

The ECR-discharge is maintained with a continous flow of nitrogen gas, which also serves to partially neutralize the space charge potential in the ionization region.

The double charge exchange takes place in a 30 mm inner diameter, 100 mm long copper channel with seven slots on the lower side as openings for Cs-vapor. The Cs is evaporated by an immersed heater in a cylindrical stainless steel chamber around the copper channel. The large openings are covered with vaned inserts in order to reduce the Cs-loss. The new Cs-channel is presently under investigation in the Lamb Shift Source.

## REFERENCES

1. W. Arnold, et al., Nucl.Instr.and Meth. 143, 441 (1977)
2. S. Jaccard, Osaka 1985, Suppl.Journ.Phys.Soc.Jpn 55,1062 (1986)
3. P.A.Schmelzbach, Paris 90, Colloque de Physique C6/545 (1990)
4. G. Lisitano, et al., Phys. Lett. 29A, 613 (1969), Appl. Phys. Lett. 16, 122 (1970)

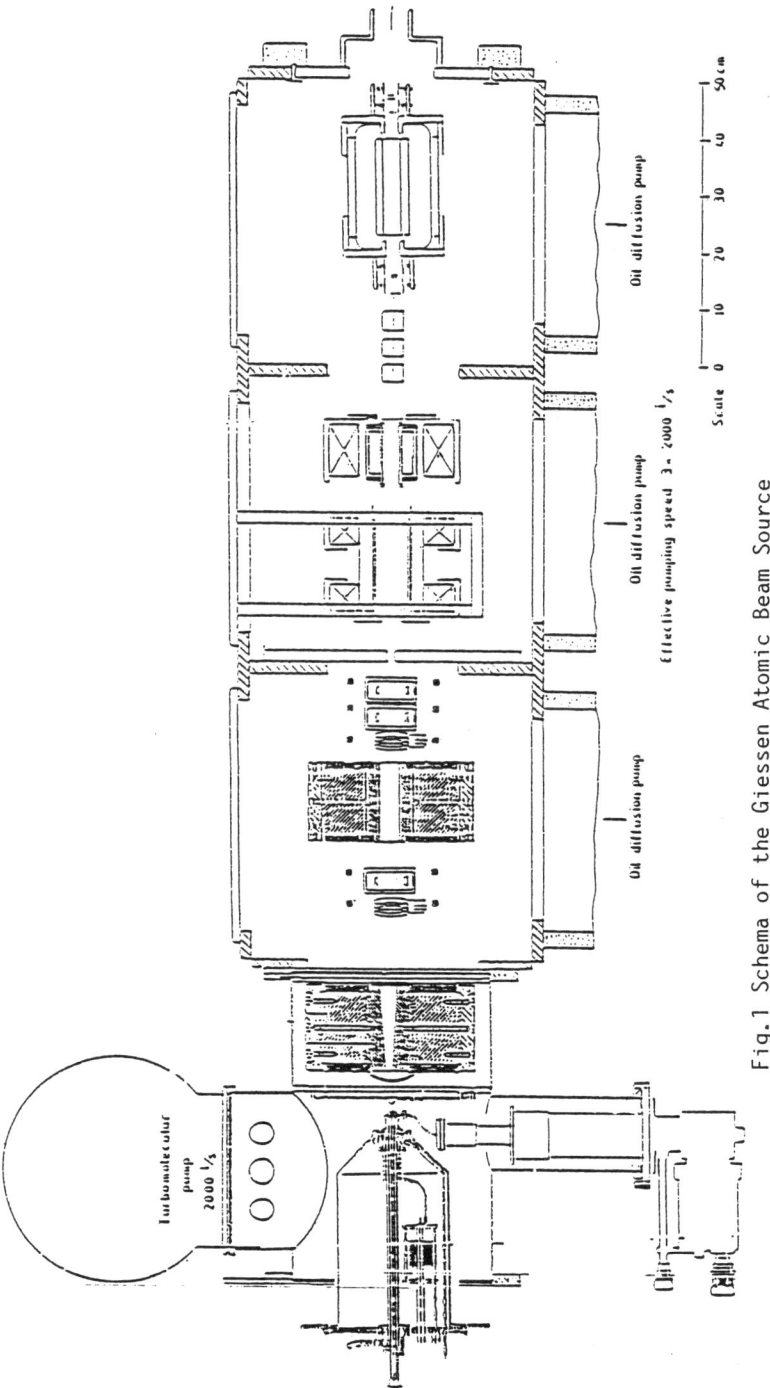

Fig.1 Schema of the Giessen Atomic Beam Source

# THE BALZERS-PFEIFFER SOURCE FOR THE SVEDBERG
# LABORATORY, UPPSALA

G. Clausnitzer, G. Keil, E. Pfaff
University of Gießen, Germany

D. Krämer, K.H. Kügler
A. Pfeiffer Vakuumtechnik, Asslar, Germany

## ABSTRACT

The commercially manufactured atomic beam source for polarized protons and deuterons has been accepted 1992 by the Svedberg Laboratory, University of Uppsala, Sweden. Hydrogen atoms emerge from a 27 MHz radiofrequency dissociator through a 30 K copper nozzle covered with nitrogen. The beam formation is achieved by a two stage differential pumping system. Two matched multipole magnets serve for an optimum spin separation; suitable radiofrequency transition units between the multipoles and before the ionizer generate the different nuclear polarization components. An electron-cyclotron-resonance (ECR) discharge driven by 2.45 GHz microwaves produces an ion beam of 40 $\mu$amps (one hyperfine state) within a phase space of 400 mm mrad at 20 keV. The polarization was estimated from measured atomic beam properties to be larger than 80%; subsequent measurements revealed lower values due to inappropriate function of the r.f.transitions.

## THE VACUUM SYSTEM

The current limitation of atomic beam polarized ion sources is predominantly due to vacuum conditions, i.e. intra beam scattering processes, which occur in the atomic beam especially in the region between the cold nozzle and the multipole magnet entrance. The optimized geometry leads to small distances and the necessary high gas flux requires the maximum conductance in order to avoid scattering losses. A special turbomolecular pump has been designed, manufactured and successfully tested especially for this application. Four rotors of the Balzers-Pfeiffer TPH 2200 are integrated in one forged aluminum housing closely around a 250 mm diameter bore located between the rotor/stator packs of each pump. This allows a compact construction of the source components "inside" the pump, a maximum conductance and a superior pumping of the beam region.
The fore line of the first stage with the largest gas load is separated from the other three combined fore lines; two small 100 l/s turbocompressors are used for the connection to one double stage 20 l/s rotary pump. The pump is running very smoothly, a pressure of $2 \cdot 10^{-4}$ mbar can be maintained in the first stage under

a 1 mbar.l/s hydrogen gas load. Appropriate inner walls separate the chambers so that pressures in the low $10^{-6}$ mbar range are measured in the third stage, where the first multipole magnet is located. The base pressures are in the low $10^{-7}$ mbar range without special outgassing procedures; the low $10^{-8}$ mbar range is reached after several days of pumping.

## THE ATOMIC BEAM FORMATION

Molecular hydrogen is dissociated in a 27 MHz gas discharge in a watercooled straight Pyrex tube of 10 mm inner diameter. Atoms emerge through a conical copper nozzle held at 30 K, which is covered by a thin solid nitrogen layer to avoid recombination. A gas flux of 0.5 mbar.l/s and a nozzle with 3 mm exit diameter are the optimum parameters for the actual installation.

The front end of the discharge tube extends approximately 10 mm into the vacuum chamber, not connected directly to the cold nozzle assembly mounted on a teflon sylphon (40 mm diameter). The nozzle is srewed into a copper plate, which is connected (flexible) to a Balzers cold head. The complete dissociator assembly including the cold head can be adjusted (3 dimensions) under operating conditions.

The common problem of a white powder (Si) on the cold surfaces occured, when new tubes were tested; it disappeared over a period of several days of operation. Smaller amounts were collected, when new tubes were cleaned with hydrofluoric acid (15 % solution) before the first operation; a conditioning with a 300 watt discharge for several hours gave a small improvement. It is known, that water vapor in the hydrogen improves the operation (dissociation degree); an addition of a small amount of oxygen ( $10^{-3}$ ) in the hydrogen feed line was used finally, to stabilize the running conditions. An increase of the extracted current by a factor of 1,5 - 2 was achieved.

A skimmer (diameter 4 mm, 10 mm high) with a 90 degree inner cone is mounted centered on the axial symmetry line and separates the first and the second chamber. The optimum distance to the nozzle was found experimentally between 10 mm and 14 mm depending on the hydrogen flux. The pressure reading under operating conditions is below $10^{-5}$ mbar. The 10 mm diameter entrance in front of the sixpole magnet concludes the atomic beam formation, which optimizes for a nozzle-sixpole distance between 50 and 60 mm.

## THE SPIN SEPARATION

A schema of the installation is shown in Fig.1. The first sixpole magnet is 120 cm long with 10 mm aperture at the entrance and 25 mm at the exit; pole cores and tips are manufactured from Hypermco 50 (similar to 2V-Permendur), the cylindrical yoke of 250 mm outer diameter is from Armco iron. A current of 250 amps through 11 windings per pole (5 x 5 with 3 mm diam.hole) produces a poletip field of 0,9 T at the entrance and 0,7 T at the exit. The second spin separation magnet has a constant aperture of 30 mm, 120 mm length and is mounted at a distance of 180 mm from the first.

The original design of the source was based on a single hyperfine state selection in order to overcome polarization losses due to too low fields in the ionizer. Therefore two radiofrequency transition units can be operated between the two multipoles in order to obtain the nuclear polarization components for protons and deuterons. In the first position a weak field unit (30 MHz for H and 10 MHz for D) is installed, which can be exchanged with a strong field unit (1,4 GHz for H) in operating conditions. A 460 MHz cavity is installed in the second position for the production of a purely tensor polarized deuteron beam. The transition units are constructed very compact with shielding plates on the sides facing the multipole magnets.

The second magnet has to fullfill the additional requirement, to deflect out the atoms, which have undergone a transition. Computer simulations of all four hydrogen components showed that a fourpole magnet produces a larger $P^2$.I-value than a sixpole. The fourpole was built from Armco iron with 22 windings per pole.

The space between the fourpole and the ECR entrance is 120 mm and contains two exchangeable radio frequency units: weak field for vectorpolarization of protons and deuterons and 330 MHz for tensorpolarized deuterons. In addition a moveable compression tube ion gauge can be shifted into the atomic beam, allowing a monitoring of the atomic flux. All radio frequency units were mounted between the two multipoles, where the correct function can be tested by a (50 %) current change of the extracted proton (deuteron) beam signal.

A carefull shielding of stray fields from the ECR-ionizer was essential especially for the weak field transition; the appropriate steel cylinder however caused axial fields below $5\mu$T, not sufficient to define the quantization axis. For an adiabatic passage of atoms of 1000 m/s the minimum field strength is around $10\mu$T.

## THE ELECTRON CYCLOTRON RESONANCE IONIZER

A 2.45 GHz microwave generator with a maximum power of 300 watts drives the discharge in a minimum B-configuration (10 pancake solenoids) with radial confinement by a Co-Sm sixpole. The vacuum container is a 65 mm diameter Pyrex tube; cavity and permanent magnets are outside, isolated from ground potential.

The plasma potential is defined by the first electrode with a 4o mm opening, positioned at the second field maximum; a slow extraction is achieved with a second elctrode operating at a very critical potential near (above) that of the plasma elextrode. The accelaration to ground potential takes place approximately 100 mm behind the field maximum.

The ECR-discharge is maintained with a continous flow of nitrogen gas into the extraction chamber, controlled by a thermo-valve on ground potential; optimum conditions are obtained for a pressure increase of 2-3 x $10^{-6}$ mbar in the extraction chamber. The nitrogen also serves to partially neutralize the space charge potential in the ionization region resulting in a low emittance ion beam.

The extracted beam is focussed by an einzel lens to the object point of a double focussing 90 degree magnet into the horizontal direction, transported to the cyclotron center, where another 90 degree deflection takes place.

A movable faraday cup can be located (electrically) in the beam after the 90 degree deflection; the current change of the mass separated beam was close to 100 $\mu$amps for two hyperfine-states, 50 $\mu$amps for the single state selection, from which 80 % were contained in a phase space of 400 mm mrad at 20 keV.

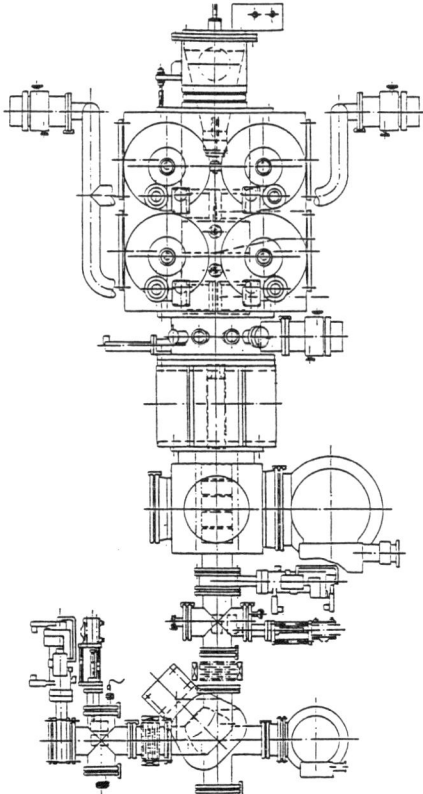

Figure 1: Schema of the Balzers Pfeiffer Source

# DEVELOPMENT OF THE RIKEN POLARIZED ION SOURCE

H. Okamura[†], H. Sakai[†], N. Sakamoto[†], T. Uesaka[†],
S. Ishida[†], H. Otsu[†], T. Wakasa[†], K. Hatanaka[‡],
T. Kubo, N. Inabe, K. Ikegami, J. Fujita, M. Kase, A. Goto and Y. Yano

[†] University of Tokyo, Hongo 7-3-1, Bunkyo-ku, Tokyo 113, Japan
[‡] Research Center for Nuclear Physics, Osaka University,
Mihogaoka 10-1, Ibaraki, Osaka 567, Japan
The Institute of Physical and Chemical Research (RIKEN),
Hirosawa 2-1, Wako-shi, Saitama 351-01, Japan

ABSTRACT    The RIKEN polarized ion source has been assembled and
is operational. The present level of performance is 140 $\mu$A with 50-60%
polarization of the ideal value. Results from operation will be described.

## INTRODUCTION

The RIKEN ring cyclotron can accelerate protons and deuterons up to 210 and
270 MeV, respectively. Use of a polarized beam will provide unique opportunities
to study the spin dependent physics at intermediate energies. The program of constructing the polarized ion source started in 1990. Since we had little experience
in this field, the decision was made to copy the reliably operating, modern and
high-performance ion source. The one developed at TUNL[1] met our requirements
and the fact that IUCF had started to construct its modified version, HIPIOS,[2]
induced us to make the same choice as IUCF. We obtained a lot of helpful information both from TUNL and IUCF and also mechanical drawings from IUCF
which allowed us rapid construction.[3,4]

The RIKEN polarized ion source of the atomic beam type is schematically
shown in figure 1. It is very similar to HIPIOS of IUCF except that the beam
traverses vertically. The assembling was completed in May 1992. The first acceleration test by the injector AVF cyclotron was performed in June, and by the
ring cyclotron in October of the same year.

## DISSOCIATOR

The performance of the dissociator was studied by using a compression tube
installed downstream of the second sextupole magnet. A cooled atomic beam is
formed by the copper nozzle with a 3 mm orifice. As the nozzle temperature $T$
falls down, the flux of the atomic beam increases approximately in proportion to
$1/T$ (figure 2). The intensity rapidly drops below 60 K because of the strong
recombination at the copper nozzle. It is recovered by the known technique of
forming a $N_2$ layer on the nozzle surface.[5,6] $N_2$ is fed in near the nozzle through the
MACOR support. The maximum beam flux into the 25 mm aperture is routinely
observed to be $2.6 \times 10^{16}$ atoms/sec.

The major difficulty we met is the short lifetime of the nozzle. During operation, the white powder, which was analyzed to be $SiO_2$, accumulates on the wall
of nozzle. It reduces the beam intensity rapidly, e.g. down to 50% within a few
hours, and sometimes plugs the nozzle orifice after a few days operation. This was
remedied by making the nozzle longer, 28 mm to 61 mm, following the suggestion
by Dr. Schmeltzbach.[6] In this way the white deposits accumulate far away from
the nozzle and the surface near the orifice will be kept clean. Another effective

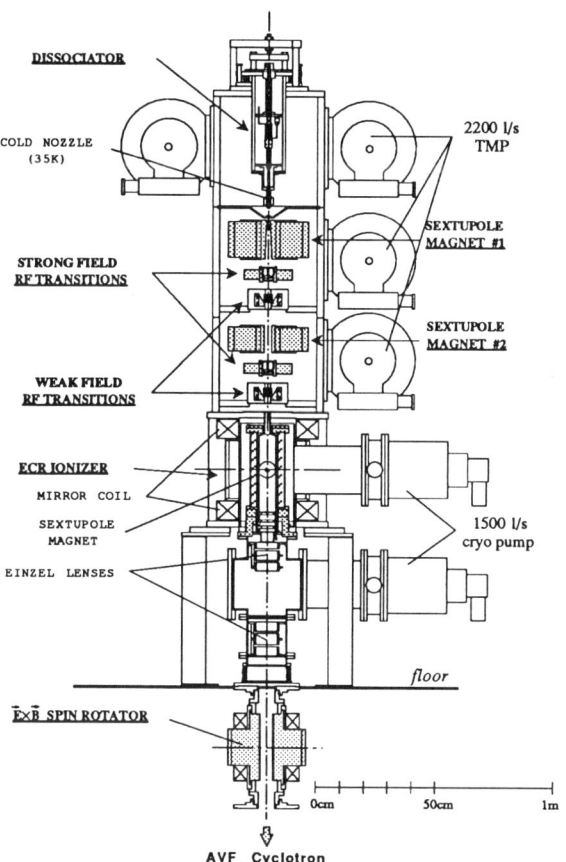

Fig. 1. Side view of the polarized ion source.

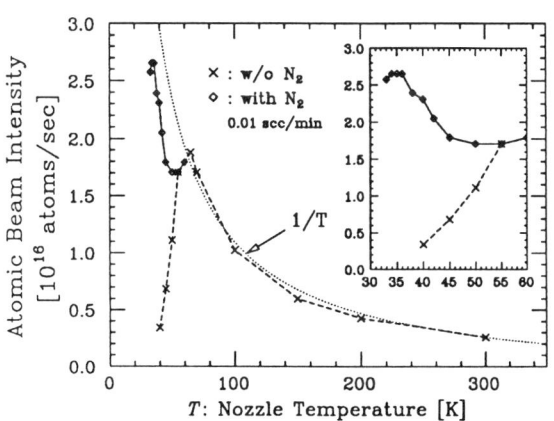

Fig. 2. Dependence of atomic beam intensity on the nozzle temperature.

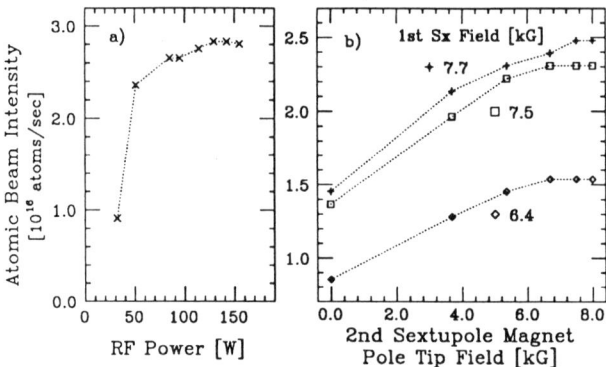

Fig. 3. Dependence of atomic beam intensity (a) on the dissociator RF power and (b) on the pole-tip field of the sextupole magnets.

procedure is to keep the RF power as low as possible. As shown in figure 3a, the power of 60 W will be enough to obtain 90% of the saturated intensity.

The currently best operating parameters are as follows: 20-30 scc/min $D_2$, 0.01 scc/min $N_2$, 37 K nozzle temperature and 60-80 W RF power with $\leq 3$ W reflected. Under these conditions no reduction of intensity is observed for at least one week. It should be noted that we have never operated the ion source continuously for more than one week. The $N_2$-layer technique works very reliably.

## SEXTUPOLE MAGNETS

The length of the first magnet was increased by 50% as recommended by TUNL.[3] Figure 3b shows the atomic beam intensity for various fields of the first sextupole magnet as a function of fields of the second sextupole magnet. The intensity does not pass the peak but seems to reach the plateau.

## ECR IONIZER

The ECR ionizer is operated with the 2.45 GHz microwave, typically at a power of 30-40 W. The low magnetic field (87.5 mT) which will reduce the proton polarization is not a problem since we are mainly interested in producing deuteron beams.

The early design of the ionizer also followed that of HIPIOS. The sextupole magnet consists of SmCo permanent magnets encapsulated in copper cans and was electrically isolated and biased to +7.5 kV operation voltage, while the vacuum chamber remained at the ground potential. The aperture of the extraction electrode was 1 cm in diameter. Good pumping was provided radially through the gap of the sextupole.

The problem that the ions were extracted axially to the upstream direction as well as radially through the gap of the sextupole magnet was immediately noticed. As a result the ions sputtered the RF transition units and the wall of vacuum chamber, inhibiting a long-term operation. We installed a mesh and an electrode to suppress the undesired extraction and the problem seemed to be settled. However, when the beam was injected into the cyclotron, we noticed that the suppressor was causing some instability to the ECR plasma. The best performance of the early version of the ionizer was 20 $\mu$A with 60% polarization of the ideal value.[7]

We have modified the design of the ionizer close to the one at PSI.[6] A Pyrex plasma chamber is installed inside the sextupole magnet which is kept at the ground potential. The plasma region is pumped through the 5 cm aperture of three-stage extraction electrodes. The stable and long-term operation of the new ionizer allowed us extensive searches of optimum conditions. Currently the ion intensity of 140 $\mu$A is routinely observed.[7] The polarization, however, is reduced to 50% of the ideal value. We believe the low polarization is primarily due to insufficient pumping which causes an increase of the unpolarized background. We are installing a Ti sublimation pump and will continue to investigate the polarization reduction problem.

SPIN ROTATING SYSTEM

The polarization is measured at the exit of the AVF cyclotron ($E_d = 14$ MeV) by using the $^{12}C(d,p)^{13}C_{gnd}$ reaction and at the exit of the ring cyclotron ($E_d = 270$ MeV) by using the $d + p$ elastic scattering. The spin direction is freely controlled by using the Wien filter downstream of the ion source. While the beam is injected and accelerated with its spin axis inclined to the cyclotron magnetic field, a single turn extraction does not lead to a loss in the amplitude of polarization. According to this principle, we have succeeded in March 1993 in rotating the spin direction without reducing the polarization or the intensity.

CONCLUSION

Owing to the newly developed but well proven technology, the RIKEN polarized ion source has been successfully constructed and is reliably operating. The method of controlling the spin direction has been established incorporated with the acceleration technique. While we will still continue to try to improve the degree of polarization, the system is ready for experimental use.

ACKNOWLEDGEMENT

We are very indebted to T.B. Clegg and the staff of HIPIOS who generously allowed us to copy their ion sources and provided us with helpful information. We are also grateful to P.A. Schmeltzbach for giving us valuable suggestions.

REFERENCES

1) T.B. Clegg, Proc. of the High Energy Spin Physics 8th Int. Symp., Minneapolis, MN, ed. K.J. Heller, A.I.P., New York, 1989, Vol. 187, 1227.
2) V.P. Derenchuk, contribution to this workshop
3) T.B. Clegg, private communication
4) M. Wedekind *et al.*, private communication
5) D. Singy *et al.*, Nucl. Instr. and Meth. B47 (1990) 167.
6) P.A. Schmeltzbach, private communication
7) The cited ion intensity is the difference between the beam current with sextupoles switched on and the one with sextupoles switched off. We do not directly cite the mass-analyzed current because that may include the current due to singly-charged molecule $H_2^+$ which will vary depending on vacuum conditions.

# POLARIZED ION SOURCE FOR LOW ENERGY NUCLEAR FUSION RESEARCH

K.Maehata, H.Arima, T.Hirose, N.Shichida and Y. Wakuta
Department of Nuclear Engineering, Kyushu University, Fukuoka 812, Japan

M.Tanifuji
Department of Physics and Research Center of Ion Beam Technology,
Hosei University, Tokyo 102, Japan

## ABSTRACT

A compact polarized ion source is under construction at Kyushu university. The ion source is designed to produce the polarized deuteron beam for measurements of the fusion reaction rate up to the deuteron enetgy of Ed =150keV. Components of the ion source such as the dissociator, the high field permanent sextupole magnets, the rf-transition units and the electron bombardment type ionizer had been fabricated. In this paper, performance and present status of the ion source are summarized and a future prospect of experiment of a polarized nuclear fusion reaction is described.

## INTRODUCTION

The fusion reaction of $d+^3He \rightarrow {}^4He+p$ is expected to have the advantage of large energy yields without neutron production. However, reactions of $d+d \rightarrow {}^3He+n$ or $d+d \rightarrow {}^3H+p$ are associated with the $d+^3He$ reaction. The d+d reaction produces neutrons or tritiums which are cumbersome in the application to fusion reactors. Therefore, in the $d+^3He$ fusion, it is important to suppress the d+d reaction.

Kulsrud et al. predicted an enhancement or a suppression of the fusion reaction rate by polarization of the reacting nuclei[1]. This prediction triggered studies on the polarized deuteron fusion. The suppression of the polarized deuteron $(\vec{d}+\vec{d})$ reaction was not confirmed in the evaluation based on R-matrix method[2]. On the other hand, Shuy et al. predicted the suppression of the $\vec{d}+\vec{d}$ reaction by means of the DWBA method[3]. Since these works depend on their own models, there has been no exact decision on the cross section modification of the $\vec{d}+\vec{d}$ reaction.

Tanifuji derived the vector and tensor analyzing powers and cross sections of the $\vec{d}+d \rightarrow {}^3H+p$ reaction by employing the invariant-amplitude method[4]. The calculated results agreed with the experimental data very well. He has also shown the large suppression of the $\vec{d}+\vec{d}$ reaction rate from his model independent analysis[5]. Hence, it is necessary to carry out the precise measurement of the $\vec{d}+\vec{d}$ reaction cross section as a function of incident deuteron energy and also by changing the degree of polarization. A compact polarized ion source is under construction to produce the polarized deuteron beam for an experiment of $\vec{d}+\vec{d}$ reaction. In this paper, the performance of the ion source and our future prospect of the $\vec{d}+\vec{d}$ reaction experiment are described.

## POLARIZED DEUTERON SOURCE

A schematic diagram of an apparatus for the $\vec{d}+\vec{d}$ reaction experiments is shown in Fig.1. The conventional dissociator operates at a frequency of 30MHz. To produce the high intensity beam, the Al nozzle attached to the Macol accommodator is cooled down to 30K via a heat link made from flexible copper-braid. Three high field permanent sextupole magnets are employed for the Stern-Gerlach separation of the atomic deuterium beam. Profile of the magnetic fields in these magnets were measured precisely and good field quality had been confirmed by analyzing the results[6].

The weak field transiton unit (WFT) and two strong field transiton units (SFT) allow various polarization states of deuteron beam ionized by the strong magnetic field ionizer. The WFT consists of a rf-coil which operates at a frequency of 10MHz in the static field of 10G. The SFT is operated at a frequency of 430MHz by applying the static field of 68G for $2 \rightarrow 6$ transition, while for $3 \rightarrow 5$ transition, the static field of 146G is applied. The WFT was tested by placing between two sextupole magnets as shown in Fig.2. The beam intensity was measured at the outlet of the second sextupole magnet. The ratio of the beam intensity during the WFT turned on and off ($I_{on}/I_{off}$) indicates the efficiency of the WFT. Fig.3 shows an example of $I_{on}/I_{off}$ in the case of the atomic hydrogen beam. The value of $I_{on}/I_{off}$ is expected to be 0.5 in the ideal WFT. The WFT can be simply optimized by minimizing the $I_{on}/I_{off}$ value.

The electron bambardment type (EB) ionizer[7] is employed as a strong magnetic field ionizer. A calculation code of the electron and ion trajectories in the EB ionizer was developed considering the electric and magnetic field distribution inside the ionizer[8]. The construction of electrodes and the field potential was optimized by simulating the trajectories of electrons and ions. Ionized current was measured to be several μA at the outlet of the ionizer. A compact Wien filter is installed before accelerating the polarized ion beam to obtain arbitrary spin orientations. The polarized deuteron beam, which is extracted from the ionizer, is accecelated by the 150kV small accelerator.

As shown in Fig.1, the polarized deuteron beam of 150keV will be transported into the storage cell[9] in a scattering chamber and collides with a polarized gas jet target. The source of the polarized deuterium gas jet target was designed in the same manner as that of the polarized deuteron source described above. The intensity of polarized deuterium gas jet which enter the cell (30cm long,1cm diameter) was estimated to be $1 \times 10^{15}$ atom/sec in our design.The target thickness in the cell was evaluated to be $2 \times 10^{12}$ atoms/cm$^2$. Charged particles emmited by $\vec{d}+\vec{d}$ reaction are detected by the Si-surface barrier detectors which are arranged around the cell.

## CONCLUSION

A compact polarized deuteron source is designed to produce the polarized deuteron beam for the experiments of the $\vec{d}+\vec{d}$ fusion reaction up to Ed =150keV. Components of the ion source have been fabricated. Performance and present status of the ion source are summarized and a future prospect of the experiments of a polarized nuclear fusion reaction is described.

## REFERENCES

(1)  R.M.Kursrud, et al., Phys. Rev. Lett. 49(1982)1248.
(2)  H.M.Hofmann, D.Fick, Phys. Rev. C55(1985)1650.
(3)  G.W.Shuy, et al., Phys. Rev. C55(1985)1649.
(4)  M.Tanifuji, Phys. Lett. B289(1992)233.
(5)  M.Tanifuji, Private Communication. May 1993.
(6)  F.Noda, et al., Tech.Reports of Kyushu University 65(1992)557.
(7)  H.F.Glavish, Nucl. Instr. and Meth. 65(1968)1.
(8)  J.Sumita, et al., Tech.Reports of Kyushu University 66(1993)121.
(9)  W.Haeberli, T.Wise, J. Phys. Soc. Jpn. Suppl. 55(1986)483.

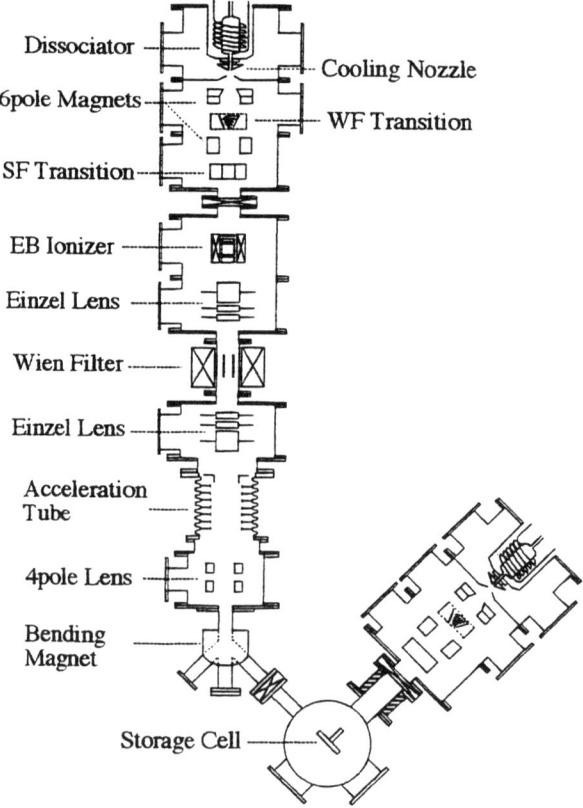

Fig.1 Apparatus for polarized fusion reaction research at low energy.
(under construction)

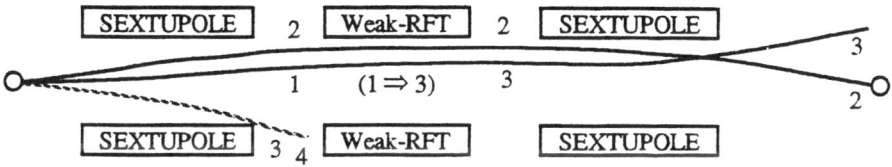

Fig.2.Schematic illustration of method for optimizing weak field rf transition unit.

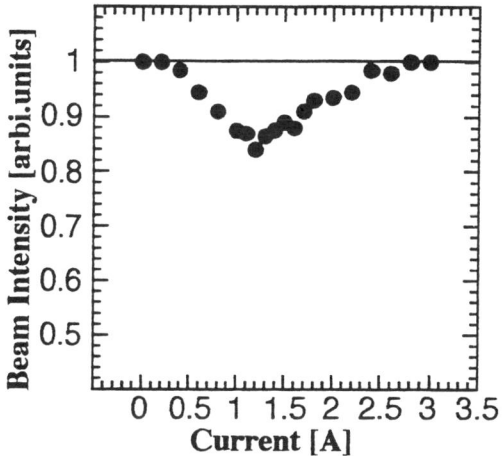

Fig.3. Beam intensity vs the current to the static magnet of weak field RFT.

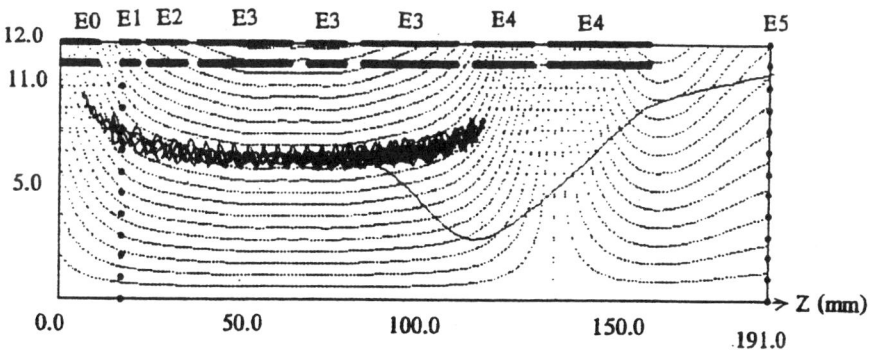

Fig.4. An example for the trajectory of the electrons and the proton in the ionizer.

# STATUS OF THE POLARIZED SOURCE FOR THE COOLER SYNCHROTRON COSY-JÜLICH

P.D. Eversheim, M. Altmeier, O. Felden, R. Gebel, M. Krekel, M. Schak
Institut für Strahlen- und Kernphysik, Univ. Bonn, 53115 Bonn, Germany

W. Kretschmer, A. Glombik, K. Mümmler, P. Nebert, G. Suft, R. Weidmann
Physikalisches Institut, Univ. Nürnberg-Erlangen, 91058 Erlangen, Germany

H. Paetz gen. Schieck, M. Eggert, S. Lemaître, H. Patberg,
R. Reckenfelderbäumer, C. Schneider
Institut für Kernphysik, Univ. zu Köln, 50937 Köln, Germany

## ABSTRACT

A polarized ion source of the colliding-beams type is presently installed at the cooler synchrotron COSY-Jülich. The source provides pulsed $\vec{H}^-$ - or $\vec{D}^-$ -beams. The working scheme, the relations which control the resulting intensity, emittance, and polarization of the source and its status is discussed.

## INTRODUCTION

Stacking injection of $H^+$ or $D^+$ into the COSY-ring is controlled by conservative forces. As a consequence, Liouville's theorem holds and injected particles have to be placed in phase-space volumes that are not occupied by particles circulating in the ring already. Since stripping $H^-$ or $D^-$ during injection is a non-conservative process this injection scheme is about an order of magnitude more efficient than stacking injection. Therefore, a polarized source for negatively charged $\vec{H}^-$ and $\vec{D}^-$ was designed and built by university groups from Bonn (atomic-beam production and handling), Erlangen (charge exchange and extraction), and Cologne (cesium-beam production and handling).

## FEATURES OF THE POLARIZED SOURCE

The negatively charged ions are produced in a charge exchange process of colliding-beams, namely e.g. a neutral nuclear polarized hydrogen-beam and a fast neutral cesium-beam according to:

$$\vec{H}^\circ + Cs^\circ = \vec{H}^- + Cs^+ \qquad (1)$$

The scheme of the source is shown in fig. 1.

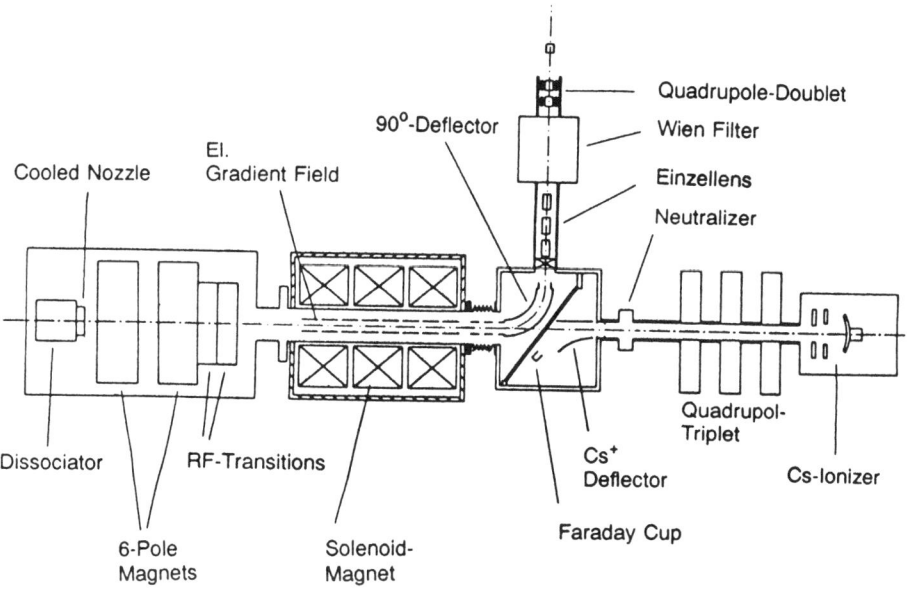

**Fig. 1** The polarized ion-source for COSY

In the atomic-beam source the molecules are dissociated in an inductively coupled 350 W rf-discharge. The atoms passing the aluminum nozzle channel of 20 mm length and 3.5 mm diameter are cooled to about 30 K. Thus, the atoms are slowed down[1] with the consequences that i) the first tapered sextupole magnet has an increased solid angle of acceptance in proportion to the beam temperature, and ii) the dwell time of the atoms in the charge exchange region increases with decreasing beam velocity. These beneficial effects are in part offset by gas scattering in– and outside the nozzle[2]. In order to reduce chromatic aberrations the atomic-beam transport system comprises a second sextupole magnet, the compressor magnet. Between these magnets or behind the compressor magnet up to two rf-transition units can produce transitions between the hyperfine states.

As an example fig. 2 shows the efficiency of the weak-field transition to produce a polarized $\vec{H}$-beam. The transition unit was tested in the Bonn polarized ion source in the presence of a 30 Gauss longitudinal magnetic field originating from the ECR-ionizer. Only after this perturbing field had been actively compensated by a set of Helmholtz coils polarization could be verified in a nuclear scattering experiment.

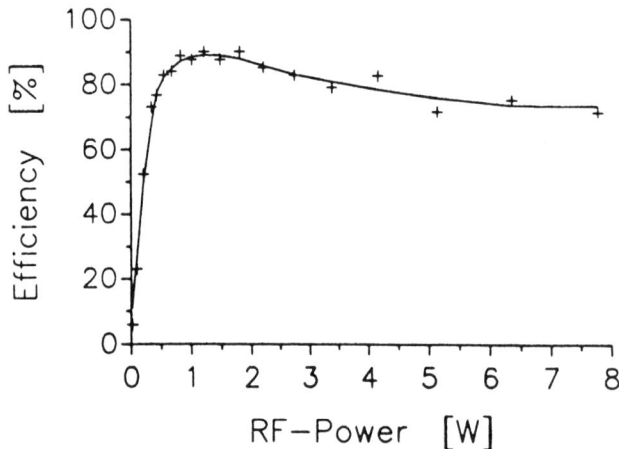

**Fig. 2** The efficiency of the weak-field rf-transition

In the polarized source for COSY these kinds of perturbing longitudinal fields are of the order of 3 Gauss.

In the charge exchange solenoid the polarization is preserved by the longitudinal field B with its vector potential $\vec{A}$. Table I shows how much polarization can be expected for a given solenoidal field.

In addition, the solenoid field defines the emittance of the beam.

**Table I** Calculated maximum polarization[3]

| Solenoid field [Gauss] | Max. polarization [%] |
|---|---|
| 0 | 50 |
| 500 | 85 |
| 1000 | 94,5 |

As a consequence of the conservation of total angular momentum $\vec{M}$ the

$$\vec{M} = e\vec{A}r + \vec{L} \qquad (2)$$

"spatial part" $\vec{L}$ of the total angular momentum will grow to the extent that the ion leaves the solenoid field and the vector potential becomes zero. At this point the "magnetic part" $e\vec{A}r$ of the total angular momentum has been converted to a macroscopic spatial angular momentum of the beam, usual beam handling devices can not cope with. The resulting transversal emittance for protons can be calculated from equ. (3)[4].

For the nominal geometric acceptance $\varepsilon$ of the injector cyclotron of:

$$\epsilon = 34{,}6 \cdot \pi \cdot B r^2 \cdot 1/\sqrt{U} \qquad (3)$$

B [gauss]; r [cm]; U [V]; $\epsilon$ [mm mrad]

$\epsilon$ = 500 mm mrad, a $\vec{H}^-$-beam potential of U = 4 kV, and a radius r = 0.5 cm where the ionization takes place, B must not exceed 1000 Gauss. This example together with the data of table I shows how the transversal emittance can be traded for polarization.

The longitudinal phase space (the energy spread of the beam) is adjusted by the electrical drift field inside the solenoid. This field accelerates the $\vec{H}^-$ ions toward the extraction end.

The fast $Cs^0$-beam for the charge exchange reaction in the solenoid is produced in two steps. First Cs vapor is thermally ionized on a hot porous tungsten surface at an appropriate beam potential of about 40-60 kV. Since it is difficult[5] to transport this $Cs^+$-beam further than some 10 cm, this beam has to be focused into the charge exchange solenoid by means of a magnetic quadrupole triplet. Normally this charged beam is electrically deflected in front of the solenoid into a Faraday cup. Just before the injection into the COSY-ring is due, the neutralizer placed between the quadrupoles and the solenoid is filled with Cs vapor. The charged $Cs^+$-beam becomes neutralized, and enters the 40 cm long charge exchange region inside the solenoid where the charge exchange process takes place selectively between atoms only. Therefore, unpolarized molecular background remains neutral and a high degree of polarization can be expected.

## STATUS AND RESULTS

The atomic-beam source and the charge exchange and extraction region have been set up at COSY. The Cs-beam source will follow. In the meantime the wiring for the remote control of the source will be completed.

Up to now degrees of dissociation in excess of 85% have been observed after some oxygen has been added to the hydrogen, providing an $H_2O$-coating which reduces the recombination of the hydrogen at the dissociator glass wall. The electrical and mechanical setting of the sextupole magnets behaves according to our Monte-Carlo simulation giving the maximum intensity for a 35 cm long drift-space between the magnets. The intensity measured with a compression tube at the position of the middle of the solenoid is presently $3 \cdot 10^{16}$ atoms per second in two substates.

The charge exchange region allows to ionize the $\vec{H}^o$ atoms at a potential of almost 30 keV.

With a quadrupole dublet a $Cs^+$-beam of 4 mA at 33 kV has been already delivered to a calorimeter spot of 0.75 cm radius[6]. A quadrupole triplet will be installed soon and the $Cs^+$ potential will be increased to accomodate a higher emission current from the tungsten surface. It is expected that this will lead to higher output currents of the Cs source at the ionizer location.

## ACKNOWLEDGEMENTS

The authors like to thank W. Grüebler and P. Schmelzbach for their technical support in building the sextupole magnets, J. Alessi for his support in building the neutralizer and his advice on the Cs source, E. Huttel for providing us with the solenoid and C. Gossett and the Seattle Nucl. Phys. Lab. for providing the opportunity to obtain first-hand colliding-beams experience.

This work was supported by the BMFT and the KFA-Jülich, Germany

## REFERENCES

1)  H.G. Mathews; Ph.D. Thesis University of Bonn, 1979
    H.G. Mathews, A. Kruger, S. Penselin, A. Weinig;
    Nucl. Instr. Meth. **213** (1983) 155

2)  W. Haeberli; Helv. Phys. Acta **59** (1986) 513

3)  W. Haeberli et al.; Nucl. Instr. Meth. **196** (1982) 319

4)  G.G. Ohlsen, J.L. McKibben, R.R. Stevens Jr., G.P. Lawrence;
    Nucl. Instr. Meth. **73** (1969) 45

5)  T. Wise, W. Haeberli; Nucl. Instr. Meth. **B6** (1985) 566

6)  H. Paetz gen. Schieck et al., contribution to this workshop

# Cs BEAM FOR THE NEW COSY-JÜLICH POLARIZED ION SOURCE

H. Paetz gen. Schieck, M. Eggert, S. Lemaître, H. Patberg,
R. Reckenfelderbäumer, and C. Schneider[1]

Institut für Kernphysik, Universität zu Köln, D-50937 Köln, Germany

## ABSTRACT

The status of the Cs part of the colliding-beams polarized ion source which is being built for the cooler synchrotron COSY-Jülich, is discussed.

## INTRODUCTION

A collaboration between the Universities at Köln, Bonn and Erlangen has developed a colliding-beams polarized ion source which is already installed in parts at the cooler synchrotron COSY[1]. The Cs part of this source is still under development.

## SETUP OF THE Cs SOURCE

As shown in fig. 1 the design follows that of ref. 2 with some modifications. These concern mainly the possibility of making large and precise remotely-controlled geometrical adjustments (extraction gap in z-, button electrode position in x- and y- and the entire source relative to the ionizer in x-, y-, $\phi_x$- and $\phi_y$-directions) under running conditions to optimize the output. The possibility of applying button voltages up to 60 kV is provided. Care has been taken to shield all insulators against Cs deposition. The necessity of using magnetic quadrupole lenses for optimal $Cs^+$ beam transport to the neutralizer and ionizer regions is investigated.

## PERFORMANCE ON THE TEST BENCH

The diagnostic tools for the investigation of the $Cs^+$ beam consist of beam-profile monitors, an emittance scanner and a carbon-covered calorimeter which will be used later on for the $Cs^0$ beam and also allows measurements of the beam current via secondary electron emission.

Profile monitor and emittance scanner results (see figs. 2.1 and 2.2) show that the rms emittance of the beam is $\epsilon = 65 \, \pi \, \mathrm{mm \, mrad}$ at 39 keV and 6.3 mA.

$Cs^+$ beams of up to 4 mA at 33 kV have been focussed (total transmission of 75%) at a distance of 165 cm using a quadrupole doublet. The necessary acceleration voltages for optimum transport of a given emission-limited $Cs^+$ beam follow

---

[1] Part of the collaboration between the Universities of Köln, Bonn and Erlangen, founded for building the COSY polarized-ion source.

an $I^{2/3}$ law as shown in fig. 3 revealing that the transport within the extraction gap is governed by space-charge limitation.

The source and beam-transport performance have been compared to numerical calculations using the code EGUN and show good agreement for the extraction region. This allows conclusions about the role of secondary electrons in the beam and the degree of space-charge neutralization. Results of these calculations are shown in figs. 4.1 and 4.2.

## FUTURE DEVELOPMENT SCHEDULE

The encouraging results obtained so far prompt us to attempt improving the performance of the source. It is planned to use a strong quadrupole triplet for focussing the $Cs^+$ beam. This together with higher Cs oven temperatures yielding higher beam currents and accordingly higher acceleration voltage will provide higher beam currents in the ionizer region. There its emittance will be carefully matched to the ionizer acceptance. Stable long-term operation will be tested and the final assembly of the complete source will take place in 1993 still. Simultaneously the parameters of the porous tungsten button will be investigated in order to improve its performance as to output and reliability.

## REFERENCES

1) D. Eversheim et al., contribution to this workshop
2) W. Haeberli et al., Nucl. Instr. and Meth. **196** (1982) 319

## ACKNOWLEDGMENTS

This work has been supported by the BMFT, Bonn and the KFA, Jülich. The help of many collaborators and of L. Fruh and the machine shop crew at Köln is gratefully acknowledged. Thanks go to J. Alessi and C. Gossett (and the Nucl. Physics Lab. at Seattle) for much advice and the possibility of hands-on experience on colliding beams.

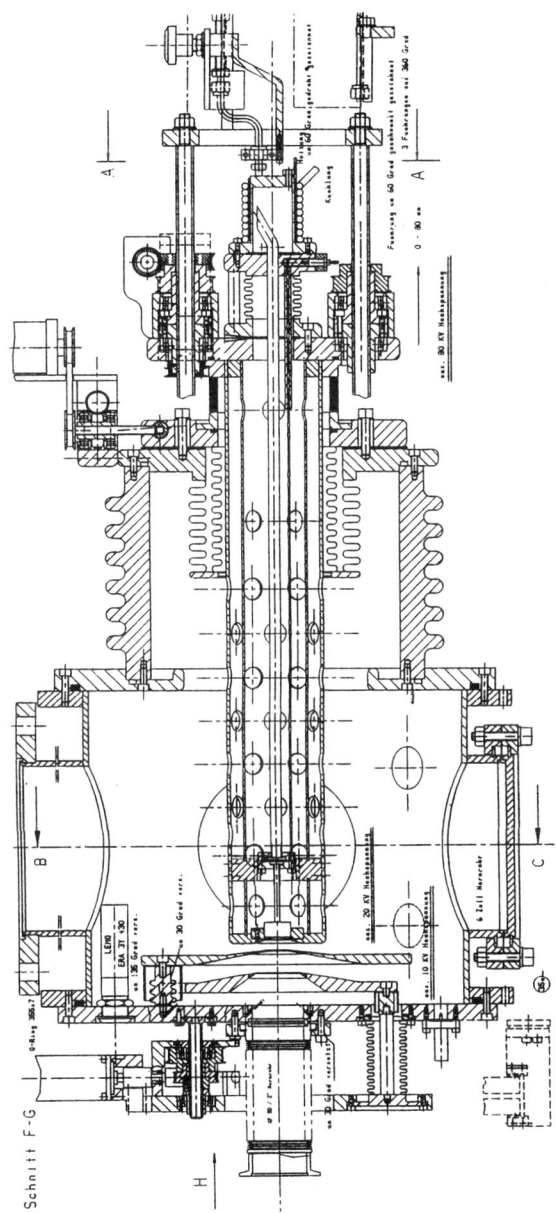

Fig. 1 Cs gun assembly

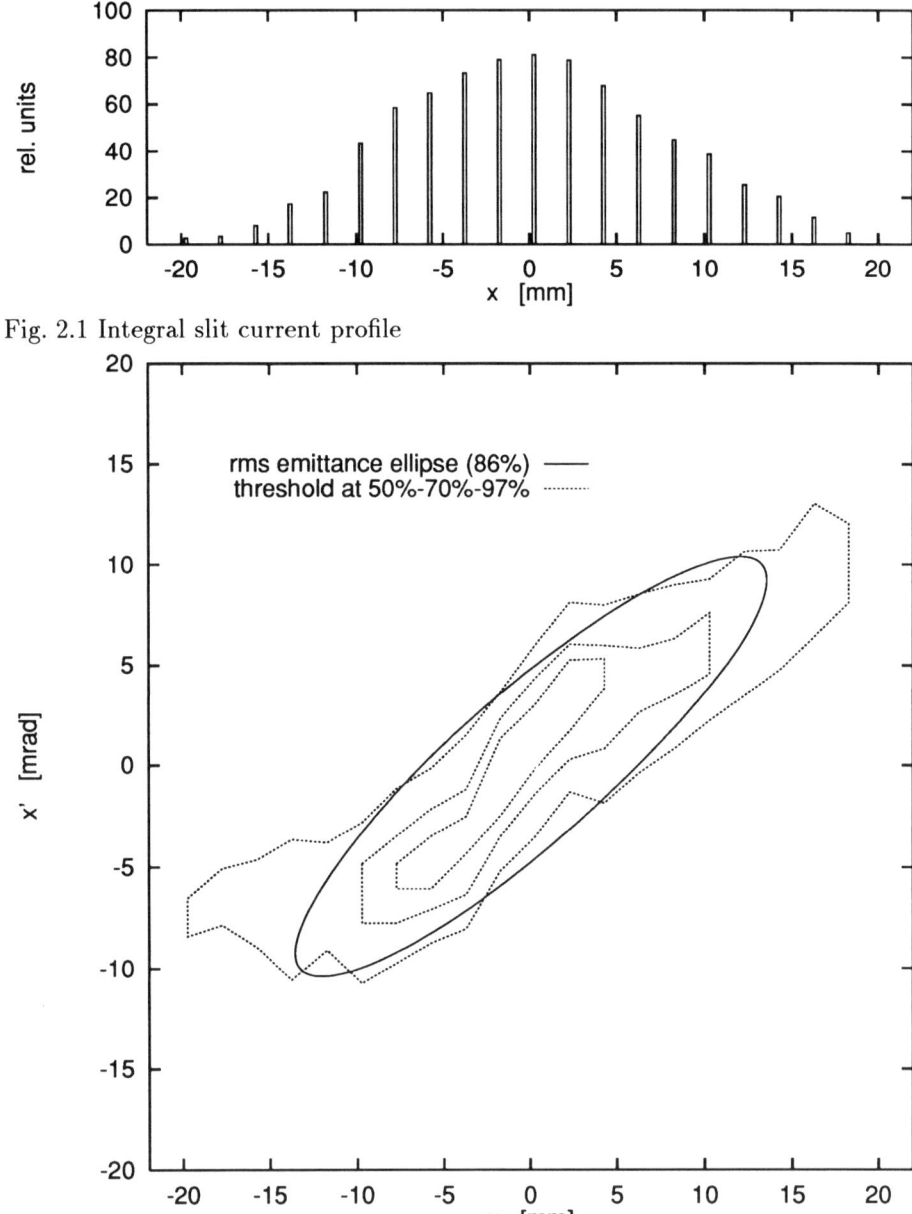

Fig. 2.1 Integral slit current profile

Fig. 2.2 Emittance plot of a $6.3\,\text{mA}$ $Cs^+$ beam at $39\,\text{keV}$, $\epsilon = 65\,\pi\text{mm}\,\text{mrad}$

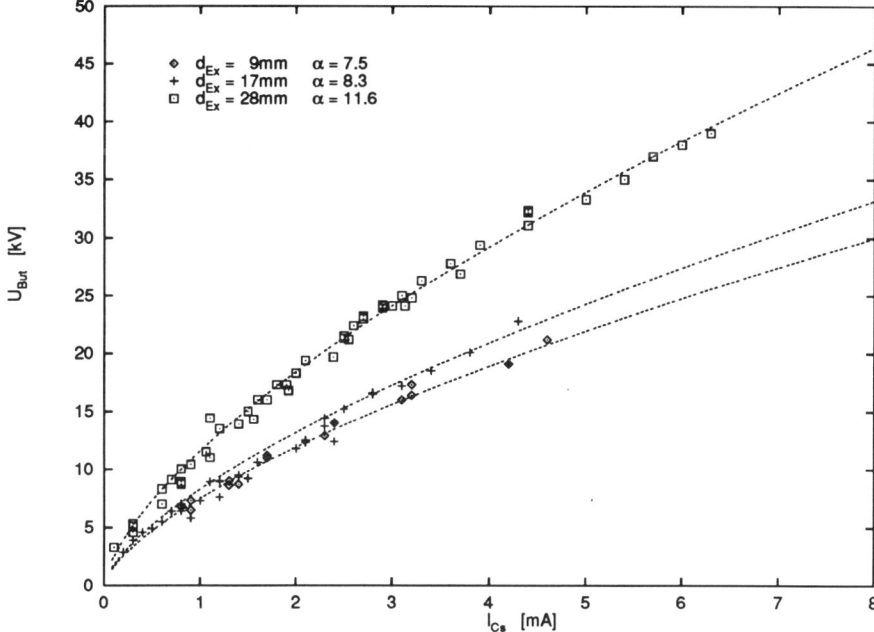

Fig. 3 $U_{But} = \alpha \cdot I^{2/3}$ dependence for optimum beam in calorimeter and different extraction gaps

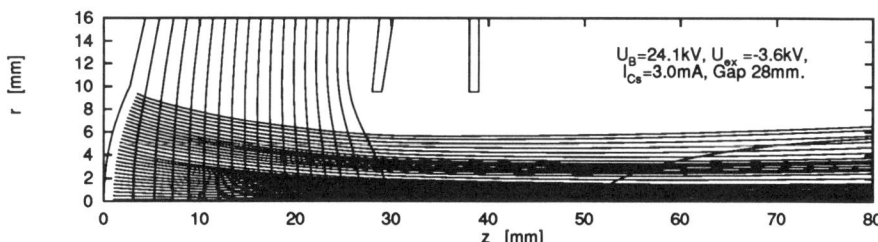

Fig. 4.1 Numerical calculation of $Cs^+$ flow using the EGUN code. The figure shows electrodes, equipotential lines $(5,10,...,85\%$ of $U_B$-$U_{Ex})$ and computed rays. Parameters are from experiment

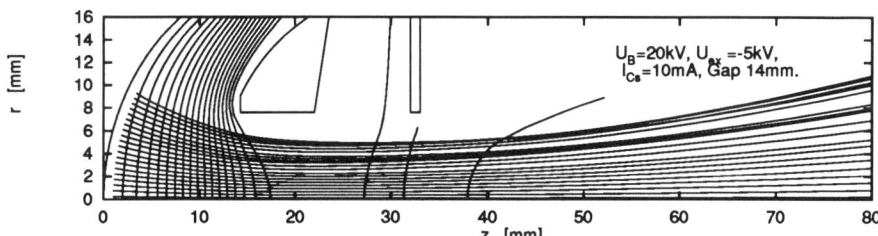

Fig. 4.2 Numerical calculation as above. Enhancement of beam current at smaller diameters with spherical geometry. This design is being tested

# A SOURCE OF POLARIZED NEGATIVE HYDROGEN IONS WITH DEUTERIUM PLASMA IONIZER

A. S. Belov, V. E. Kuzik, L. P. Nechaeva, G. A. Vasil'ev
Yu. V. Plokhinski and V. P. Yakushev
Institute for Nuclear Research of Russian Academy of
Sciences, Moscow, 117312, Russia.


V. G. Dudnikov

Budker Institute for Nuclear Physics of Russian
Academy of Sciences, Novosibirsk, 630090, Russia.

## ABSTRACT

An atomic beam – type source of polarized negative hydrogen ions with deuterium plasma ionizer has been developed. A charge-exchange reaction between thermal polarized hydrogen atoms and negative deuterium ions is used for production of polarized hydrogen negative ions. A polarized beam with peak current up to 150 $\mu$A and pulse duration 100 $\mu$s has been obtained from the source at repetition rate 5 Hz . The results of experimental test of the source developed and polarized beam emittance measurement are presented.

## INTRODUCTION

One of the methods of producing polarized negative hydrogen ions $(H^-)$ is using the charge-exchange reaction between polarized hydrogen atoms $(H^o)$ with thermal energy and unpolarized negative deuterium ions $(D^-)$. The method had been proposed by W. Haeberli[1].

The charge-exchange reaction between $H^o$ atoms and $D^+$ ions has been used successfully in the pulsed high-intensity polarized proton source developed at the Moscow Institute for Nuclear Research[2].

It has been suggested that the deuterium plasma ionizer of this source be used for production of $H^-$ ions as well[2]. This suggestion has been tested, but owing to the low density of $D^-$ ions in the plasma jet produced by an arc – discharge plasma source, only a 0.2 $\mu$A peak current of $H^-$ could be extracted from the ionizer[3]. In order to increase $H^-$ current a new deuterium plasma source has been developed which produces plasma enriched by $D^-$ ions. The source is surface – plasma type. The description of the source as well as results of its test are presented in this report.

## GENERAL DESCRIPTION OF THE SOURCE

The pulsed atomic beam - type polarized proton source with deuterium plasma ionizer that has been described in detail in ref.[2] was used to study the possibility of producing a beam of $H^-$ ions from the same source.

The schematic diagram of the source illustrating the method tested is shown in fig.1.

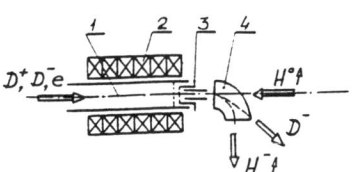

Fig.1. Schematic diagram of the polarized source.
1: charge-exchange region, 2: solenoid, 3: ion optical system,4: deflecting magnet.

The polarized atomic hydrogen beam and the deuterium plasma jet are injected from opposite directions into the charge-exchange region (1) inside the solenoid (2) that creates a longitudinal magnetic field of about 1 kG. The $H^-$ ions produced in the charge-exchange region are moved to the ion - optical system (3) and are accelerated to the energy 10 keV. The $H^-$ beam is deflected and is separated from the $D^-$ and electron beams in the bending magnet (4).

The parameters of the atomic hydrogen beam have been measured earlier.[4] The atomic beam is pulsed. The peak intensity of the atomic hydrogen beam entering the charge-exchange region is $2 \cdot 10^{17}$ atoms/s and the most probable velocity of the atoms is $2 \cdot 10^5$ cm/s.

## DEUTERIUM PLASMA SOURCE

In order to increase the density of $D^-$ ions in the deuterium plasma jet produced by the pulsed arc-discharge source (see ref.[2]) a plasma - surface converter has been installed in the region where the plasma jet expands. The schematic diagram of this new source is shown in fig.2. Molecular deuterium is injected into the discharge region by a fast electromagnetic valve (1). The typical gas flow rate is $3 \cdot 10^{17}$ mol.$(D_2)$/pulse. The arc discharge between cathode (2) and anode (4) is ignited by a 10 $\mu$s long 5 kV pulse applied to the gas valve relative to the cathode. Simultaneously, a voltage pulse with duration 200 $\mu$s is applied between the cathode and the anode, resulting in a low pressure arc discharge in hydrogen.

The diaphragms (3) between the cathode and the anode are at a floating potential. The coil (5) creates longitudinal magnetic field which increases the plasma output from the source.

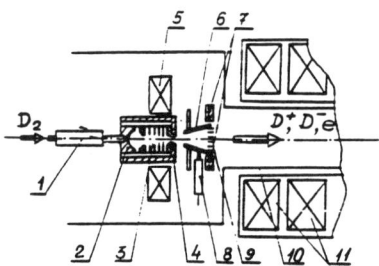

Fig.2. Schematic diagram of the D⁻ plasma source.
1: gas valve, 2: cathode, 3: diaphragms, 4: anode, 5: coil, 6: converter cone, 7: permanent magnets, 8: cesium oven, 9: grid, 10: electrostatic screen, 11: solenoid.

Deuterium plasma flows out of the discharge volume through a $\emptyset$4 mm hole in the anode and is injected into the plasma – surface converter (6), where part of the $D_+$ ions are converted into $D^-$ ions and initiates the secondary emission of negative ions from the surface.

The internal conical surface layer of the converter is made of molybdenum. A grid (9) made of molybdenum sheets is installed at the exit from the converter cone. The sheets are 0.1 mm wide and 3 mm high. Cesium is fed into the converter cone from the cesium oven (8) to catalyze the production of negative deuterium ions at the surface of the converter$^5$. Permanent magnets (7) create a transversal magnetic field in the region of the exit grid. The magnetic field is 250 G at the axis between the magnets. The transversal magnetic field is used to decrease the electron density in plasma and thereby to decrease the stripping of negative ions in the plasma jet downstream the converter.

The whole assembly of the plasma source is under a voltage of –10 kV, which is also applied to the screen (10) surrounding the plasma source and the ionization region.

## RESULTS

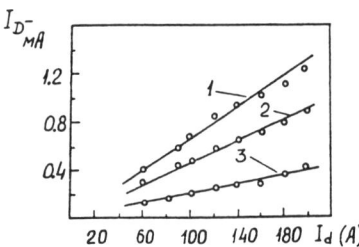

Fig.3. Peak current of the D⁻ ion pulse versus the discharge current in the plasma source for different currents in the source coil.
1: $I_c$ = 3.8 A, 2: $I_c$ = 2 A, 3: $I_c$ = 1 A.

The experimental results for the D⁻ current extracted from the ionizer are shown in fig. 3 versus the magnitude of the discharge current in the plasma source for different values of magnetic field created by the source coil. The D⁻ current was recorded by a Faraday cup after the bending magnet. The D⁻ current is increased with the discharge current and with the magnetic field in the plasma source. It is almost linearly proportional to the discharge current. The largest magnitude of the discharge current was 200 A. This limit was set by the existing power supply. At this value of the discharge current, the peak current of the D⁻ beam extracted from the ionizer was 1.2 mA.

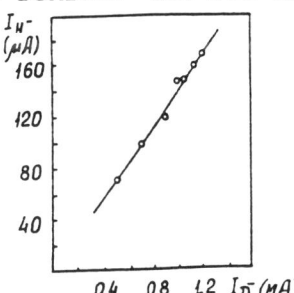

Fig. 4. Peak current of the H⁻ pulse versus the D⁻ ion peak current.

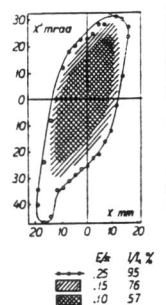

Fig. 5  Phase space area occupied by the beam of polarized negative hydrogen ions.

The experimental dependence of H⁻ ion current on the D⁻ ion current is shown in fig. 4. The H⁻ ion current grows linearly with the D⁻ current. The peak H⁻ current of more than 150 μA is obtained at D⁻ peak current 1.2 mA. The linear dependence allows one to suppose that it is possible to increase H⁻ beam current generated by such source futher by increasing D⁻ density in deuterium plasma.

Emittance of the H⁻ beam extracted from the source has been measured by a two slits method. The phase space area occupied by the beam is shown in fig. 5. The most part of the beam (76%) is inside normalized phase space area 0.15π cm mrad.

The authors are grateful to S. K. Esin for useful discussions and support of this work and to G. E. Derevyankin and V. S. Klenov for useful discussions.

## REFERENCES

1. Haeberli W. Nucl Instr. Meth., 1968, vol. 62, N 3, p. 355-357.
2. Belov A. S., Esin S. K., Kubalov S. A., Kuzik V. E., Yakushev V. P., Nucl. Instr. Meth., 1987, A 255, N 3, p. 442-459.
3. Belov A. S., Kuzik V. E., Yakushev V. P., Journal of Tech. Physics, 1990, vol. 60, N 2, p. 179-181.
4. Belov A. S., Kubalov S. A., Kuzik V. E., Yakushev V. P., Nucl. Instr. Meth., 1985, A 239, p. 443-454.
5. Dudnikov V. G. Proc. of the 4th International Conf. on Ion Sources 1991, Bensheim, Germany, Rev. Sci. Instrum., 63, 1992, p. 2660-2668.

# INVESTIGATION ON THE LIMITATION OF POLARIZED ION BEAMS PRODUCED BY AN ELECTRON BEAM IONIZER

Jean-Louis LEMAIRE
Laboratoire National Saturne 91191- Gif sur Yvette cedex France

## ABSTRACT

For several years, the Saturne National Laboratory polarized proton and deuteron ion source h a s produced $H^+$ and $D^+$ beam intensities of $700 \mu A$, that allow to accelerate $3. 10^{11}$ particles at 2.9 Gev in the Saturne synchrotron.
Developments were carried out on the atomic beam /1/, ionization process and sextupole selection. We present the last results concerning the improvements on the Electron Beam Ionizer (Anac type), that have led to an intense beam current of 2.5 mA.

## I/ STATUS OF THE OPERATING ATOMIC BEAM SOURCE

### 1) The polarized proton and deuteron beam source

The polarized ion source, which has been already described in reports /2/, /3/ is routinely operated at the Saturne National Laboratory for the production of positive polarized hydrogen and deuterium beams. It delivers currently polarized beam of 500 $\mu A$. More precise tuning allows beam intensity of 700 $\mu A$.

We recall that the atomic beam is pulsed for the purpose of the injection process in the accumulator ring Mimas. the pulse duration is then 1.5 ms . Since 1989 the ionizer is also pulsed, allowing better stability of the beam and higher extracted beam current as mentionned before.

So it is very uncomfortable to make major modifications without being sure of a net improvement on the Saturne beam performances and we took use of the shut down periods to make some improvements. Indeed, these modifications are related to the results obtained with the test bench and from analytical calculations.

### 2) The test bench results

Fig 1 shows the experimental arrangement of the test bench. Systematic studies of the atomic jet have been carried out through the whole temperature range from 6 Kelvin to room temperature. Liquid helium is fed from a liquid helium external tank to a reservoir setted close to the nozzle when one runs at 6K. An additional circuit using a separate closed loop helium refrigerator allows to cover the temperature region from 10K up to 300K.

We measured mainly two characteristic parameters of the beam:

The first parameter is the velocity : it is measured using a time of flight technique. A fast chopper rotating at 100 Hz, produces short atomic beam pulses which are detected after a flight path of 1.52m on an ionization open gauge. This gauge has been calibrated before. The chopper consisted of a rotating blade or a disk with a hole and could be positioned either after the nozzle (TdV1) or after the skimmer (TdV2). A serie of measurements have been made on $H_0$ and $H_2$ for all these configurations.They have given the same results.

106     

We then decided to use the configuration where the chopper is a rotating blade positioned after the nozzle (between the nozzle and the skimmer).

The second parameter is the dissociation degree.This quantity, defined as follows, give the relative populations of $H_0$ and $H_2$ in the beam. We use a mass spectrometer designed according to reference /4/, but built by ourselves. Incoming hydrogen jet, composed of $H_0$ and $H_2$ is ionized by a well defined in size electron beam. $H^-$ and $H_2^+$ ions are produced, then extracted through a 300V pulsed voltage and are identified, depending upon their time of flight through this apparatus.

We called this device a mass spectrometer. It is worth to notice that this device can be located anywhere downstream the atomic jet. So it can give the dissociation-recombination rate of the beam (dissociation degree) any place. So far, we have only made these measurements after the skimmer.

$$\alpha = \frac{H_0}{H_0 + 2 H_2} \approx \frac{H_0^+}{H_0^+ + H_2^+}$$

with the following ionization cross sections
$$\sigma_0 = 3.2 \; 10^{-17} \, cm^2 \quad and \quad \sigma_2 = 5.8 \; 10^{-17} \, cm^2$$

At the moment the measurements are made without a selecting sextupole magnet or a limiting aperture, simulating a sextupole . The nozzle was made out of copper, gold plated .

As results, we retained 3 regions:

**region 1** - 80K to 100K (precision in temperature is critical)
        dissociation degree 45%
        atom velocity 1700 m/s
        Mach number ~ 7

**region 2** - 35K without nitrogen
        dissociation degree 8%
        atom velocity 1150 m/s

        - 35K with nitrogen  (temperature has to be adjusted very precisely under 34K)
        dissociation degree 45%
        atom velocity 1300 m/s
        Mach number ~ 6

**region 3** - 6K
        dissociation degree very small

To our understanding, unless there is a very strong density dependence recombination effect at 6K which could be overcome in some future new design taking use of new ideas, we consider that region # 1 and region # 2 give comparable results. As a matter of fact, from our experience we normally run in operation the source at the region # 1 temperature. Nevertheless, region # 2 has a real interest if nitrogen is added; the $H_0$ density is ~10% higher and the atom velocity is smaller. Then the sextupole magnet should be easier to design and build. But in this case the temperature must be controled precisely and the nitrogen gas fed directly into the nozzle and not into the bottle were the dissociation takes place, in order to keep a beam stable (our results concern a pulsed atomic jet).
We have confirmed with the test bench that region # 1 is the best at present for our operating source. We are now confident that the atom velocity is very closed to 1700 m/s and corresponds to

a supersonic jet.

### 3) Improvements on the operating source

a- <u>sextupole:</u> Initially, the sextupole magnet was tapered on the first half part and had a constant aperture on the second half part. It was also designed to separate atoms produced at 300K. As the sextupole was supposed to be saturated, the pole tip field was constant along the magnet. We have made modifications of the bore radius aperture in order to take advantage of the lower velocity atoms that were produced when the nozzle became cooled down to 80K. A new tapering of the first part and an enlargement of the second part as shown on fig 2, has been decided after the last results have been obtained on the test bench and monte-carlo simulations optimizing the acceptance of the sextupole. ( Fig 3 shows the result of the monte-carlo simulation convoluted with a gaussian beam having a fwhm = 500m/s). Consequently to this modification, it became necessary to power the coil with 300A instead of the usual 200A in order to get the proper focusing magnetic field. The net improvement closed to 20% was encouraging, but the power supply had to be pulsed slowly to deal with prohibitive thermal heating of the sextupole magnet coils.

After several months of satisfactory operation, we slowly lost the polarization of the proton beam during experiments. Investigations showed that the ferrites which provide the DC magnetic field in the transitions were demagnetized because of the pulsed magnetic fringing field leaking from the sextupole -varying magnetic flux in time- (fig 4). Finally, it was decided to run again at 200A DC after having remagnetized the ferrites. We had missed an improvement but experienced that no permanent magnet should be located closed to a pulsed magnetic field.

b- <u>nozzle:</u> - For many years the nozzles we used, were made of copper, gold coated in order to be chemically clean. They gave satisfaction for stability except that they had to be replaced many times a year due to a white deposit that choked the extremity of the nozzle. This deposit was analyzed and turned to be $SiO_2$ coming certainly from the pyrex bottle, eroded by the RF discharge (quartz is worse). We now use aluminum nozzle, coated with a silicone film RTV141 from Rhône Poulenc that works as good as gold coated copper, as long as recombination and stability are concerned but is cheaper and is a lot better to avoid the deposit at the extremity of the nozzle.

## II/ INVESTIGATIONS ON THE ELECTRON BEAM IONIZER

### 1) Operating modes

a- <u>DC mode:</u> - The original EB Ionizer was design to run with DC voltages on the electrodes. $E_1$ extracts the electrons from a tantalum filament which has in our case an omega shape. An electronic density is built up in a reflex configuration obtained with electrode voltages surrounded by solenoïdal magnets. These electrodes are named $E_2$, $E_3$, $E_4$, $E_5$ /3/.
$E_3$ is the unipotential electrode where the atoms become ionized.

b- <u>pulsed mode:</u> - In order to lengthen the life time of the filament we pulsed $E_1$ .This led to a more stable beam because of smaller electron losses on the electrodes. Then we pulsed also $E_2$ after we realized that this electrode was contributing to the electron extraction of the filament. The pulse duration is set to 1.5 ms and determines the beam pulse length. $E_3$ is always kept DC: we call this mode, the pulsed mode.
This modification was actually a great improvement because it allowed better stability of the electron beam and consequently better stability of the ion beam In this case higher voltages on the electrodes

led to the result of 700 $\mu$A beam. The present limitation is given by the thermal heating of the coils of the solenoïdes. The source is routinaly run on this mode.

c- accumulated mode: - We experimentally saw that pulsed mode operation permited new voltage configurations and by systematically measuring the energy spread of the ion beams we observed that the energy spread was dependant on the filament heating and on these voltages set up. There is no doubt that the energy spread is related to the transverse potential well into which the ionization process of the atomic beam occurs, but up to now it was very difficult to tell if, once produced the ions fill up the potential well before being extracted. This is a crucial point because it defines the neutralization state of the electron beam, the time constant and the efficiency of the ionization process which are unknown on this type of ionizer.

As shown on fig 5, the $E_3$ electrode has been cut in two parts for which, voltages can be adjusted separately. If the second part called $E'_3$ is set to a voltage equal to $E_3 + \mathcal{E}$, the positive ions should not be extracted. This is exactly what happened, except that $E'_3$ had to be equal to $E_3 + V_{SC}$ where $V_{SC}$ is a positive value voltage which corresponds to the potential well into which the ions are created. Depending upon the configuration sets up , we have a new direct experimental way to measure the ion beam energy spread or the potential well or the electron beam current.

Fig 6, shows the voltages configuration. As can be seen, two parameters are optimized: $t_{min}$ is the delay after which the beam is extracted and $t_{dev}$ is the duration of the extraction. Both can be adjusted. In order to get the maximum of beam, $t_{min}$ is optimized experimentally between 40 $\mu$s to 60 $\mu$s. For higher values of $t_{min}$ no gain is obtained. $t_{dev}$ has a minimum value ~ 30 $\mu$s (short spilling) which corresponds to the peak current of 2.5 mA as shown on fig 6 and 7: in this case the beam pulse has a triangular shape. If the value of $t_{dev}$ is increased from 40 $\mu$s to 1.5 ms ( longer spilling), a plateau appears on the signal and this value is the same as for the pulsed mode current. No particular other tuning of the ionizer is required.

## 2) Emittance measurements

The emittance is then measured about 1 meter downstream after the beam is extracted This apparatus consists of a sampling hole and a multiwire profiler collecting the beam current passing through the hole. This device is steped accross the beam section and emittance is reconstructed in the wellknown usual way. This device is very precise : the hole diameter is 0.5 mm and the spacing of the collecting wires is 300 $\mu$m /5/.

We have measured the emittance of the beam for three configurations; the first one corresponds to the pulsed mode, the second one corresponds to the accumulated mode where the emittance is measured in the peak current value ( fast electronic amplifiers were used in this case) and the third one corresponds to the " longer $t_{dev}$ accumulated beam"

All the emittances are the same . Not in terms of matching parameters but in terms of phase space volume (fig 8).

pulsed mode :        total beam current on a faraday cup is 650 $\mu$A

                    normalized emittance is $E_n / \pi$ = 1.2 $10^{-6}$ m . rad        for 160 $\mu$A

accumulated mode:    total beam current on a faraday cup is 2. mA

                    normalized emittance is $E_n / \pi$ = 1.2 $10^{-6}$ m . rad        for 510 $\mu$A

That shows that the intensity of the beam when it is produced in the accumulated mode is 4 times higher in the same emittance volume than in the other modes (the brillance has strongly increased). It also tends to prove that in this case the ionisation region approaches the space charge neutralization.

### 3) Conclusions

We have obtained for the first time a peak current of positive " polarized " proton beam of 2.5 mA from an Electron Beam Ionizer by running the ionizer differently as one normally does it. We called this mode, the accumulated mode.
Space charge neutralization seems to be reached and allows to get smaller emittances as being measured. But we still have to check if the polarization of this high intensity beam is correct.
In the future, this mode of operation which produces ion beams in a space charge neutralized region will be more looked into in order to reach the real limits of this kind of ionizer.

Aknowledgments: I want to thank P.A Chamouard for supporting this work, for constant interest and for encouraging theses new ideas. I am indebted to P.Y Beauvais and R. Ferdinand for helpfull discussions and for developping the pulsed power supplies of the ionizer, the instrumentation,the monte-carlo calculation for the design of the sextupole magnet and the emittance measurement device. I also thank J.P Penicaud for usefull discussions and magnetic calculations on the sextupole.

### references:

/1/ Present results on the polarized proton and deuteron source at Saturne: J-L Lemaire
Proceedings of the international workshop on polarized ion sources and polarized gas jets
Tsukuba, Japan 1990

/2/ Saturne latest results for polarized beams: P-A Chamouard, J-M Lagniel, J-L Lemaire,
A Tkatchenko, Y Yonnet
Proceedings of the 7th international conference on polarization phenomena in nuclear physics
Paris, France 1990

/3/ The Saturne polarized particle source: J-L Lemaire, R Vienet, P-Y Beauvais
International workshop on polarized sources and targets
Montana, Switzerland 1986

/4/ Study of the velocity distribution in an intense pulsed hydrogen beam
A.S Belov, S.A Kubalov, V.E Kuzik,V.P Yakushev
Nuclear Instruments Methods A 239 (1985) 443-454

/5/ Mesure fine de l'emittance d'un faisceau d'ions, application aux particules polarisées et aux ions
lourds: R Ferdinand, P-Y Beauvais, J-L Lemaire
Internal report  LNS - SM 93/02

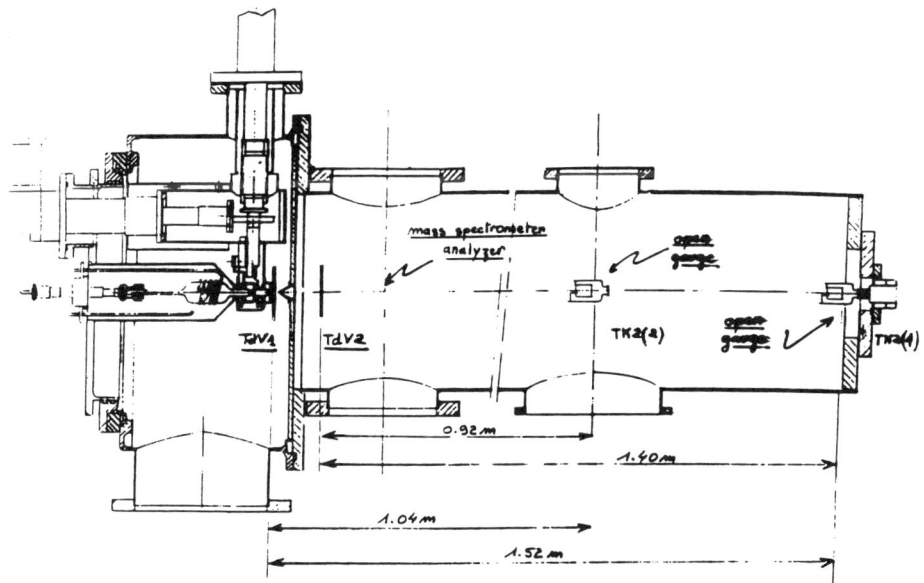

FIG 1 : TEST BENCH SET UP

FIG 2 : MODIFIED SEXTUPOLE MAGNET APERTURE

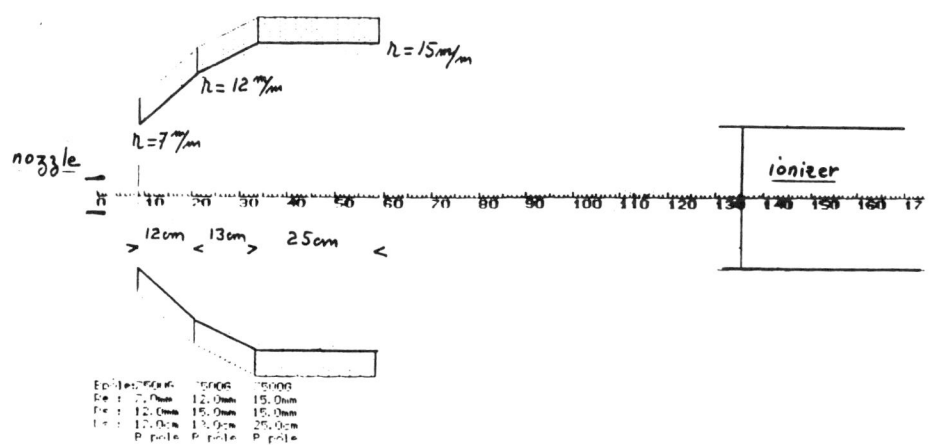

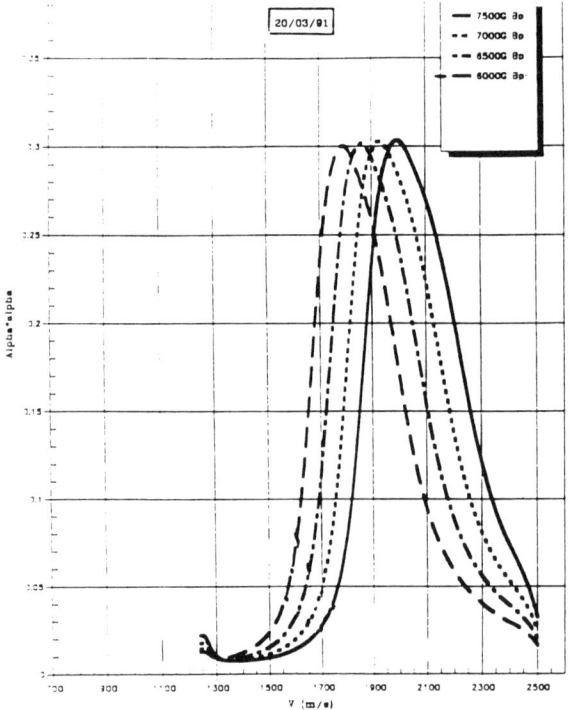

FIG 3 : SEXTUPOLE ACCEPTANCE FROM MONTE CARLO SIMULATION

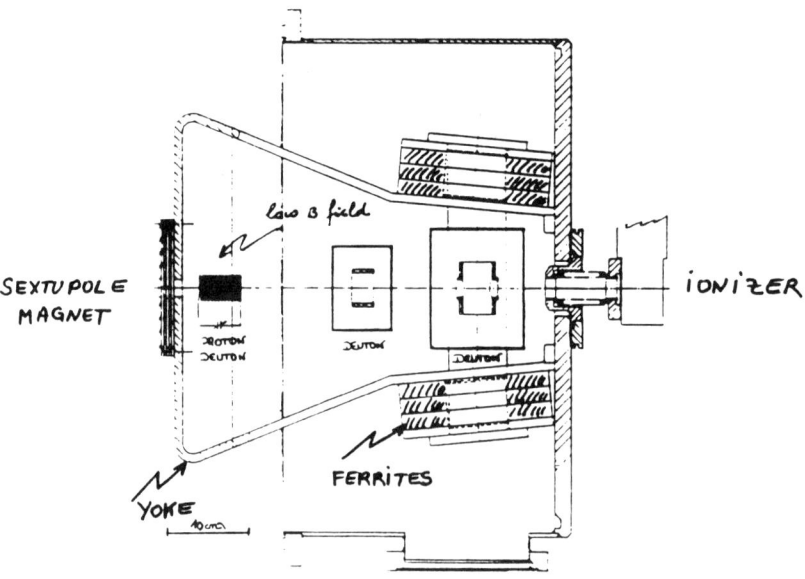

FIG 4 : TRANSITIONS REGION

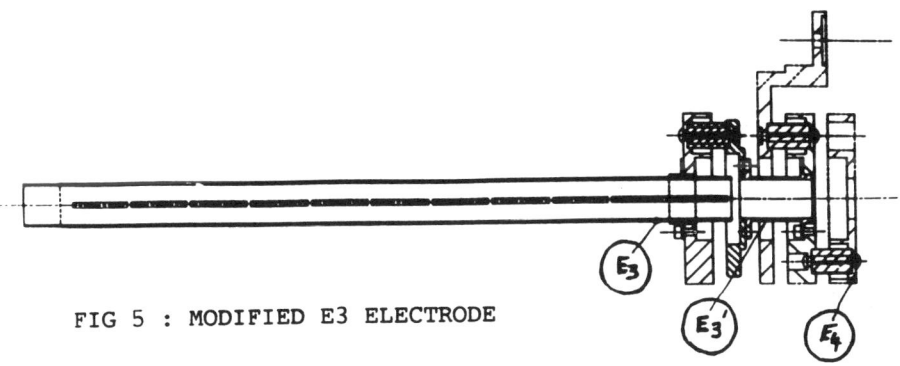

FIG 5 : MODIFIED E3 ELECTRODE

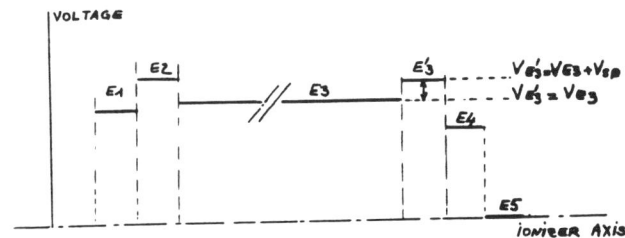

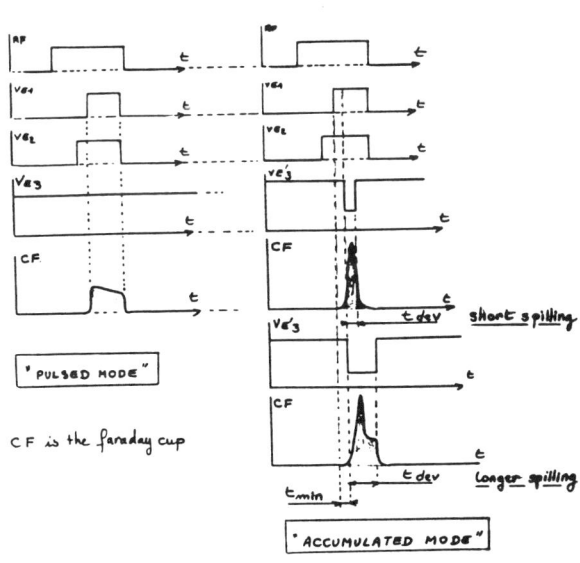

FIG 6 : VOLTAGE CONFIGURATIONS FOR THE ACCUMULATED MODE

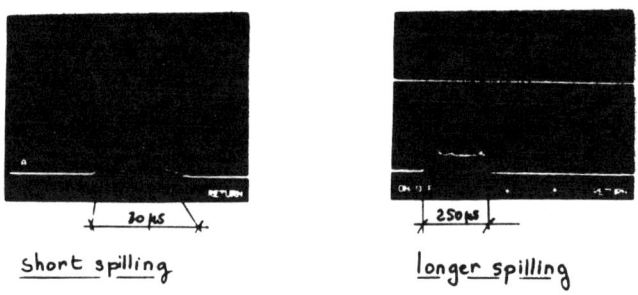

Short spilling          longer spilling

FIG 7 : BEAM SIGNALS ON A FARADAY CUP

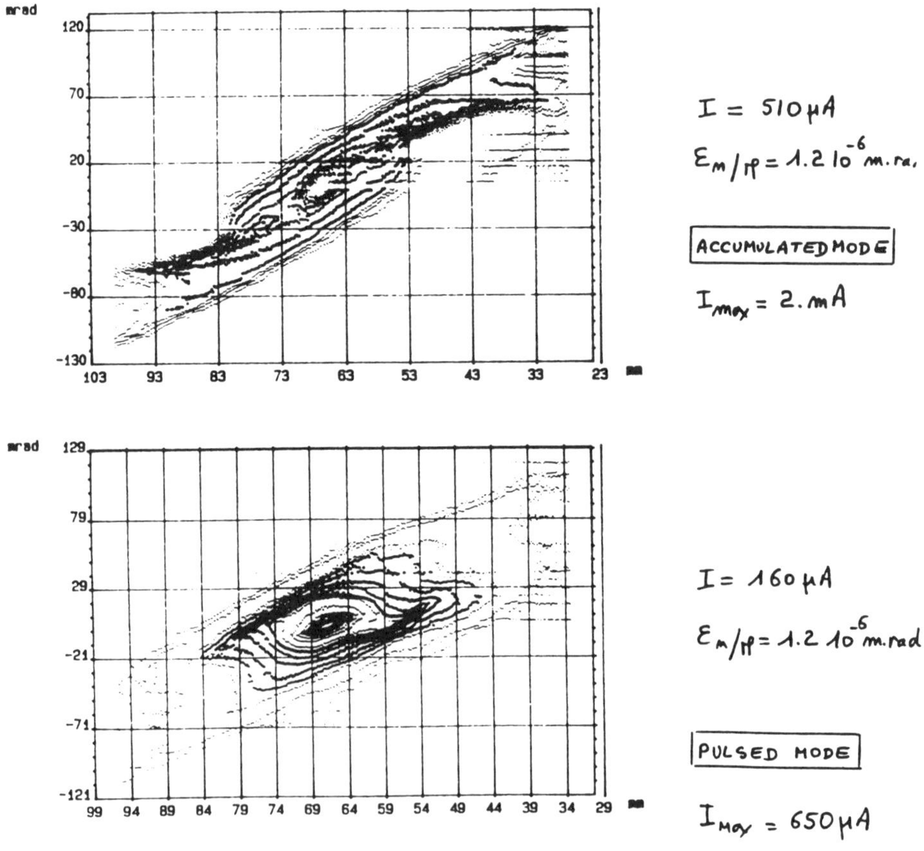

$I = 510\,\mu A$

$\varepsilon_{m/\pi} = 1.2\ 10^{-6}\ m.ra.$

ACCUMULATED MODE

$I_{max} = 2.\,mA$

$I = 160\,\mu A$

$\varepsilon_{m/\pi} = 1.2\ 10^{-6}\ m.rad$

PULSED MODE

$I_{Max} = 650\,\mu A$

FIG 8 : EMITTANCE RESULTS

# ON-LINE, EFFICIENT, LOW-ENERGY PROTON AND DEUTERON POLARIMETRY

E. R. Crosson, C. D. Roper , and T. B. Clegg
Triangle Universities Nuclear Laboratory, Durham, NC. 27708-0308*

## ABSTRACT

A polarimeter based on a three-level interaction in the n=2 states of hydrogen-like atoms has been developed to reveal rapidly the relative hyperfine state population of H or D beams. Initial tests results with this polarimeter are briefly described, and plans for installing this device permanently on an atomic beam polarized ion source are discussed.

## INTRODUCTION

Though beam polarimetry is necessary for any practical use of spin polarized proton or deuteron beams, polarimeter development often lags development of intense sources of these polarized particles. As a result, laboratories can lose valuable accelerator and experimental time determining whether a ion source is providing beams appropriately spin polarized for an experiment. In addition, once an experiment is underway, continual beam polarization monitoring is essential to assure accuracy of the data collected.

Traditional proton and deuteron polarimeters have long been based on nuclear scattering[1]. Though they are well-understood and can be calibrated absolutely, they suffer from low efficiency and often low analyzing power at the beam energy of the primary experiment. Modern polarimetry increasingly employs devices based on atomic interactions. These are favored because of the increased efficiency associated with the larger underlying atomic cross sections and because of the intrinsically high analyzing powers associated with hyperfine state selectivity of these interactions.

Atomic polarimetry for hydrogen-like atoms utilizes unique features of their excited n=2 states[2]. Such excited states can easily be created by charge-exchange of a low-energy (~0.5-10 keV) polarized ion beam in an alkali vapor. Atoms in the $2S_{1/2}$ state are metastable, with a lifetime of ~1/7 s in the absence of applied electromagnetic fields. Inside a magnetic field of ~57.5 mT, the $2S_{1/2}$ atoms with electron spin projection $m_J = -1/2$ have an energy which becomes degenerate with $2P_{1/2}$ atoms having $m_J = +1/2$. Thus, if a small electric field is simultaneously applied to induce Stark mixing of these levels, the $2S_{1/2}$ atoms having $m_J = -1/2$ are selectively caused to decay to the ground state. The remaining excited $2S_{1/2}$ state atoms with $m_J = +1/2$ have hyperfine state populations reflecting the nuclear polarization of the incident ion beam. In the most common method of atomic polarimetry used[3], these atoms then emerge from the magnetic field and are intentionally caused to decay while measuring the resulting flux of Lyman-$\alpha$ radiation. This photon flux carries information about the incident ion beam's polarization, so intensity ratios with the incident ion beam polarized and then unpolarized reveal this polarization clearly, conveniently, and accurately.

This technique is far less convenient and accurate for deuteron beam polarimetry because of the need to monitor *both* its vector *and* tensor polarization. Tests recently in our laboratory developed an old polarimetry idea[4] which works with both H *and* D

*Work supported in part by the U.S. Department of Energy under Grants No. DE-FG05-91-ER40619 and DE-FG05-88ER40442.

beams. That scheme utilizes the nuclear "spin filter" developed in the late 1960's at Los Alamos for the Lamb-shift polarized source[5]. The spin filter cavity, placed inside a highly uniform axial magnetic field, produces a carefully designed axial distribution of dc and rf electromagnetic fields. These enable, for $2S_{1/2}$ H or D atoms traveling down its axis, an atomic three-level interaction between $2S_{1/2}$ states having $m_J = +1/2$ and $-1/2$ and $2P_{1/2}$ states with $m_J = +1/2$. When metastable H or D atomic beams are allowed to enter the spin filter, $2S_{1/2}$ atoms having unique nuclear spin projection $m_I$ may be selected and allowed to pass. Only the dc B-field must be changed to pass atoms with an alternate $m_I$.

## RESULTS OF INITIAL SPIN FILTER POLARIMETER TESTS

The spin-filter from our laboratory's old Lamb-shift source was temporarily installed downstream from our atomic beam polarized ion source[6]. Its polarized ion beam was converted into a similarly polarized metastable beam by charge-exchange in cesium vapor inside a ~0.2 T magnetic field. These nuclear-spin-polarized metastable atoms then entered the spin filter. It selected those having unique $m_I$ and allowed them to pass into a downstream chamber where they were caused to decay in a region of high electric field viewed by an ultraviolet-sensitive phototube. By scanning the magnetic field inside the spin filter to pass successively those atoms with different $m_I$, the photon flux revealed the relative populations of the individual hyperfine states of the metastable beam. Figure 1 shows typical plots obtained for H and D. No change of the spin filter operating parameters was required between the measurements for H and D. From the relative peak sizes shown for different hyperfine states, one can immediately calculate the incident beam's nuclear polarization.

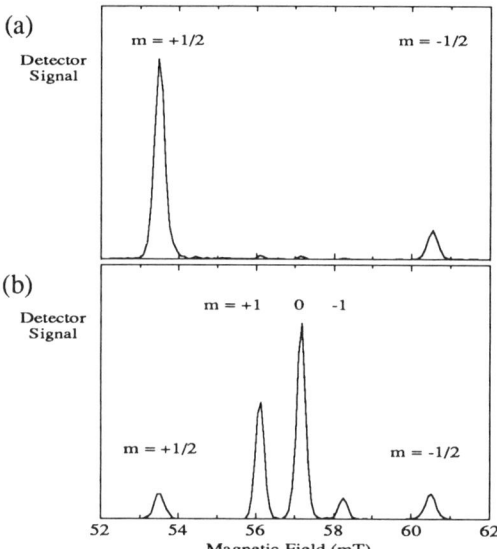

Fig. 1. Spectra generated for polarized beams of (a) hydrogen with $p_z = 0.75 \pm 0.02$ and (b) deuterium with $p_z = 0.29 \pm 0.02$ and $p_{zz} = -0.71 \pm 0.03$.

Computer spectra like those shown in Figure 1 were easily obtained in ~10 s. They have immediate appeal, and users can learn quickly how to interpret relative peak sizes in terms of their own beam polarization requirements.

Even more convenient for tuning and optimizing the beam's polarization, however, was connecting the phototube output current directly to an electrometer. When the spin filter was tuned to pass atoms in an individual hyperfine state, a typical electrometer signal was ~300nA. This could then be peaked up *in real time* by tuning transition unit parameters affecting the beam polarization. Alternatively, the phototube output could be sent through a current-to-voltage amplifier and into a storage oscilloscope. There spectra for H or D similar to those in Figure 1 could then be displayed. With a 1Hz refresh rate of the oscilloscope screen, this method too could be used to gain immediate insight into ion source transition unit performance. Both these latter techniques were user friendly and extremely efficient.

Further tests revealed that the same information about the primary beam's polarization could be obtained whether the cesium density was adjusted for optimum production (1) of metastable atoms, or for optimized (2) positive or (3) negative output polarized ion current. In the latter two cases, the photon flux was reduced to 7% and 40%, respectively, of that in case (1). Values of the beam polarization obtained under these three conditions agreed to within our measurement precision of <2%. These results led us to begin planning a permanent installation of the spin filter polarimeter on our polarized ion source as a diagnostic device for all users.

## PLANS FOR PERMANENT SPIN-FILTER POLARIMETRY

Since metastable atoms used by the polarimeter are quenched to the ground state in even modest electric fields, and since correct polarimeter operation requires a uniform B-field of better than 0.5 G over its spin filter cavity, its most practical installation on our atomic beam polarized source[7] inserts it into the beam axis immediately downstream of the cesium charge-exchange canal. If the canal and the polarimeter cavity are at the same dc potential and ion beam acceleration occurs only after the cavity, then the electric field of this first ion beam accelerating gap can serve also to quench the "spin-filtered" metastable atoms. Reflection the resulting Lyman-α decay radiation from that acceleration gap into an ultraviolet-sensitive phototube then can provide continual polarimetry for all ion source users.

We are working now on engineering details of this design scheme. The B-field needed will be provided by additional external coils similar to those which have until now provided the B-field for the cesium charge-exchange region of our source. We have used the computer code POISSON[8] from Los Alamos to design the number and placement of these coils and the size and dimensions of an iron flux return to achieve the B-field uniformity required. Output of a typical calculation shows that we can expect to achieve the 0.5 G uniformity needed over 20 cm by adding only 26.5 cm to the overall length of the source. We include in the flux return system an iron plate between the cesium canal and the polarimeter cavity. This enables maintaining this uniformity while allowing some minor tuning of the coil currents for the ECR ionizer and the cesium regions to achieve optimum output beams from the ion source. Nevertheless, we foresee providing for on-line, computer-controlled measurement of the axial B-field in the region of the cavity and computer adjustment of the coil currents as necessary.

The scheme shown here may also provide for efficient polarimetry of ions emerging from polarized gas jet targets[1] if it can be shown to have adequate sensitivity to molecular components of this ion flux. This is really a question of determining the relative production rate of 2S metastable atoms from charge exchange of molecular ions in cesium and whether the polarization is preserved in the charge-exchange process.

## REFERENCES

1. J.S. Price and W. Haeberli, Nucl. Instrum. Meth. A326, 416 (1993).
2. W.E. Lamb, Jr. and R.C. Retherford, Phys. Rev. 81, 222 (1951) .
3. A.N. Zelenski, S.A. Kokhanovskii, V.M. Lobashev, and V.G. Polushkin, Nucl. Instrum. Meth. A245, 223 (1986); A.S. Belov, S.K. Yesin, S.A. Kubalov, V.E. Kuzik, A.A. Stepanov, and Y.P. Yakushev, Nucl. Instrum. Meth. A255, 442 (1987).
4. J.E. Brolley, G.P. Lawrence, and G.G. Ohlsen, in Proc. of 3rd Int. Symp. on Polarization Phenomena in Nuclear Reactions, Eds. H.H. Barschall and W.

Haeberli, (Univ. of Wisconsin Press, Madison, 1971) p. 848.
5. J.L. McKibben, G.P. Lawrence, and G.G. Ohlsen, Phys. Rev. Lett. 20, 1180 (1968).
6. S.K. Lemieux, T.B. Clegg, H.J. Karwowski, W.J. Thompson, and E.R. Crosson, accepted for publication in Nucl. Instrum. and Meth. A.
7. T.B. Clegg, H.J. Karwowski, S.K. Lemieux, R.W. Sayer, E.R. Crosson, W.M. Hooke C.R. Howell, H.W. Lewis, A.W. Lovette, H.J. Pfutzner, K.A. Sweeton, and W.S. Wilburn, submitted to Nucl. Instrum. Meth. A.
8. POISSON (LAACG), The Los Alamos Accelerator Code Group, AT-6, Mail Stop H829, AT-Division, LANL, Los Alamos, NM 87545, U.S.A.

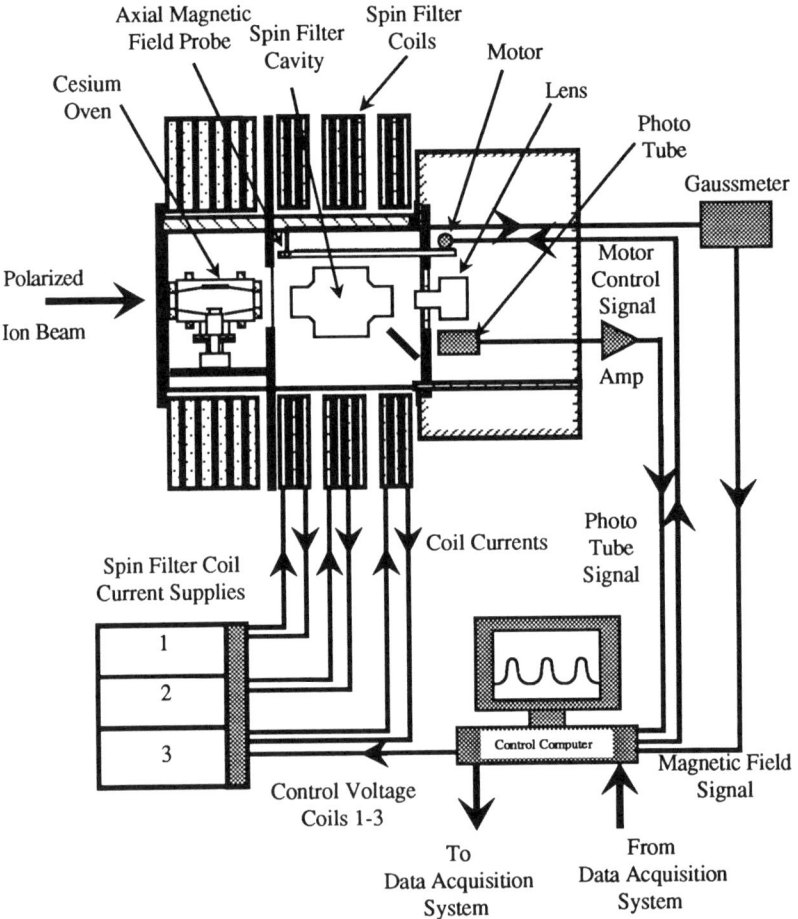

Fig. 2. Schematic showing the components of the planned spin filter polarimeter installation on the TUNL atomic beam polarized ion source.

# ROUNDTABLE DISCUSSION ON ATOMIC BEAM ION SOURCES

Thomas B. Clegg

Department of Physics and Astronomy, University of North Carolina, Chapel Hill, NC
27599* and Triangle Universities Nuclear Laboratory, Durham, NC 27708-0308

## ABSTRACT

Topics covered during a freewheeling discussion about the optimum designs, best
performance, and future potential of atomic beam polarized sources are summarized.

## ATOMIC BEAM SYSTEMS

The best polarized atomic beam fluxes reported to date have come from the
University of Wisconsin[1] and at MPI-Heidelberg[2], where similar systems built to feed
polarized gas storage cells for targets yield similar results. The Wisconsin system
provides $6.7 \times 10^{16}$ atoms/s into a 10 mm diameter, 13 cm long tube with its entrance 26
cm beyond the last focusing sextupole. They have also shown that at a position 17 cm
beyond the last sextupole, the total beam is included inside a 2 cm diameter and the
FWHM of the beam is 7 mm, inferred from measurements with a smaller 5 mm
diameter compression tube. In these systems the nozzle temperature is typically
operated at ~85 K, and the dissociation degree is near 80%. Both the Wisconsin and
Heidelberg systems utilize massive pumping systems with differential pumping
employing two skimmer apertures between the cooled dissociator nozzle and the
entrance of a multiple element, permanent magnet sextupole focusing system.

Are these results directly applicable to ion sources? Can one assume that the
sextupole geometries which are best for gas storage cell targets provide also the
optimum performance for ion sources? The effective ionizer volume into which the
atomic beam must be focused for an ion source may be different than that needed for a
target. An electron-cyclotron-resonance (ECR), cesium beam, or $D^{\pm}$-plasma ionizer
provides an active ionization volume 10 to 15 cm long and 2.5 cm diameter placed ~30
cm beyond the last sextupole. This differs in shape from the cylindrical ionization
volume of an electron-beam ionizer which is longer and smaller in diameter and nearer
in shape to that of the gas target for which the Wisconsin and Heidelberg systems were
optimized.

If the sextupole systems were redesigned to provide an atomic beam with larger
diameter, how much of any beam at larger radii would really be used? Could ions
produced at larger radii be extracted effectively as part of a usable beam? This question
led to a discussion of beam emittance growth upon extraction from any ionization
volume where an axial B-field is present. Ohlsen et al.[3] showed that if a beam of
particles of mass M were extracted initially from an emission surface of maximum
radius $R_{max}$ inside an axial field of magnitude $B_z$ and transported into a B-field-free
region downstream, the minimum transverse beam emittance there would be given by

$$\varepsilon_t = 3.46\pi B_z R_{max}^2 / M^{1/2} \text{ cm·rad·(eV)}^{1/2} \text{ (with } B_z \text{ in kG, } R_{max} \text{ in cm, and M in amu).}$$

First beam emittance measurements reported[4] for the new polarized source at the
Indiana University Cyclotron Facility (IUCF) were combined with knowledge of the
B-field at the beam extraction surface inside the ionizer to determine that the effective
emission surface radius is $\approx$ 6 mm in their ECR ionizer. This result, if upheld by
further measurements, implies that only the near-axis ECR plasma region is effective

*Work supported in part by the U.S. DOE under Grant No. DE--FG05-88ER40442.

in producing the beam. Thus, the optimum atomic beam profile needed for ECR ionizers may also be very similar to that provided by the Wisconsin and Heidelberg sextupole assemblies.

There is also the question of whether the best atomic beam *flux*, which the Wisconsin and Heidelberg systems were designed to provide, will also provide the best output ion current. Most ionizers are not highly efficient; better measurements of this efficiency are needed. Derenchuk reports[4] a measurement showing that the ratio of ion beam extracted to atomic beam injected for the ECR system at IUCF is ~8%. For such systems, the *density* of the gas in the ionizer seems a more important parameter to optimize, implying that lower velocity atomic beams are more important for these ion source applications. For $D^{\pm}$ plasma ionizers like that developed by Belov and his colleagues at Moscow[5], the efficiency becomes high enough that it is probably more important to maximize the atomic beam *flux* than *density*.

Though there is yet no firm experimental proof, some participants felt that when sextupole systems are designed to match the most probable atomic beam velocity, the optimum dissociator nozzle temperature will ultimately be found between 50 and 100 K. Finding this optimum operating temperature experimentally could be facilitated by a properly designed combination of permanent and electromagnet sextupoles so the effective focal length could be "tuned" as the nozzle temperature is adjusted. If the optimum temperature were > 50 K, then laboratories would not need to add $N_2$ gas to the dissociator discharge to reduce recombination at the cold dissociator nozzle[6]. This would have two added benefits: reduced technical complication for maintaining the proper $N_2$ flow and nozzle temperature over many days of continuous operation, and reduced maintenance for removing the nozzle and cleaning its interior of slight surface contamination each time the dissociator is switched off after use.

What diagnostic measurements or calculations for atomic beam systems are needed? As mentioned above, the optimum dissociator nozzle temperature is not yet known. As this temperature is lowered, the velocity of the exiting atomic beam decreases. This causes an increase in interbeam gas scattering because the scattering cross section(s) increase. These cross sections are not well known; better measurements would be helpful, both for atomic and molecular scattering. Relevant Monte Carlo gas dynamic calculations now seem possible, as discussed by Boyd[7], but their accuracy will ultimately depend to some extent on the accuracy of such cross section data. Also problematic in these calculations is modeling the reemergence of atoms and molecules from surfaces in the dissociator-skimmer-sextupole regions into the dynamical flow. Some felt it important that sources of error be included in the calculations. There was no final consensus, but if calculations could reveal whether increased dissociator gas flows will improve the atomic beam flux, or if accurate calculations could reduce the need for expensive and time-consuming experimental tests, they could surely be beneficial.

It is not even clear how this modeling of gas flows and scattering could best be accurately tested by comparison with measurements. Electron beam scattering from the atoms and molecules emerging from the dissociator nozzle was suggested as a possibility for probing the gas dynamics in this region. However, the pressure in the dissociator region might not be low enough to allow such measurements to be made with confidence. Some participants in the discussion felt this would be an important goal. Others believed that the extensive empirical optimization of the Heidelberg and Wisconsin systems left relatively little room for further improvement. In their opinion, effort to make such calculations and experimental tests may not be justified because, for example, tests at Wisconsin show that only a ~10% increase in their atomic beam flux would be possible if *infinitely* large pumping speed were provided. Such a small gain could not justify the cost of ever-larger pumping systems.

Another gain in performance would be possible if the rf transitions needed to introduce nuclear polarization to the atomic beam could be reduced in axial length. This would allow shortening the distance between the last sextupole and the ionizer volume, allowing a larger flux of atoms to be focused into the ionizer. However, the problem is not as simple for ion sources as for target applications. All ionizers for hydrogen require a strong (> 100 mT) axial B-field to optimize the output beam's nuclear polarization. Such strong axial fields are, however, extremely difficult to shield from the nearby transitions where well-defined, transverse and small (~1 to 10 mT) fields are needed. In almost all present rf transition systems, either significant magnetic shielding is inserted or significant axial separation is allowed to minimize this stray field interference with transition unit performance. Clausnitzer suggested turning this problem into a benefit. He proposed reorientating the rf field in the transition unit(s) by 90 degrees so it would be perpendicular to the atomic beam. This then would allow the static axial field in the transition unit to be parallel to the beam axis and parallel to the problem ionizer stray field. This should greatly facilitate providing, in the shortest possible axial distance, the carefully shaped axial field gradient needed for proper transition unit operation for all particles in the beam.

## IONIZER SYSTEMS

The primary question about ionizers was whether the performance of ECR systems now in use made them a better choice than the traditional electron-bombardment systems developed much earlier. Pierre Schmelzbach reported that direct comparison of these two types of ionizers on one atomic beam system at PSI showed that the ECR ionizer provided beam with considerably lower energy spread. This led to much improved beam bunching efficiency into the phase acceptance of their cyclotron. The ECR ionizer is also less difficult to operate experimentally and is less affected by stray magnetic fields from the cyclotron.

The ECR ionizer does not provide beams with quite as large polarization as did the older electron-bombardment ionizer. Typical reductions are ~5%. Presumably this comes from the fact that the active ionization region of the ECR plasma extends to larger radii, beyond the region where the polarized atomic beam exists. There the ions produced come from unpolarized background gas, diluting the overall output beam polarization.

Another possible contribution to the polarization reduction is that of having off-axis components of the B-field inside the ECR ionizer coming from the sextupole field and the axial B-field gradient needed to confine the ECR-heated plasma. Given that recent IUCF measurements show that the effective ion beam comes from very near the axis, and given that these plasma confining B-fields provide transverse components which are zero on axis and grow with radius, it is not believed that their contributions to the depolarization are large.

Schmelzbach's comments about having regions of possible high and low ionization efficiency in the ECR ionizer because of standing electromagnetic waves inside was noted as a point to be studied further. Also deserving thought are differences in performance between the ECR ionizers at IUCF and PSI, where there is a quartz liner inserted inside the sextupole magnet, and systems like ours at TUNL which have no quartz liner. The former design seems to facilitate extraction of beams of higher intensity, for reasons not yet fully understood.

Neither the ECR nor the electron-beam ionizer yet aproaches the efficiency of the Belov-style[5] ionizer, for which the underlying ionization cross sections are larger. Belov suggested that the emittance of the extracted beam from his system might be further improved by radial rather than axial ion beam extraction, across the lines of B-field confining the $D^\pm$ plasma.

## FUTURE POTENTIAL AND GOALS

Discussion near the end of the session turned to whether the atomic beam-style source can be competitive with the optically pumped source for providing the beams needed for high-energy, low-duty-cycle accelerators. A decision was made to collect data at the conference describing the performance of all operating atomic beam *and* optically pumped sources so valid performance comparisons could be made. In addition, it was pointed out that the goal for both polarized source communities should be the same, to provide intensities equal to those of unpolarized beams needed at each laboratory. A tabulation of these unpolarized intensities at principal laboratories was also collected and is presented in this volume by Y. Mori.

## REFERENCES

1.  A. Roberts, Report at this conference.
2.  K. Zapfe, Report at this conference.
3.  G.G. Ohlsen, J.L. McKibben, R.R.Stevens, Jr, and G.P. Lawrence, Nucl. Instrum. Meth. 73, 45 (1969).
4.  V. P. Derenchuk, Private communication at this conference.
5.  A.S. Belov, S.K. Yesin, S.A. Kubalov, V.E. Kuzik, A.A. Stepanov, and Y.P. Yakushev, Nucl. Instrum. Meth. A255, 442 (1987). source paper
6.  D. Singy, P.A. Schmelzbach, W. Grüebler, and W.Z. Zhang, Nucl. Instrum. Meth. A278, 349 (1989).
7.  I. Boyd, Report at this conference.

# III. OPTICALLY PUMPED POLARIZED TARGETS AND ION SOURCES (EXCEPT $^3$He)

# A LASER-DRIVEN SOURCE OF SPIN POLARIZED ATOMIC HYDROGEN AND DEUTERIUM

M. Poelker, K. P. Coulter, R. J. Holt, C. E. Jones,
R. S. Kowalczyk, L. Young
Argonne National Laboratory, Argonne, IL 60439, USA

D. Toporkov
Budker Institute for Nuclear Physics, Novosibirsk, Russia

## ABSTRACT

Recent results from a laser-driven source of polarized hydrogen (H) and deuterium (D) are presented. The performance of the source is described as a function of atomic flow rate and magnetic field. The data suggest that because atomic densities in the source are high, the system can approach spin-temperature equilibrium although applied magnetic fields are much larger than the critical field of the atoms. We also observe that potassium contamination in the source emittance can be reduced to a negligible amount using a teflon-lined transport tube.

## INTRODUCTION

Experiments have been performed and have been proposed that use stored internal targets of H and D atoms within electron storage rings [1]. Sources of polarized H and D, namely the ultra cold source and the atomic beam source, are described in these proceedings [2]. The laser driven source of polarized H and D described herein relies on the technique of spin-exchange optical pumping where photon angular momentum is transferred to target nuclei via spin-exchange collisions with polarized potassium atoms. The principle of spin-exchange optical pumping has long been known [3], however, early attempts to develop this idea into a practical source have met with limited success as a result of radiation trapping [4]. Recently it was reported that high D polarization could be obtained by performing spin-exchange optical pumping in a high magnetic field thereby alleviating the ill effects of radiation trapping [5]. In this paper, we report the first polarized H results from the Argonne laser-driven source. Source performance is described as a function of H and D atomic flow rate and magnetic field. Also, we report greatly reduced potassium contamination in source emittance using a teflon-lined transport tube.

## EXPERIMENTAL TECHNIQUE

The laser-driven source apparatus is described in refs. [5,6]; only a short description of the source is given here. Molecular H (or D) is dissociated in an rf inductive discharge and sent into a drifilm-coated spin-exchange cell containing potassium vapor. The potassium is optically pumped and polarized with a single-frequency titanium sapphire laser in the presence of a high magnetic field. Electron spin is transferred to H (D) atoms during collisions with polarized potassium atoms;

$$H(\uparrow) + K(\downarrow) = H(\downarrow) + K(\uparrow) \qquad (1)$$

H (D) atoms leave the spin-exchange cell and travel through a transport tube to a vacuum chamber that contains an electron-spin polarimeter. The polarimeter consists of a quadrupole mass analyzer (QMA) and a permanent sextupole magnet which focuses electron "spin-up" atoms and defocuses electron "spin-down" atoms. Atomic polarization is determined by comparing QMA signals when the pump laser is blocked and then directed into the spin-exchange cell [5]. The QMA is also used to measure degree of dissociation.

## FLOW RATE AND MAGNETIC FIELD RESULTS

The performance of the source has been investigated as a function of flow rate. Figure 1 shows measured H and D atomic polarization values as a function of total intensity of atoms. Polarization values were obtained over similar dissociator region pressures where degree of dissociation (Df) was high and nearly constant ($\approx 75\%$). Maximum H flow rate exceeds that of D by a factor of approximately $\sqrt{2}$ as a result of increased dissociator conductance for H. The data were obtained with a magnetic field of 4.4 kGauss and with approximately 3 Watts of pump laser power incident upon the spin-exchange cell. The potassium sidearm was maintained at 188°C resulting in a potassium density within the spin-exchange cell of approximately $4x10^{11}$ atoms/cm$^3$. Higher H and D polarization values are obtained by increasing the potassium density within the spin-exchange cell [6], however higher sidearm temperatures are undesirable as that would require maintaining the spin-exchange cell and transport tube at temperatures $> 250°C$ [7]. High atomic polarization ($P_e \approx 80\%$) is measured for both H and D at a flow rate of $2x10^{17}$ atoms/second. Polarization values decrease with increasing flow, however D polarization is observed to decrease more rapidly than H polarization.

The performance of the source was also investigated as a function of magnetic field. A simple rate

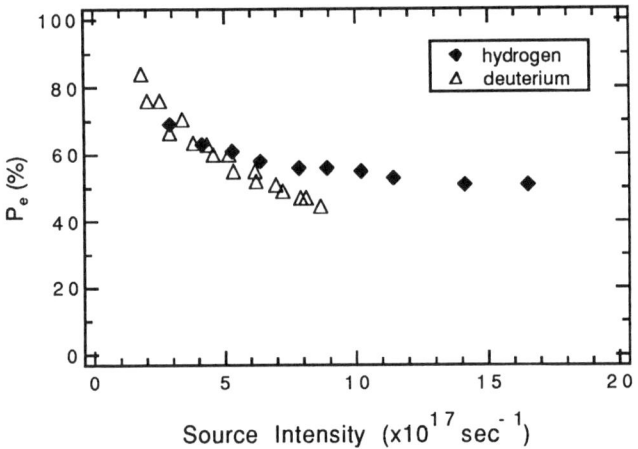

Figure 1.  H and D atomic polarization as a function of total source intensity.

equation model that accounts only for optical pumping of potassium atoms and spin-exchange collisions between potassium and H (D) atoms predicts that H (D) polarization should not vary with changing magnetic field [8]. Figure 2 shows measured H and D atomic polarization values as a function of flow rate for three different magnetic field conditions in the spin-exchange cell. The data were obtained with approximately 3 Watts of pump laser power incident upon the spin-exchange cell. The potassium density within the spin-exchange cell was approximately equal to $4x10^{11}$ atoms/cm$^3$. The degree of dissociation over the range of flow rates tested was approximately 75%. H polarization is observed to be relatively insensitive to changing magnetic field; D polarization however depends significantly on magnetic field. The field dependence of the D data cannot be explained with the simple rate equation optical-pumping spin-exchange model.

Recently, it has been proposed that spin-exchange optical pumping in a high magnetic field can yield direct polarization of the nucleus without actively performing rf hyperfine transitions [9]. Although high magnetic fields generally weaken the influence of the hyperfine coupling between the nucleus and the electron, frequent H·H (D·D) collisions increase the total probability for a hyperfine interaction to occur. Under such a condition, atoms approach spin-temperature equilibrium and nuclei become polarized through hyperfine interaction with polarized electrons. A system approaches spin-temperature equilibrium in a time given by;

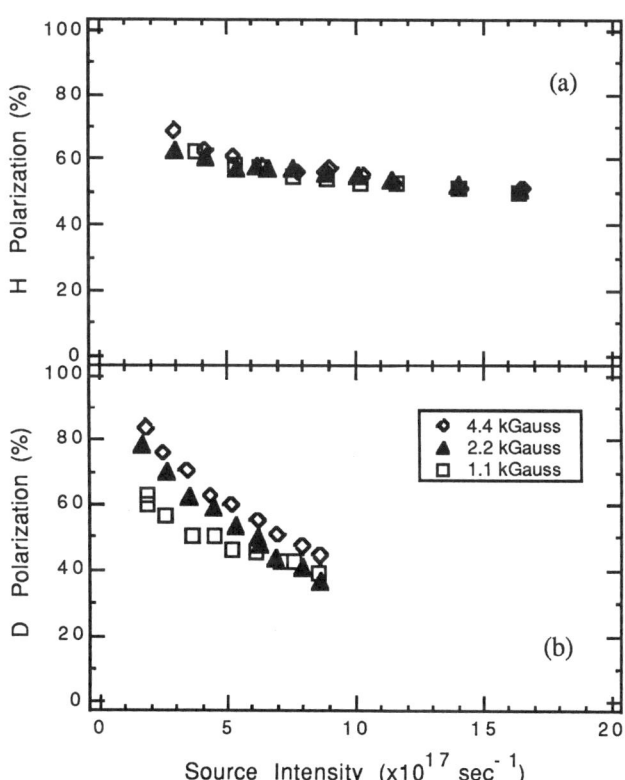

Figure 2 (a) H polarization as a function of total H intensity for three magnetic field conditions. (b) D polarization as a function of total D intensity for three magnetic field conditions.

$$T_{st} = (1 + (\frac{g_s \mu_B B}{\delta v_{hfs}})^2) T_H$$

(2)

where $\delta\nu_{hfs}$ is the ground-state hyperfine splitting in zero magnetic field, $g_s\mu_B B$ is the energy shift of the electron in a magnetic field B, and $T_H^{-1} = n_H < \sigma_{se}(HH)\cdot v >$ is the thermally averaged H·H spin-exchange rate, where $n_H$ is the hydrogen density, $\sigma_{se}(HH)$ is the spin-exchange cross section equal to $2x10^{-15}cm^{-2}$ [10] and v is the thermally averaged velocity.

H and D polarization data obtained with the Argonne laser-driven source as a function of flow rate and magnetic field provide an excellent test of this novel feature of the spin-exchange process since densities within the source are high and because the large difference in the magnetic moments of the nuclei will give very different rates for approach to spin-temperature equilibrium. For comparable flow rates, H will approach spin-temperature equilibrium more quickly than D because of the comparatively larger hyperfine splitting for H ( $\delta\nu_{hfs}(H)$=1420 MHz, $\delta\nu_{hfs}(D)$=327 MHz). This statement is evident when calculating $T_{st}$ values for flow rate and magnetic field conditions parameterized by Figure 2 and comparing these values to $t_{dwell}$, the dwell time of an atom in the spin exchange cell. Atoms approach spin-temperature equilibrium when $T_{st} << t_{dwell}$. For H atoms, $T_{st}$ values are less than $t_{dwell}$ for both high and low field conditions ($0.03 < T_{st} < 3.0$ msec; $t_{dwell}(H) = 5$ msec). As a result, H polarization is relatively insensitive to changing field in Figure 2a. Thus, H atoms approach spin-temperature equilibrium for each of the three magnetic field conditions tested. For D atoms, $T_{st}$ values can be made both less than and greater than $t_{dwell}$ ($0.43 < T_{st} < 67$ msec; $t_{dwell}(D) = 7$ msec). As a result, D polarization in Figure 2b is observed to be highly sensitive to changing magnetic field. D atoms approach spin-temperature equilibrium only for low field conditions.

## POTASSIUM GETTER RESULTS

We observe that when a teflon sleeve is inserted into the transport tube of the laser-driven source, potassium contamination in the source emittance is reduced to a negligible amount. Potassium contamination is undesirable for internal target experiments described in ref.[1] because it dilutes the target with unpolarized potassium nuclei and necessitates the heating of the target cell to prevent potassium condensation and subsequent molecular recombination on the target cell wall. To investigate the extent to which the teflon sleeve acts as a potassium getter, a 10 cm long piece of teflon tubing (19 mm o.d., 17 mm i.d.) was inserted into the bore of the transport tube and the ratio of potassium atoms to hydrogen atoms in the source emittance was measured with the QMA described above. Also, the effect of the teflon sleeve on source polarization was monitored by measuring H polarization throughout the duration of the test.

Figure 3a shows the ratio of potassium atoms to H atoms in the source emittance measured during 140 hours of source operation. (Source operation was not continuous. Typically, the source would operate for six or seven hours followed by overnight shut-down and continued operation the next day.) Throughout the test, the total source intensity was constant to within 5% at $5x10^{17}$ atoms/second. The data were obtained with a magnetic field of 4.4 kGauss and with approximately 3 Watts of

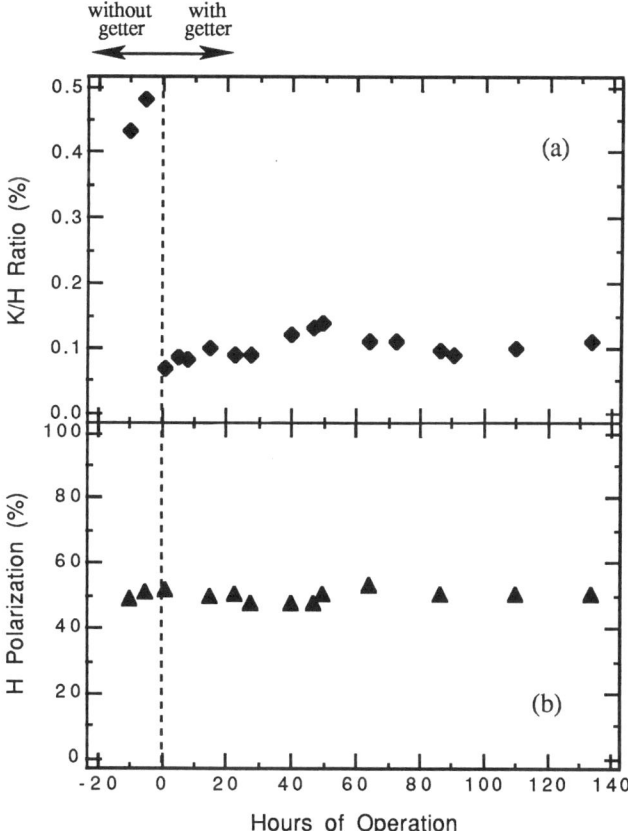

Figure 3 (a) Measured K/H ratio as a function source operation time. (b) H polarization as a function of source operation time.

pump laser power incident upon the spin-exchange cell. Before installation of the getter, approximately one potassium atom was detected in the source emittance for every 200 H atoms, corresponding to a K/H ratio of 0.5%. After installation of the getter, the K/H ratio was reduced by approximately a factor of five to a value of 0.1% . Monte Carlo simulations predict that 10% of the atoms detected with the electron-spin polarimeter travel from the source without hitting the transport tube wall. It is possible that the K/H ratio of 0.1% could be reduced further if attempts were made to ensure that all potassium atoms hit the teflon surface (e.g. future getter designs could incorporate a small bend). Figure 3b gives evidence that the teflon sleeve is not detrimental to maintaining high H polarization. Throughout the duration of the test, H polarization remained at a value of 50%.

## CONCLUSIONS

The flow rate and field dependence of the D polarization data suggest that H and D atoms approach spin-temperature equilibrium in the source. This implies that nuclei become polarized without actively performing rf hyperfine transitions. For H atoms, high vector polarization at a flow rate of $1.6x10^{18}$ atoms/second (with a 75% degree of dissociation) can be inferred ($P_z = 51\%$), since vector polarization is equal to electron polarization ($P_e = P_z$) for H atoms in spin-temperature equilibrium. For D atoms in spin-temperature equilibrium (i.e. under low magnetic field conditions), a measured electron polarization of 50% at a flow rate of $6x10^{17}$ atoms/second (with a 75% degree of dissociation) would correspond to vector and tensor polarization of $P_z = 57\%$ and $P_{zz} = 27\%$, respectively. We are currently working towards measuring $P_{zz}$ directly; our efforts are described in these proceedings [11].

Potassium contamination is reduced to a negligible amount using a teflon-lined transport tube and suggests the possibility of source operation with a cold target cell. These factors greatly enhance the figure of merit of a laser-driven target.

We thank Joe Gregar for his expert glass blowing of the spin-exchange cell apparatus. Also, we appreciate the assistance offered from students Germaline Calagday, Taliver Heath, Jamie Kustak, David Flory and Jennifer Gerbi.

This work was supported by the U.S. Department of Energy, Nuclear Physics Division and the Office of Basic Sciences, under contract No. W-31-109-ENG-38

## REFERENCES

1.   K. Zapfe, "Tests of a High-Density Polarized Gas Target Storage Cell and Spin Filter for Stored Ion Beams," A. D. Roberts, "Polarized H and D Target Experiments in the Indiana Cooler," and D. Toporkov, "Experiments with Polarized Deuterium Target at VEPP-3 Storage Ring: Status and Perspective," these proceedings and references therein.

2.   R. S. Raymond, "Status of the Ultra-Cold Polarized Jet for Neptun-UNK," and F. Stock, "Optimization of the FILTEX-HERMES Atomic-Beam Source", these proceedings and references therein.

3.   W. Happer, Rev. Mod. Phys. **44**, 169 (1972).

4    L. Young et al., Nucl. Phys. **A497**, 529c (1989).

5.   K. P. Coulter et al., Phys. Rev. Lett. **68**, 174 (1992).

6.   M. Poelker et al., to be published.

7.   D. Swenson at LANL observes "drifilm"-coating failure at temperatures $\approx 300°C$. See also D. R. Swenson and L. W. Anderson, Nucl. Instrum. Methods  Phys. Res., Sect. **B 29**, 627 (1988).

8.   T. E. Chupp et al., Phys. Rev. C **36**, 2244 (1987).

9.   T. Walker and L. W. Anderson, submitted for publication Nucl. Instrum. Meth., see also T. Walker, "Constraints on Optically-Pumped Spin-Exchange Targets," these proceedings.

10.  H. R. Cole and R. E. Olson, Phys. Rev. **A 31**, 2137 (1985).

11.  C. E. Jones, "Measurement of $P_{zz}$ of the Laser-Driven Polarized Deuterium Target," these proceedings.

# Measurement of $p_{zz}$ of the Laser-Driven Polarized Deuterium Target

C. E. Jones, K. P. Coulter[†], R. J. Holt, M. Poelker,
D. H. Potterveld, R. S. Kowalczyk

*Argonne National Laboratory, Argonne, IL 60439*

M. Buchholz, J. Neal, J. F. J. van den Brand

*Univ. of Wisconsin, Madison, WI 53706*

## Abstract

The question of whether nuclei are polarized as a result of H−H (D−D) spin-exchange collisions within the relatively dense gas of a laser-driven source of polarized hydrogen (deuterium) can be addressed directly by measuring the nuclear polarization of atoms from the source. The feasibility of using a polarimeter based on the D + T → n + $^4$He reaction to measure the tensor polarization of deuterium in an internal target fed by the laser-driven source has been tested. The device and the measurements necessary to test the spin-exchange polarization theory are described.

## Introduction

The laser-driven polarized hydrogen/deuterium source (LDS), which is described elsewhere[1], has been demonstrated recently[2] at flow rates of polarized hydrogen up to $1.7 \times 10^{18}$/s with $\geq$ 50% electron polarization. Deuterium polarizations of $\geq$ 40% have also been obtained at flow rates up to $0.9 \times 10^{18}$/s. Compared to conventional atomic beam sources, which have not exceeded a hydrogen flow rate of $8 \times 10^{16}$/s (with 95% polarization), the optical pumping technique can polarize rapidly relatively dense samples of the hydrogen isotopes. Such a source is of general interest for atomic, plasma, nuclear, and high energy physics. One specific use of the LDS which has recently received much attention is as the source of polarized nuclei for an internal target in an electron storage ring. The Argonne laser-driven polarized deuterium internal target will be installed in the VEPP-3 electron storage ring (Novosibirsk, Russia) in 1994 for a measurement of the tensor analyzing power in electron-deuteron scattering at high momentum transfer.

For this application one must polarize the target nuclei, and to date only the electron polarization of the hydrogen and deuterium atoms from the laser-driven source has been measured. Given the recent progress in the development of the source[2], the most important outstanding question concerning its viability as a polarized target for nuclear physics experiments is the nuclear polarization achievable with the system. Recently Walker and Anderson[3] suggested that in a dense polarized gas of any of the hydrogen isotopes spin-exchange collisions between the hydrogen atoms drive the system to a spin temperature distribution of the magnetic substates, effectively transferring polarization from the electrons to the nuclei. If this suggestion is correct, the nuclei can be polarized without inducing RF transitions, a significant technical advantage. Measurements of the electronic polarization of both hydrogen and deuterium as a function of density and magnetic holding field indicate that the calculation of Walker and Anderson may be correct[2], but the definitive test will be a direct measurement of the nuclear polarization. To this end, a polarimeter has been set up at Argonne National Laboratory to measure the tensor polarization of deuterium from an internal target fed by the LDS. The device is described here, along with the series of measurements that will be performed to test the predictions of Walker and Anderson.

## Determination of $p_{zz}$

The polarimeter is based upon the technique developed by Price and Haeberli[4]. As the analyzer it employs the low energy $D + T \rightarrow n + {}^4He$ reaction, which displays an anisotropy in the angular distribution of the outgoing neutrons if the incident deuterium ions are tensor polarized. At low energy the $D(T,n)^4He$ reaction can proceed through both $J^\pi = 3/2^+$ and $J^\pi = 1/2^+$ resonances in ${}^5He$. The unpolarized cross section has a broad resonance peak centered at a deuteron energy of 110 keV, corresponding to the $J^\pi = 3/2^+$ channel, which dominates the cross section at low energy. The expression for the cross section for a tensor-polarized incident deuteron beam is

$$\sigma(\theta) = \sigma_0(\theta) \left( 1 - \frac{f}{4} p_{zz} (3 \cos^2 \theta - 1) \right).$$  [1]

$\sigma_0$ is the unpolarized cross section, which is isotropic, $\theta$ is the angle between the direction of the outgoing neutron and the spin of the deuteron in the center-of-mass system, $p_{zz}$ is the tensor polarization of the deuteron, and the dilution factor $f$ accounts for the small admixture of the $J^\pi = 1/2^+$ reaction channel which has no tensor analyzing power. The dilution factor has been measured and found to be near unity[5] ($f \approx 0.96$) for ion energies in the range used for the polarimeter. The neutron anisotropy R, defined as

$$R = \frac{\sigma(0°)}{\sigma(90°)} = \frac{1 - \frac{f}{2} p_{zz}}{1 + \frac{f}{4} p_{zz}},$$  [2]

is plotted as a function of $p_{zz}$ in Figure 1.

The apparatus is shown schematically in Figure 2. The polarized deuterium flows from the optical pumping cell through a glass transport tube and a teflon getter, which removes the potassium, into the storage cell. A

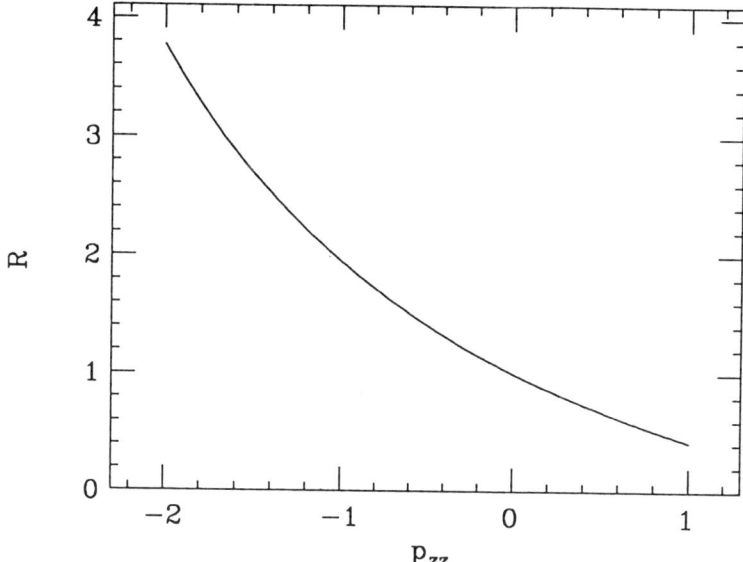

**Figure 1.** Anisotropy in the neutron yield as a function of tensor polarization.

solenoid surrounding the storage cell creates the magnetic holding field that serves both to decouple the electronic and nuclear spins and as a guide field for the ions to prevent them from hitting the cell walls. The storage cell is 48 cm long and 22 mm in diameter. The longitudinal holding field is 330 Gauss over most of the length of the storage cell; the value at the center of the cell where the atoms enter from the LDS is 270 Gauss. A small hole opposite the entrance port allows $\sim 10\%$ of the atoms to leak out of the cell to be analyzed by a sextupole magnet and quadrupole mass spectrometer, which serve as an electron polarimeter. The storage cell is coated with Drifilm to reduce recombination at the surface and is heated to $140°$ C to prevent any potassium entering the cell from condensing on the surface where it would cause molecular recombination.

Electrons from an electron gun are accelerated to $\sim 2$ kV and pass through the storage cell, ionizing the deuterium. The ions follow the magnetic field lines to the end of the storage cell, where they are either reflected by a positive potential at the electron gun end or extracted and accelerated toward a tritiated titanium foil target at the other end. A Wein filter $(\vec{E} \times \vec{B})$ between the storage cell and the tritium target acts as a velocity selector to separate deuterium atoms and molecules. The spin direction, which is parallel to the solenoidal holding field in the storage cell, precesses in the $B$-field

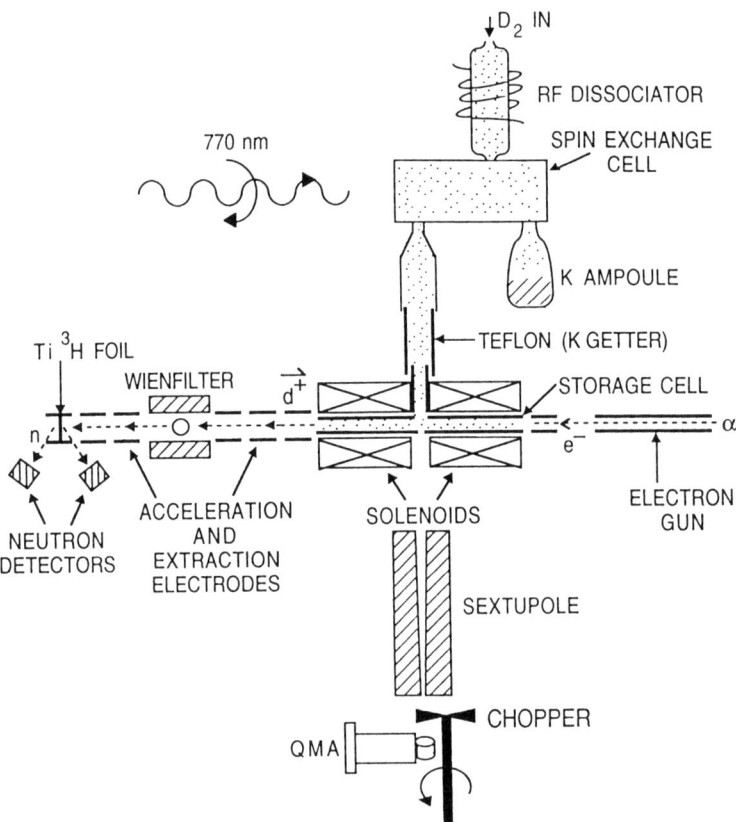

**Figure 2.** Schematic of the laser-driven source, the deuteron nuclear polarimeter, and the electron polarimeter.

of the Wein filter according to the relationship

$$\frac{d\vec{S}}{dt} = \frac{g\mu_N}{\hbar}\vec{S} \times \vec{B}, \qquad [3]$$

where $g = 0.857$ is the Landé $g$-factor for the deuteron. The spin rotates $\sim 40°$ in the field used in the measurements described here. The tritium foil is mounted inside an electrostatic lens maintained at 30 kV. Outgoing neutrons are detected in two NE110 scintillators placed outside the vacuum chamber at $\theta = 0°$ and $\theta = 90°$. Initial tests of the polarimeter with unpolarized deuterium show that the measurements are reproducible to better than $\Delta R/R = 0.01$. Corrections for the finite detector size are less than 0.5%. The largest systematic uncertainty is expected to be the determination of the deuteron spin orientation at the tritium foil. This can be found by moving the detectors to find the maximum value of $R$.

## Tests of the Spin Temperature Prediction

According to Walker and Anderson, in a dense sample of polarized hydrogen, deuterium, or tritium, spin-exchange collisions drive the system to a spin temperature distribution of the magnetic substates where the relative population density of a sublevel with total angular momentum magnetic quantum number $m_F$ is

$$\rho(m_F) = e^{\beta m_F}/N. \qquad [4]$$

Both $N$ and $\beta$ depend upon the total angular momentum stored in the system. Even in a relatively large magnetic field the system can be driven toward a spin temperature distribution if the density is sufficiently high.

The experimental consequences of this prediction are marked. The theory predicts that both the vector and tensor polarization of the deuterons will be nonzero. If D−D spin-exchange collisions redistribute the angular momentum between the atoms and nuclei, then optical pumping with either $\sigma_+$ or $\sigma_-$ light will tend to deplete the $m_I = 0$ state, resulting in $p_{zz} > 0$ for either helicity pumping light. Figure 3 shows the values of the electron, vector, and tensor polarizations of deuterium in a population that has a spin temperature distribution with temperature parameter $\beta$, in the limit of $B \gg B_{crit}$, where the critical field $B_{crit}$ for deuterium is 117 Gauss. Since we observe an electronic polarization of $\sim 50\%$ at high density currently with the LDS, we should measure a tensor polarization of $\sim +0.3$ if the system is in spin temperature equilibrium. The limit for the tensor polarization achievable with a spin temperature distribution is $+1.0$. In Figure 3 one also sees the interesting feature that the vector polarization of the deuteron is larger than the electron polarization for a spin temperature distribution, so an electron polarization of 50% corresponds to a vector polarization of $p_z = 65\%$.

Whether the polarized atoms from the LDS have a spin temperature distribution depends upon the equilibration rate, which is given in Reference 5 as

$$\frac{1}{T_{ST}} = \frac{(h\nu_{hfs})^2}{(h\nu_{hfs})^2 + (g_s\mu_B B)^2}n < \sigma_{se}v > . \qquad [5]$$

$h\nu_{hfs}$ is the ground state hyperfine splitting in zero field, $g_s\mu_B B$ is the field-dependent electron energy shift, and $n < \sigma_{se}v >$ is the thermally-averaged H−H (D−D) spin exchange rate, which depends upon the density $n$, the velocity $v$, and the spin-exchange cross section[6] $(\sigma_{se} = 2 \times 10^{-15}\text{cm}^2)$. The hyperfine splitting for hydrogen is 1420 MHz, for deuterium is 327 MHz, and for tritium is 1517 MHz, so spin temperature equilibrium is reached much

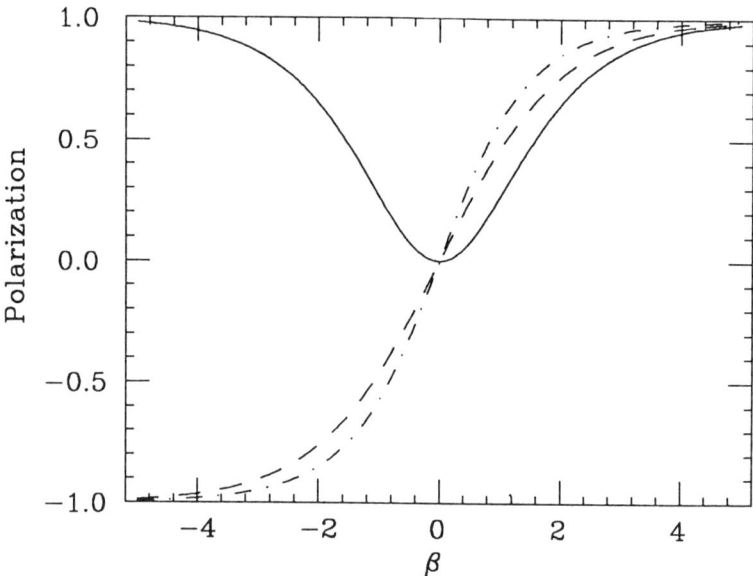

**Figure 3.** The electron (dash), vector (dotdash), and tensor (solid) polarizations for deuterium atoms in spin temperature equilibrium as a function of the temperature parameter $\beta$.

faster in hydrogen and tritium than in deuterium. From equation 5 it is clear that $T_{ST}$ decreases with increasing density and increases with decreasing $B$-field. For the LDS, the dwell time of deuterium atoms in the spin-exchange cell is 7 ms, and the time for deuterium to reach spin temperature equilibrium in the spin-exchange cell is estimated to vary from $0.4 < T_{ST} < 70$ ms for the range of densities and magnetic fields over which the LDS is operated. This is an interesting region to study because it should be possible to observe the transition to a spin temperature distribution.

A series of measurements are planned to ascertain whether the deuterium target can be polarized without the use of RF transition units. The target polarization $p_{zz}$ will be measured for both $\sigma_+$ and $\sigma_-$ pumping light as a function of flow $(2 - 10 \times 10^{17}/\text{s})$ and magnetic holding field in the spin-exchange cell $(0.8 - 4 \text{ kG})$. In addition, RF transitions can be induced and the value of $p_{zz}$ measured. If the system is driven to spin temperature equilibrium in the transport tube then the transitions should not result in a higher value of $p_{zz}$.

Perhaps the most important outcome of a verification of the predictions of Walker and Anderson would be the demonstration that the laser-driven source is already producing high flows of vector polarized hydrogen, deuterium, and, potentially, tritium with polarizations $\gtrsim 50\%$. In addition, it becomes feasible to polarize hydrogen, deuterium, or tritium nuclei in a sealed system. An experiment which makes use of this technique to construct a sealed polarized tritium target was recently proposed[7] for CEBAF. Finally, these tests permit us to optimize the achievable polarization figure of merit, $p_{zz}^2 f$, where $f$ is the flow rate, for experiments requiring tensor polarized deuterium.

## Acknowledgements

We thank Scott Price for advice on the design and operation of the polarimeter. We are grateful to Joe Gregar for the glassware used for the source. We appreciate the assistance of students David Flory, Jennifer Gerbi, Taliver Heath, and William Weintraub. This work is supported by the U.S. Department of Energy, Nuclear Physics Division and the Office of Basic Sciences, under contract No. W-31-109-ENG-38, and by the National Science Foundation under contract No. PHY-9019983.

## References

† current address: Physics Department, University of Michigan, Ann Arbor, MI 48109.

1 K. P. Coulter *et al.*, Phys. Rev. Lett. **68**, 174 (1992).

2 M. Poelker *et al.*, "A Laser Driven Source of Spin Polarized Atomic Hydrogen and Deuterium," these proceedings.

3 T. Walker and L. W. Anderson, "Limitations of Optically-Pumped Spin-Exchange Targets," these proceedings.

4 J. S. Price and W. Haeberli, Nucl. Instrum. Meth. **A326**, 416 (1993).

5 G. G. Ohlsen, J. L. McKibben, and G. P. Lawrence in *Polarization Phenomena in Nuclear Reactions, Proceedings of the Third International Symposium, Madison, 1970*, edited by H. H. Barschall and W. Haeberli (The University of Wisconsin Press, Madison, 1971), pp. 503-505.

6 H. R. Cole and R. E. Olson, Phys. Rev. **A31**, 2137 (1985).

7 C. E. Jones, spokeswoman, CEBAF Proposal 93-016 (1993).

# LIMITATIONS OF OPTICALLY PUMPED SPIN-EXCHANGE-POLARIZED TARGETS

T. Walker and L. W. Anderson
Department of Physics, University of Wisconsin-Madison, Madison, Wisconsin 53706

## ABSTRACT

The effects of spin-exchange collisions on the polarization of dense spin-polarized samples of hydrogen and deuterium are analyzed. It is shown that even in large magnetic fields spin-exchange collisions transfer angular momentum between the electrons and the nuclei. This effect has important implications for the operation of spin-polarized targets and sources of hydrogen and deuterium. For the specific case of sources that are spin-polarized by spin-exchange collisions with optically pumped alkali atoms, spin-exchange not only polarizes the hydrogen and deuterium electron spins, but polarizes the nuclear spins as well.

## INTRODUCTION

Spin-exchange collisions have long been used to produce electron- and nuclear-spin-polarized samples of atoms that cannot be conveniently polarized by direct optical pumping [1]. While the basic origin of spin-exchange is well-understood [2], there are features of these collisions that involve the roles of nuclear spins, external magnetic fields, and time-scales for production of a spin-temperature distribution that are not widely appreciated. These important effects are not only fundamentally interesting but have significant consequences for applications of spin-exchange such as the production of a spin-polarized deuterium target [3] for scattering experiments. In this paper we summarize our recent detailed analysis [4] of hydrogen/deuterium spin-exchange collisions with the aim of developing insights into the roles the above-mentioned effects play in spin-polarized targets.

When two atoms with antiparallel electron spins collide, the probability of spin exchange is about 50% . Thus the cross-sections for spin-exchange are large, $10^{-15} - 10^{-14}$ cm$^2$. Hyperfine interactions and magnetic fields have negligible effect on the collisions at temperatures above a few Kelvin. Between collisions, however, the electron and nuclear spins precess about each other and about the external magnetic field. As a result, in the limit of rapid spin exchange the system reaches spin-temperature equilibrium [5]. This is true independent of the size of the external magnetic field. Of course, the rate at which the system approaches equilibrium is extremely important in determining whether a spin-temperature adequately describes the state of the system. For hydrogen atoms in a large magnetic field the time constant for the approach to equilibrium is

$$T_{ST} = \left(1 + \left(\frac{g_s\mu_B B}{\delta\nu_{HFS}}\right)^2\right)T_H = (1 + x^2)T_H, \tag{1}$$

where $\delta\nu_{HFS}$ is the ground-state hyperfine splitting in zero magnetic field, $g_s$ is the electron $g$-factor, $\mu_B$ is the Bohr magneton, $B$ is the magnetic field, $x = g_s\mu_B B/\delta\nu_{HFS}$ is the Breit-Rabi field parameter, and $T_H^{-1} = n_H\langle\sigma_{se}(HH)v\rangle$ is the thermally averaged hydrogen-hydrogen spin-exchange rate. Equation 1 is approximately true for deuterium as well. Thus in a large magnetic field the rate of relaxation to spin-temperature equilibrium is considerably smaller due to the reduced coupling of the electron to the nucleus. However, if many collisions occur substantial angular momentum may still be transferred between the electron and the nucleus.

## HYDROGEN-HYDROGEN SPIN-EXCHANGE IN ARBITRARY FIELDS

We consider hydrogen-hydrogen spin-exchange collisions in arbitrary magnetic fields. The results for deuterium are similar; see Ref. [4] for details. The possible states of hydrogen are $|1\rangle = |\uparrow 1/2\rangle, |2\rangle = \cos\theta|\uparrow -1/2\rangle + \sin\theta|\downarrow 1/2\rangle, |3\rangle = -\sin\theta|\uparrow -1/2\rangle + \cos\theta|\downarrow 1/2\rangle, |4\rangle =$

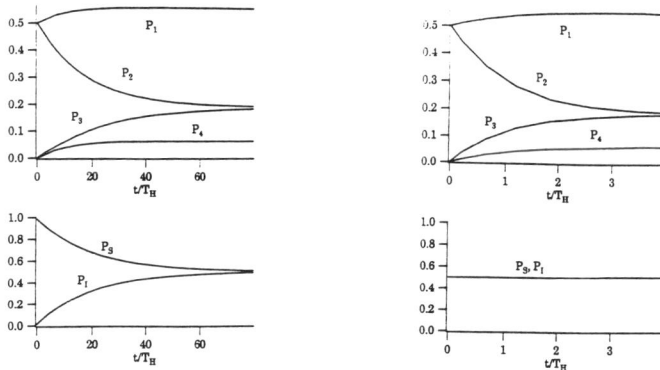

Figure 1: Evolution of states of H in 2.2kG and 0 kG fields.

$|\downarrow -1/2\rangle$. Here the arrows denote the projection of the electron spin along the direction of the magnetic field, and $\pm 1/2$ are the projections of the proton spin along the magnetic field direction. The magnetic-field dependent mixing angle $\theta$ is given by $\tan 2\theta = \delta\nu_{HFS}/g_s\mu_B B = 1/x$. Spin-exchange causes the atoms to change states consistent with conservation of the z-component of the angular momentum. An example of a possible spin-exchange reaction is $|1\rangle + |2\rangle \longrightarrow |1\rangle + |4\rangle$.

The rate equations for the populations of these states can be derived using the method of Purcell and Field[2]. Using $P_1 + P_2 + P_3 + P_4 = 1$ and $\langle F_z \rangle = P_1 - P_3$, the equations are:

$$T_H \frac{d(P_2 - P_4)}{dt} = -\sin^2 2\theta (P_2 - P_4) \tag{2}$$

$$T_H \frac{d(P_2 + P_4)}{dt} = -(P_2 + P_4) + \frac{1 - \langle F_z \rangle^2}{2} + \frac{\cos^2 2\theta}{2}(P_2 - P_4)^2 \tag{3}$$

In steady state, $P_2 - P_4 = 0$ and the system can be characterized by a spin-temperature, with $P_1 = e^\beta/N$, $P_2 = P_4 = 1/N$, $P_3 = e^{-\beta}/N$, and $N = 4\cosh^2\frac{\beta}{2}$. The constant $\beta$ is related to $\langle F_z \rangle$ by $\langle F_z \rangle = \tanh\frac{\beta}{2}$. Note that this distribution is independent of the strength of the magnetic field.

Eq. 2 shows that the system approaches spin-temperature with a time constant $T_{ST}$ given by Eq. 1. An immediate consequence of Eqn. 1 is that in a given large magnetic field deuterium ($\delta\nu_{HFS} = 327$ MHz) will reach a spin-temperature distribution more than one order of magnitude slower than hydrogen ($\delta\nu_{HFS} = 1420$ MHz). At small magnetic fields the two isotopes reach equilibrium at approximately the same rate.

In Fig. 1 we show the time evolution (in units of the spin-exchange time $T_H$) of the populations of the various states of H, for magnetic fields of 2.2 and 0 kG. We also have plotted the electron and nuclear spin-polarizations as a function of time. For the 0 G case, an important feature peculiar to H-H spin-exchange is apparent. Since states $|2\rangle$ and $|4\rangle$ have $\langle S_z \rangle = \langle I_z \rangle = 0$ at zero field, the difference $P_1 - P_3$ solely determines the spin polarizations. Since this quantity is conserved in spin-exchange collisions, the collisions do not affect the spin polarizations, even though the state populations are changed. At high fields, this is no longer true and the collisions redistribute the angular momentum between the electron and nuclear spins.

For deuterium, we show in Fig. 2 how the states (numbered in order 1...6, with 1 being the highest energy and 6 the lowest in a positive magnetic field) and polarizations evolve for the interesting case of initially tensor-polarized atoms evolving at low field. At high field, the degradation of the tensor polarization is slowed by approximately the factor $1 + x^2$.

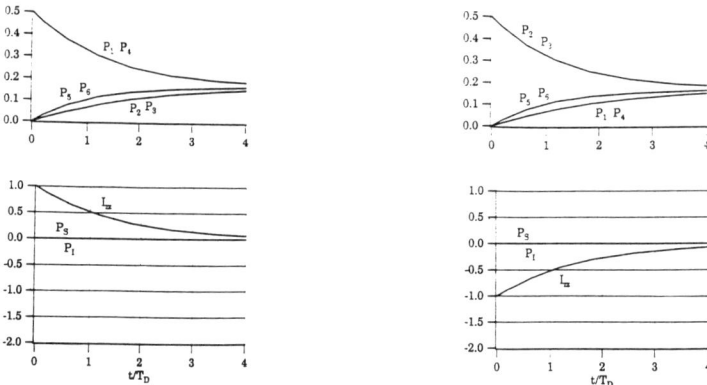

Figure 2: Evolution of D due to spin-exchange in low field.

So far we have considered only H-H or D-D spin-exchange. This analysis will apply to any polarized target in the transport or storage parts of the apparatus. We now consider spin-exchange collisions in the presence of an optically-pumped alkali(A) vapor in a large magnetic field. The combined effects of A-H and H-H collisions are to polarize the H nuclei as well as the H electrons provided the H-H collision rate is sufficiently high. Here the rate $T_{ST}^{-1}$ that the H atoms come to spin-temperature equilibrium again plays a vital role.

We perform the calculations exactly for hydrogen and make approximate extensions for deuterium. Since a large magnetic field is necessary for efficient optical pumping of the A atoms [6], we restrict the calculations to large magnetic fields. In high field, the relevant equations for A-H and H-H spin exchange can be written as

$$\frac{d\langle F_z\rangle}{dt} = \frac{1}{T_A}\left(P_A/2 - \langle F_z\rangle + \langle I_z\rangle\right) \qquad \frac{d\langle I_z\rangle}{dt} = \frac{1}{2T_{ST}}(\langle F_z\rangle - 2\langle I_z\rangle) \qquad (4)$$

where $P_A$ is the alkali electron spin polarization and where $T_A = n_A\langle\sigma_{se}(KH)v\rangle$ is the thermally averaged alkali-hydrogen spin-exchange rate.

For simplicity, we consider the limit $T_{ST} \gg T_A$. This corresponds to the situation where the hydrogen electron spin-polarization is in spin-exchange equilibrium with the alkali electron spin-polarization. Then the transient behaviors of $\langle I_z\rangle$ and $\langle F_z\rangle$ are

$$\langle F_z\rangle = \frac{P_A}{2} + \langle I_z\rangle \qquad \langle I_z\rangle = \frac{P_A}{2}\left(1 - \exp\frac{-t}{2T_{ST}}\right) \qquad (5)$$

This says that the hydrogen electron spin is quickly polarized to a value $P_A/2$, and the nuclear spin is then slowly polarized by repeated H-H spin-exchange collisions.

For deuterium we expect similar behavior, i.e.

$$\frac{d\langle F_z\rangle}{dt} = \frac{1}{T_A}\left(P_A/2 - \langle F_z\rangle + \langle I_z\rangle\right) \qquad \frac{d\langle I_z\rangle}{dt} = -\frac{1}{2T_{ST}}\left(\langle F_z\rangle - \frac{3}{2}\langle I_z\rangle\right) \qquad (6)$$

In the limit of $T_{ST} \gg T_A$, the transient solutions are

$$\langle F_z\rangle = \frac{P_A}{2} + \langle I_z\rangle \qquad \langle I_z\rangle = P_A\left(1 - \exp\frac{-t}{4T_{ST}}\right) \qquad (7)$$

In the next section we will show that these equations imply that spin-exchange collisions partially polarize the D nuclei under the conditions of the Argonne polarized target.

APPLICATION TO THE ARGONNE SPIN-POLARIZED DEUTERIUM TARGET

For information on the Argonne target, see Ref. [3]. Using the parameters given there, we make the following estimates. The value of $T_{ST}$ in the optical pumping cell is 210 s$^{-1}$. With a dwell time of about 6.9 ms, we find from Eqn. 7 that $\langle I_z \rangle = 0.3 P_K$ and $\langle F_z \rangle = .8 P_K$ so due to the large number of spin-exchange collisions the nuclear spin is substantially polarized despite the large magnetic field.

In the exit tube, the magnetic field drops off to small values, so the estimated 30 spin-exchange collisions should be sufficient to bring the atoms to spin-temperature equilibrium. Then $P_I \approx P_S \sim 0.53 P_K$. The sextupole magnet measures $P_S$, so at high potassium densities a deuterium electron spin polarization of 53% should be measured by the sextupole polarimeter. This is in reasonable agreement with recent measurements of about 60% by the Argonne group [7].

CONCLUSIONS

We have demonstrated that spin-exchange collisions are important for high density polarized targets. We now conclude by summarizing some of the ramifications of spin-exchange collisions for polarized targets in general and the Argonne target in particular.

Since H-H spin-exchange collisions at low field do not change the vector polarization, spin-exchange should not affect these targets once the nuclei are polarized.

For D targets spin-exchange affects both the vector and tensor polarizations in all fields. Tensor polarizations are especially susceptible, and in the limit of large numbers of collisions the tensor polarization cannot be negative. Magnetic fields can be used to increase the number of collisions required to degrade the tensor polarization.

Spin-exchange not only affects the nuclear polarizations in the target but may also play a role in the production of the desired polarization. This is because spin-exchange collisions that occur before the atoms reach an RF transition region can redistribute the populations of the various levels. This occurs for either hydrogen and deuterium at all fields. In addition, spin-exchange causes shifts in the magnetic resonance frequencies.

The efficiency of the spin-exchange method for polarizing hydrogen will be much greater than for deuterium in the same magnetic field. The consequences of this are discussed in another paper for this conference.

In general, it will be necessary for designs of high-density polarized H and D targets and sources to take into consideration the effects of spin-exchange collisions.

This research is supported by the NSF (Grant Numbers PHY-9005895 and PHY-9257058) and the University of Wisconsin Research Committee. T. W. is an Alfred P. Sloan Fellow.

# References

[1] L. W. Anderson, F. M. Pipkin, and J. C. Baird, Phys. Rev. Lett. 1,229 (1958).

[2] E. M. Purcell and G. B. Field, Astrophys. J. 124 542 (1956).

[3] K. Coulter , R. Holt, E. Kinney, R. Kowalczyk, D. Potterveld, L. Young, B. Zeidman, A. Zghiche, and D. Toporkov, Phys. Rev. Lett. 68, 174 (1992).

[4] T. Walker and L. W. Anderson, Nucl. Instrum. Meth. A, in press.

[5] L. W. Anderson, F. M. Pipkin, and J. C. Baird, Phys. Rev. Lett. 4, 69 (1960).

[6] L. W. Anderson and T. Walker, Nucl. Instrum. Methods Phys. Res., Sect. A 316,123 (1992).

[7] D. Toporkov, private communication.

# SPIN-EXCHANGE OPTICAL PUMPING IN A HIGH MAGNETIC FIELD

L. W. Anderson, and Thad Walker
Department of Physics, University of Wisconsin-Madison
Madison, Wisconsin 53706

## ABSTRACT

The prospects for the production of a nuclear polarized H target by direct spin exchange optical pumping is analyzed. It is shown that it is possible to produce a dense nuclear polarized H target by spin exchange optical pumping without rf transitions.

## I. INTRODUCTION

There is currently high interest in the construction of nuclear polarized atomic hydrogen or deuterium (H or D) gas targets for use in storage rings. Coulter *et al* have reported a polarized D beam produced by spin exchange optical pumping. The beam has a flux of 2.1 $\times 10^{17}$ atoms/s and an electron spin polarization of 0.73.[1] Walker and Anderson have shown that the nuclei are partially polarized by spin exchange as well.[2] This paper analyzes some of the problems associated with spin exchange optical pumping to produce directly a nuclear polarized H or D target.

A spin exchange optically pumped polarized H target can be produced as follows. An alkali (A) vapor and H atoms flow through an optical pumping cell. The cell is located in a high magnetic field. The alkali atoms are polarized by optical pumping, and the atomic hydrogen is electron spin polarized by alkali-hydrogen (A-H) spin exchange collisions. There are also hydrogen-hydrogen (H-H) spin exchange collisions and alkali-alkali (A-A) spin exchange collisions. The H nuclear spin polarization is produced by repeated H-H spin exchange collisions. In a spin exchange polarized target the alkali density is much less than the atomic hydrogen density. For this reason we ignore A-A spin exchange collisions. The polarized H atoms are lost by relaxation or by flow out of the cell. For a target to be useful the polarization must be lost primarily by flow out of the cell and hence we ignore relaxation. The polarization of the A atoms is primarily lost to the H atoms by A-H spin exchange collisions. Radiation trapping influences the optical pumping rate of the A atoms and thereby influences the polarization rate of the H atoms.

## II. RATE EQUATIONS

The rate equations for the spin exchange optical pumping of H in large magnetic fields are the following:

$$\frac{dP_A^e}{dt} = \frac{R}{N}(1 - P_A^e) - \frac{n_H}{n_A}\frac{1}{T_A}(P_A^e - P_H^e) \tag{1}$$

$$\frac{dP_H^e}{dt} = \frac{1}{T_A}(P_A^e - P_H^e) - \frac{1}{2T_{ST}}(P_H^e - P_H^n) - \frac{P_H^e}{T_D} \tag{2}$$

and

$$\frac{dP_H^n}{dt} = \frac{1}{2T_{ST}}(P_H^e - P_H^n) - \frac{P_H^n}{T_D} \tag{3}$$

where $P_A^e$ and $P_H^e$ are the alkali and hydrogen electron spin polarization, $P_H^n$ is the hydrogen nuclear spin polarization, $n_A$ and $n_H$ are the A and H number densities, R is laser absorption rate per atom in photons/s, N is the average number of photons needed to polarize an A atom, and $T_D$ is the average dwell time of an H atom in the optical pumping cell. The A-H spin exchange rate is

$$\frac{1}{T_A} = n_A < \sigma v >_{AH} \tag{4}$$

and the H-H spin-temperature equilibration rate is

$$\frac{1}{T_{ST}} = \frac{n_H < \sigma v >_{HH}}{1 + x^2} \approx \frac{n_H < \sigma v >_{HH}}{x^2} \tag{5}$$

where $< \sigma v >_{AH}$ and $< \sigma v >_{HH}$ are the averages of the spin exchange cross sections times the relative velocity over the velocity distribution function for the AH and HH collisions respectively and $x$ is the ratio of the magnetic field to the critical field for H or D. The critical field for H is 507 G, and the critical field for D is 113 G. In Eqns.(2) and (3) we use $T_D^{-1}$ as the loss rate for both $P_H^e$ and $P_H^n$. This is correct if the primary loss rate for $P_H^e$ and $P_H^n$ is flow out of the optical pumping cell. In Eqn.(1) we have ignored all A polarization loss mechanisms other than A-H spin exchange. In order to have a useful target it is necessary that A-H spin exchange is the primary mechanism by which A atoms lose polarization. We ignore the loss of A or H electron spin polarization to A nuclear polarization because there are so few A atoms compared to the number of H atoms.

## III. ANALYSIS

In the steady state where

$$\frac{dP_A^e}{dt} = \frac{dP_H^e}{dt} = \frac{dP_H^n}{dt} = 0$$

the Eqns.(1), (2), and (3) can be solved. The solutions for $P_H^e$ and $P_H^n$ are the following

$$P_H^e = \left[ 1 + 2 \left( \frac{1 + T_{ST}/T_D)}{(1 + 2T_{ST}/T_D)} \right) \left( \frac{T_A}{T_D} + \frac{n_H}{n_A T_D} \frac{N}{R} \right) \right]^{-1} \tag{6}$$

and

$$P_H^n = P_H^e \left( 1 + \frac{2T_{ST}}{T_D} \right)^{-1} \tag{7}$$

It should be noted that the term $(1+T_{ST}/T_D)/(1+2T_{ST}/T_D)$ varies only from 1/2 when $T_{ST} = \infty$ to 1 when $T_{ST}$ is 0. The electron spin polarization of the H atoms by spin exchange optical pumping in the limit where $T_{ST}/T_D$ is very large has been treated by Anderson and Walker[3]. They assumed no nuclear spin which is equivalent to assuming $T_{ST}/T_D$ is infinite. In this paper we consider the prospects for producing directly nuclear polarization of the H atoms. This requires that $T_{ST} \ll T_D$ as will be shown. In this situation $(1+T_ST/T_D)/(1+2T_{ST}/T_D) \approx 1$. In order to have a high $P_H^e$ it is necessary that $2T_A \ll T_D$ and $2n_H N/(n_A T_D R) \ll 1$. The first condition, $2T_A \ll T_D$, indicates that $n_A < \sigma v >_{AH} T_D/2 \gg 1$ so that the alkali density is high enough that a hydrogen atom makes a large number spin exchange collision with polarized alkali atoms during the dwell time in the optical pumping cell to acquire a polarization close to that of the alkali atoms. The value for $< \sigma v >_{AH}$ for K and H spin exchange collisions is $2.3 \times 10^{-9}$ $cm^3/s$ at 400K and is nearly independent of the temperature. Thus it is necessary that $n_A T_D >> 9 \times 10^8$ $s/cm^3$. If we take $n_A$ to be the same as for the Argonne target so that $n_A = 2 \times 10^{12} cm^{-3}$ then $T_D >> 4.5 \times 10^{-8} s$. If one desires $P_H^e$ to be 0.9 or greater then it is necessary that $T_D \gtrsim 4.5$ ms provided $2 n_H N/(n_A T_D R)=0$. If $2 n_H N/(n_A T_D R)$ is not zero then $T_D$ must be even longer than 4.5 ms.

The second condition

$$2 n_H N/(n_A T_D R) << 1 \qquad (8)$$

indicates that $F << P_L/(2h\nu N)$ where F is the flow rate of hydrogen atoms out of the optical pumping cell and $P_L$ is the laser power. The limitation on the flow rate follows from the definition of the flow rate $F=n_H V/T_D$ and from result that $R \lesssim P_L/(h\nu n_A V)$ where V is the volume of the optical pumping cell combined with the second condition. The result that $F << P_L/(2h\nu N)$ simply says that the flow rate of polarized H atoms from the optical pumping cell must be less than the number of pumping photons incident on the optical pumping cell per s divided by the average number of photons needed to polarized an alkali atom.

If one obtains a large $P_H^e$ then it is possible to have a large $P_H^n$ provided $2T_{ST}/T_D << 1$ which is equivalent to $2x^2/(n_H < \sigma v >_{HH} T_D) << 1$. This condition also places a limitation on the flow rate out of the cell. Substituting the result that $T_D = (n_H V)/F$ yields

$$F << \frac{n_H^2 < \sigma v >_{HH}}{2x^2} \qquad (9)$$

This limitation indicates that the flow rate must be low enough that there are enough H-H spin exchange collisions to polarize the nuclear spin. The spin exchange rate depends on the magnetic field through x. At very large fields the flow rate must be very slow. In order that the magnetic field is small enough for the nuclear polarization to be large it is necessary that

$$x << \sqrt{\frac{n_H^2 < \sigma v >_{HH}}{2x^2}} \qquad (10)$$

The value of $< \sigma v >_{HH}$ is $4 \times 10^{-10}$ $cm^3/s$ at 400K and is relatively independent of the temperature. If we take $n_H = 10^{14}$ atoms/$cm^3$, $F=2 \times 10^{17}$ atoms/s, and V=50cm then we find that $x << 22$ or $B << 22 B_c=11kG$. In order that $P_H^n$ be 0.9 of $P_H^e$ it is necessary that $B \leq 3.5kG$ for the assumed $n_H$, F, and V.

The first condition on the flow rate, $F << P_L/(2h\nu N)$, also depends on the magnetic field through the fact that the average number of photons needed to polarize an alkali atom depends on both the alkali density and the magnetic field since the radiation trapping depends on the magnetic field.[4,5] Coulter et al. have shown that it is possible to optically pump to a high polarization a K vapor with $n_A = 10^{12}$ atoms/$cm^3$ in the presence of atomic hydrogen with $n_H = 10^{14}$ atoms/$cm^3$ and for a flow rate of $2.1 \times 10^{17}$ atoms/s if the magnetic field is about 2kG.[1] This high an alkali density can not be optically pumped to a high polarization in a low magnetic field. Since the various numerical values for

$n_A, n_H, T_D$ and V chosen for our calculations are similar to the target of Coulter et al. it seems clear that one can construct a nuclear polarized H target provided the magnetic field at the target is in the range 2 to 3.5kG.

For deuterium (D) the small critical field changes the situation significantly. The ratio of the rate of polarization of D to H under the same conditions is approximation $[x_D^2 < \sigma v >_{HH}]/[x_H^2 < \sigma v >_{DD}] \approx 30$. Thus higher densities or lower magnetic fields are necessary in order to achieve high polarizations of D than are necessary for high polarization of H. In lower fields the optical pumping rate will be adversely affected, thus limiting the flow. At higher densities (at constant flow) longer dwell times are necessary, so relaxation processes will be more important. We note that in the Argonne optical pumping cell the D atoms make hundreds of wall collisions. For longer dwell times more wall collisions will occur and relaxation by wall collisions may be important. In addition other relaxation processes may begin to become important as well. For D if one satisfies the condition to obtain nuclear polarization the ground level Zeeman-hyperfine level populations will be given by those predicted by the spin temperature.[2] In this situation one will have both vector and tensor polarization.

## CONCLUSIONS

The analysis presented in this paper indicates that it is possible to produce directly a nuclear spin polarized H atom target similar to the electron spin-polarized target of Coulter et al. by spin exchange optical pumping provided the magnetic field at the optical pumping cell is between 2 and 3.5kG. It is more difficult to produce directly a nuclear polarized D target by spin exchange optical pumping. For D it is probably preferable to use a very high magnetic field at the optical pumping cell followed by rf transitions to produce a nuclear polarized D target. However, Walker and Anderson have pointed out that if spin exchange collisions occur in the low or intermediate magnetic field required for the rf transitions then spin temperature populations will be approached,[2] and this must be considered in estimating the nuclear polarization that can be obtained.

## ACKNOWLEDGMENTS

This paper is supported in part by the National Science Foundation.

## BIBLIOGRAPHY

1. K. P. Coulter, R. J. Holt, E. R. Keirney, R. S. Kowalczyk, D. H. Patterveld, L. Young, B. Zeidwan, A. Zghiche, and D. K. Toporkov, Phys. Rev. Lett. **68**, 2137 (1992).

2. T. Walker and L. W. Anderson, Nucl. Instr. and Methods, to be published.

3. L. W. Anderson and T. Walker, Nucl. Instr. and Methods **A316**, 123 (1992).

4. D. Tupa, L. W. Anderson, D. L. Huber, and J. E. Lawler, Phys. Rev. **A33**, 1045 (1986).

5. D. Tupa and L. W. Anderson, Phys. Rev. **A36**, 2142 (1987).

# LASER OPTICAL PUMPING OF POTASSIUM IN A HIGH MAGNETIC FIELD USING LINEARLY POLARIZED LIGHT

Cody Martin, T. Walker and L. W. Anderson
Dept. of Physics, University of Wisconsin, Madison, WI 53706

D. R. Swenson
Los Alamos National Laboratory, MP-5, H838, Los Alamos, New Mexico 87545

ABSTRACT

It is shown experimentally that in a high magnetic field a potassium vapor can be optically pumped to a high electron spin polarization by light polarized parallel to the magnetic field and incident normal to the magnetic field. The polarization of the K vapor is measured both by observing the fluorescence and by the Faraday effect. This method of optical pumping may be useful for spin polarized targets.

## A. INTRODUCTION

The optical pumping of an alkali metal vapor in a high magnetic field is of interest.[1-7] In a typical experimental setup $\sigma^+$ or $\sigma^-$ circularly polarized light incident parallel to the magnetic field is used for the optical pumping of an alkali vapor. Anderson and Walker have suggested that the use of light, that is polarized parallel to the magnetic field B and therefore is incident normal to B, may be advantageous for high field optical pumping.[1] The average number of photons required to polarize a K atom should be less for $P_{1/2}$ $\pi$ light incident normal to the magnetic field than for $\sigma^+$ or $\sigma^-$ light incident parallel to the magnetic field. The use of $\pi$ light may also enable one to use a simpler or more efficient experimental setup for the optical pumping. This paper reports the first experiments on the high field polarization of a K vapor by optical pumping with $\pi$ light incident normal to the magnetic field. Optical pumping using $P_{1/2}$ $\pi$ light incident normal to the magnetic field works as follows. Figure 1 shows the energy levels in a high magnetic field of an idealized K atom with no nuclear spin. The wavelength of the light is set to match one of the $\Delta m_J=0$ transitions. In order to be concrete we take the wavelength of the light to correspond to the $4^2S_{1/2}(m_J=-1/2) \rightarrow 4^2P_{1/2}(m_J=-1/2)$ transition. A K atom absorbs a photon and is excited to the $4^2P_{1/2}(m_J=-1/2)$ state. This level undergoes spontaneous radiative decay to both the $4^2S_{1/2}(m_J=-1/2)$ and $4^2S_{1/2}$ $(m_J=1/2)$ states with transition probabilities of $1/3$ and $2/3$ respectively. Since some of the radiative decays take atoms to the $m_J=1/2$ state and since no absorption out of this level occurs the K atoms are optically pumped out of the $m_J=-1/2$ state and into the $m_J=1/2$ state.

## B. APPARATUS

An Ar$^+$ laser pumped Ti:Sapphire laser beam is chopped by a rotating wheel at a waist. The chopped beam turns on or off in less than $1\mu s$. The laser beam is steered through a crystal polarizer and a $\lambda/2$ plate. The linearly polarized light is incident in a direction normal to the magnetic field on a cell containing K vapor plus 100 torr of He. The cell is located between the poles of an electromagnet. The pole faces have a 1.27 cm diam. hole bored through their center. The crystal polarizer in the absence of the $\lambda/2$ plate is adjusted to transmit light polarized parallel to the pole face of the magnet and thus perpendicular to the magnetic field. The $\lambda/2$ plate is adjusted to change the polarization of the laser light from

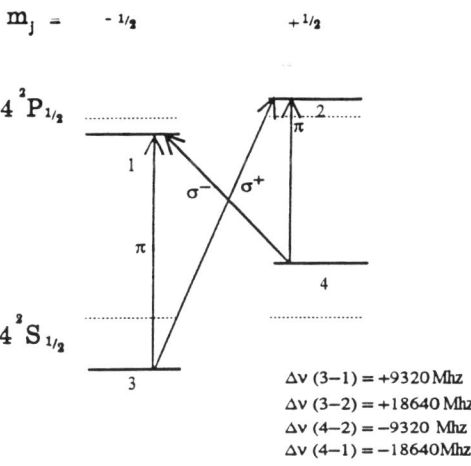

$m_j = -1/2 \quad\quad +1/2$

Figure 1. Energy levels of $^{39}$K in ignoring nuclear spin in a magnetic field of 10,000 G, showing both the $\pi$ ($\Delta m_J = 0$), and $\sigma^{+/-}$ ($\Delta m_J = +/- 1$) transitions.

$\Delta v$ (3-1) = +9320 Mhz
$\Delta v$ (3-2) = +18640 Mhz
$\Delta v$ (4-2) = -9320 Mhz
$\Delta v$ (4-1) = -18640Mhz

perpendicular to the magnetic field to parallel to the magnetic field. The magnetic field is set to 1 Tesla ($10^4$ Gauss).

We measure the polarization of the optically pumped K vapor in two different ways. The first way is to measure the polarization of the K vapor by the Faraday effect and the second way is to observe the resonance fluorescence from the vapor as it is pumped. The Faraday effect measurements are carried out as follows. The beam from a Liconix diode laser is used as a probe. The probe beam is linearly polarized, passes through the pole faces of the magnet and is incident on the K vapor cell parallel to the magnetic field. The polarization of the probe beam is analyzed after passage through the K vapor cell, and the beam is then focused by a lens onto a fast photodiode. If the K vapor is polarized the induced optical activity rotates the plane of polarization of the probe beam. The analyzer is adjusted to pass the minimum intensity when the K vapor is unpolarized. The time dependent intensity is recorded using a Le Croy transient digitizer. The time dependence of polarization of the K vapor can be obtained from the rotation of the polarization of the diode laser light.[9] The rotation of the polarization is obtained from the measured intensity at the photodiode.

The polarization of the K vapor is measured using resonance fluorescence as follows. The resonance fluorescence of the K cell is detected using a lens that images the fluorescence from the center of the K vapor cell onto a photomultiplier via a fiber optic cable. The K cell temperature is adjusted to produce about 50% absorption of the pump beam from the Ti:Sapphire laser. The time dependence of the fluorescence signal is recorded by a LeCroy transient digitizer. The polarization of the K vapor is obtained from the fluorescence as a function of the time.

## C. EXPERIMENTAL PROCEDURES AND RESULTS

There are three naturally occurring isotopes of K. The abundances are 93.1% for $^{39}$K, 0.0119% for $^{40}$K, and 6.9% for $^{41}$K. Because of its abundance we have primarily optically pumped $^{39}$K. Since the isotope shift in K is very small we also pump the other isotopes. The nuclear spin of $^{39}$K is 3/2. The ground level hyperfine separation in zero field is 461.7 Mhz. In a high magnetic field the selection rules for an allowed dipole transition with light linearly polarized parallel

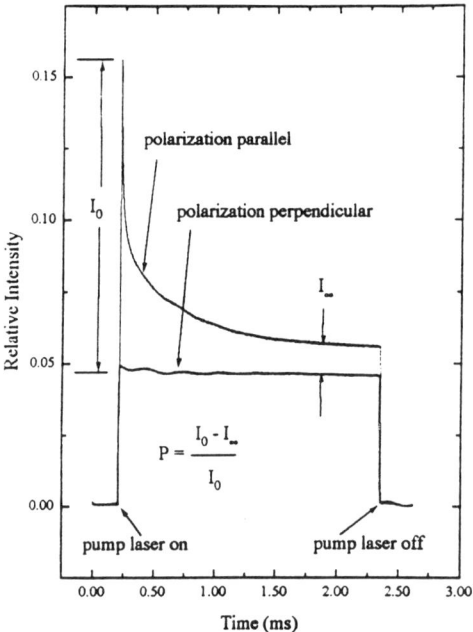

Figure 2. The resonance fluorescence signal when the pump beam turns on with the pump beam polarized perpendicular to the magnetic field or polarized parallel to the magnetic field.

to an external magnetic field and incident at right angles to the magnetic field are $\Delta m_J=0$ and $\Delta m_I=0$. Figure 2 shows the resonance fluorescence signal when the K vapor is pumped on the $4^2S_{1/2}(m_J = -1/2) \rightarrow 4^2P_{1/2}(m_J = -1/2)$ transition. Data is shown for both the pump beam polarized parallel to the magnetic field and polarized perpendicular to the magnetic field in Fig. 2. The chopping of the pump beam achieves turn on and turn off times less than 1 $\mu$s. The K vapor relaxes in the dark. The $4^2S_{1/2}(m_J = -1/2) \rightarrow 4^2P_{1/2}(m_J = -1/2)$ transition can only absorb light polarized parallel to the magnetic field. Thus we expect to absorb light and to optically pump the vapor when the light is polarized parallel to the magnetic field, whereas when the light is polarized perpendicular to the magnetic field we expect no absorbed light although there will be some scattered light for this polarization. As is seen in Fig. 2 there is almost no optical pumping when the pump light is polarized perpendicular to the magnetic field. The signal with the light polarized perpendicular to the magnetic field is almost independent of wavelength, and we believe it represents primarily scattered light. The signal in Fig. 2 shows a large initial fluorescence immediately after the pump beam is turned on and then a rapid decrease in the fluorescence as the K is optically pumped to a high polarization. The electron spin polarization of the K vapor is taken as $P = (I_0 - I_\infty)/I_0$. The analysis yields a steady state polarization for the K vapor of 0.9. The pumping rate , $\tau_p^{-1}$, is determined both by the pump laser intensity $I_\nu$ and the electron polarization loss rate of the K vapor $\tau_L^{-1}$. The pumping rate $\tau_p^{-1}$ is given by

$$\tau_p^{-1} = \tau_L^{-1} + \frac{ph\nu}{\sigma_{lu}I_\nu}$$

where $\sigma_{lu}$ is the absorption cross section at the pump wavelength and p is the average number of photons required to polarize a K atom. As shown in Fig. 2 the

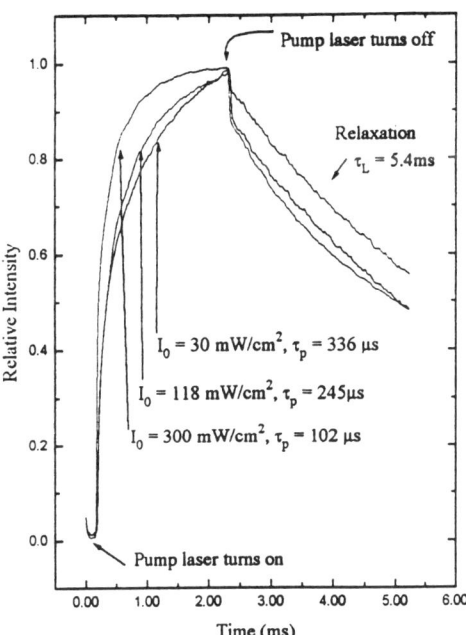

Figure 3. The Faraday rotation signal for pump laser light intensities of 300, 118 and 30 mw/cm². The traces show three distinct regions. The first displays the increase of the rotation signal as the pump beam turns on and $\theta$ increases with the increasing polarization of the K vapor. The time to polarize the vapor is seen to decrease with increasing pump light intensity. The second region shows a sharp transient feature of 50-60 $\mu$s duration. The third region shows a slow loss of polarization with a time constant $\tau_L = 5.4$ ms.

time to pump the K vapor cell to a steady state polarization is equal to 69 $\mu$s for a laser intensity of 300 mW/cm². Because the Ti:Sapphire laser lases on several modes and the exact frequencies of the pumping light are not known we can not calculate $\tau_p^{-1}$.

The average number of photons required to polarize a K vapor using $P_{1/2}$ $\pi$ polarized light incident normal to the high magnetic field is 1.5 photons per K atom at low buffer gas density. As the buffer gas density increases the K atoms excited to the $4^2P_{1/2}$ level begin to make collisions with the buffer gas while in the $4^2P_{1/2}$ level. These collisions reorient the angular momentum of the excited K atom. At high buffer gas densities the atoms excited to the $4^2P_{1/2}$ level are completely reoriented and the probability that an atom decays back to a given $m_J$ state in the ground level is the same for all $m_J$ states. In this situation the average number of photons required to polarize a K vapor is 2 photons per K atom. The average number of photons required to polarize a K vapor using $\sigma^+$ or $\sigma^-$ polarized light incident parallel to the high magnetic field is 3 photons per K atom at a low buffer gas density and 2 photons per K atom at high buffer gas density. At the He buffer gas density in our K cell the reorientation in the excited $4^2P_{1/2}$ level is complete and p=2. The optically pumped alkali targets used for the optically pumped H⁻ ion sources contain no buffer gas and p will be 1.5 for $\pi$ polarized pump light. Thus there may be a major advantage to use $\pi$ polarized pump light for the ion source. Likewise there is an advantage in the use of $\pi$ polarized pump light for the Argonne optically pumped target.

Figure 3 shows a typical Faraday rotation signal showing both the pumping and the relaxation in the dark for three different pump laser intensities. The Faraday rotation angle $\theta$ is given by $\theta = VBnl$, V is the verdet constant, B is the magnetic field intensity, and nl is the product of the K number density times the cell length. The Verdet constant depends both upon the population densities of the magnetic sublevels and upon the transition frequencies of the respective polarization components of the light.[9] Therefore it depends on the magnetic field strength through the Zeeman splitting of the energy levels and upon the polarization of the vapor by optical pumping. The transmitted intensity through the analyzer is governed by $I(\theta) = (I_{max} - I_{min})\cos^2(\theta - \phi) + I_{min}$. For our experiment $I_{min}$ is very near zero and $\phi$ is chosen to be $\pi/2$. Figure 3 shows $I(\theta)$ as a function of time for three different pump laser intensities. We do not know the diode laser frequency with sufficient accuracy to obtain absolute values for the K vapor polarization and number density. We can, however, analyze the Faraday effect data to obtain pumping and relaxation times. The pumping times and the relaxation time are given in Fig. 3. As expected the pumping rate increases as the pump laser intensity increases.

The electron spin relaxation in these traces has an interesting unexplained dependence on time. The polarization loss rate is very high for the first 60 $\mu$s and then decreases to a much slower rate. The polarization decreases in the dark due to various effects. For example diffusion of polarized atoms out of and the diffusion of unpolarized atoms into the probe laser beam causes polarization loss. Collisions of a polarized alkali with the He buffer gas or the cell walls also causes polarization loss. Other effects can also cause a loss of polarization. We can not quantitatively explain the time dependence of the polarization loss in the dark.

## ACKNOWLEDGEMENT

This research is supported in part by the NSF (PHY-9005895). T. W. is an Alfred P. Sloan Research Fellow.

## BIBLIOGRAPHY

1. L. W. Anderson, Nucl. Instr. and Methods 167, (1979) 583.

2. Y. Mori, K. Ito, A. Takagi, and S. Fukamoto, Proc. of the Workshop on Polarized Proton Ion Sources, A.I.P. Conference Proc. No. 80, Edited by A. D. Krisch and A. T. M. Lin, New York (1982), pg. 201.

3. R. L. York, O. B. Van Dyck, D. R. Swenson and D. Tupa, Proc. of the International Workshop on Polarized Ion Sources and Polarized Gas Jets, edited by Y. Mori, Tsukuba, Japan (1990), pg. 170.

4. A. N. Zelenski, S. A. Kokhanovskii, V. G. Polushkin and K. N. Vishnevskii, Proc. of The International Workshop on Polarized Ion Sources and Polarized Gas Jets, edited by Y. Mori, Tsukuba, Japan (1990), pg. 154.

5. R. L. York, Proc. of The International Workshop on Polarized Ion Sources and Polarized Gas Jets, edited by Y. Mori, Tsukuba, Japan (1990), pg. 142.

6. L. Buchman, C.D.P. Levy, M. McDonald, R. Ruegg, and P. W. Schmor, Proc. of The International Workshop on Polarized Ion Sources and Polarized Gas Jets, edited by Y. Mori, Tsukuba, Japan (1990), pg. 161.

7. K. P. Coulter, R. J. Holt, E. R. Kinney, R. S. Kowalcyk, D. H. Patterveld, L. Young, B. Zeidman, A. Zgheche, and D. K. Toporkov, Phys. Rev. Lett. 68 (1992) 174.

8. L. W. Anderson and Thad Walker, Nucl. Instr. and Methods A316 (1992) 123.

9. Z. Wa, M. Kitano, W. Happer, M. Hou. and J. Daniels, Appl. Opt. 25 (1986) 4483. See also M. Dulick. D. R. Swenson, D. Tupa, R. L. York, W. A. Cornelius, O. Van Dyck, unpublished.

# PRESENT STATUS AND FUTURE PROSPECTS
# OF OPTICALLY PUMPED POLARIZED ION SOURCE

Yoshiharu Mori

*National Laboratory for High Energy Physics(KEK)*
Oho 1-1, Tsukuba-shi, Ibaraki-ken 305, Japan

abstract

Present status and future prospect of optically pumped polarized ion source(OPPIS) are described.

## 1. Introduction

An idea of the optically pumped polarized ion source (OPPIS) for producing polarized negative hydrogen ion source , which is based on electron-capture reactions of negative hydrogen ions from the optically pumped sodium atoms, was originally proposed by Anderson[1]. The first OPPIS based on this idea was successfully developed at KEK in 1983.[2] Since then, OPPIS has been developed at various laboratories, LAMPF,TRIUMF and INR.[3][4][5] Recently, it was found at KEK that, with dual-optically pumped scheme, OPPIS was also able to generate highly polarized negative deuterium ions which have been thought difficult to make so far.[6][7] In order to increase the beam intensity further in future, a new type of OPPIS with spin-exchange scheme is being investigated. It may not be a dream to obtain intense polarized ion beams with OPPIS as well as ordinary unpolarized ion beams.

## 2. Present status of OPPIS for polarized negative hydrogen ions

A schematic diagram of the optically pumped polarized ion source is shown in Fig. 1. Low-energy (several keV) protons pick up polarized electrons from the optically pumped alkali atoms and form electron-spin polarized hydrogen atoms. In order to avoid depolarization due to a spin-orbital angular momentum coupling in the excited state (2P state) hydrogen atoms, which are formed in electron pick-up reactions in this proton beam energy range, a high magnetic field of more than 1 T is

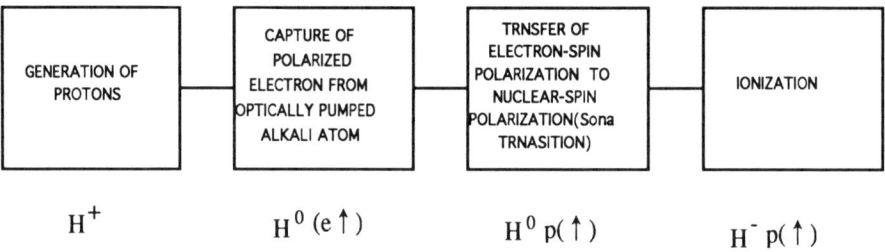

Fig. 1  Schematic diagram of OPPIS for polarized negative hydrogen ions.

necessary.  The effect of the depolarization is calculated as a function of the magnetic field strength by several groups.[8,9]  The measured polarization transfer was
10 to 15 % lower than the theoretically expected value.   This might be partly due to
the unpolarized atomic hydrogen beam which is formed by charge-exchange reactions with residual hydrogen gas molecules flowing from the proton source.  The
electron-spin polarized hydrogen atoms pass through a zero-crossing magnetic field
and the electron-spin polarization transfers to nuclear-spin polarization by a nonadiabatic transition(Sona transition).   Ionizing the nuclear-spin polarized hydrogen
atoms to negative hydrogen ions can be done by the charge-exchange reaction with
another alkali atoms such as sodium.

The optical pumping of the alkali atoms in the neutralizer was previously performed with a dye laser because of its wavelength tunability to match the absorption

Table 1  Characterisitcs and performance of OPPIS in the world.

|  | KEK | LAMPF | TRIUMF | INR |
|---|---|---|---|---|
| intensity |  |  |  |  |
| H⁻ | 100μA(H ) | 50μA(2μA) | 25μA | 400μA |
| H⁺ | 1mA(H ) | - | - | 4mA |
| polarization | .65 | .64(.77) | .80 | .65 |
|  |  |  |  |  |
| emittance | 2.0 | 1.0 | 1.0 | 1.0 |
| (πmm.mrad) |  |  |  |  |
| duty factor | 0.001 | 0.1 | DC | 0,0002 |

resonance lines concerning to the optical pumping.   However, recently, tunable solid state laser such as Ti-sapphire or Alexandrite laser has been widely used because of its relatively large power and ease for maintenance.

Characteristics and performance of OPPIS at various laboratories are summarized in table 1.   The beam intensity from OPPIS reaches more than 100 $\mu$ A for polarized negative hydrogen ion beam and 1 mA for polarized proton beam with relatively small beam emittance, respectively, in pulsed mode operation and 25$\mu$ A for negative hydrogen ion beam in DC mode operation.   The beam polarization depends very much on the external magnetic field strength in the electron-capture process.   When the magnetic field strength is about 2.5T, the beam polarization of 80 % was realized in TRIUMF.

### 3. OPPIS for polarized negative deuterium ions

Polarized protons have been successfully accelerated in the KEK 12-GeV proton synchrotron (KEK-PS) since 1985.   Many experiments have been carried out with polarized proton beams so far.   Recently, several proposals for the

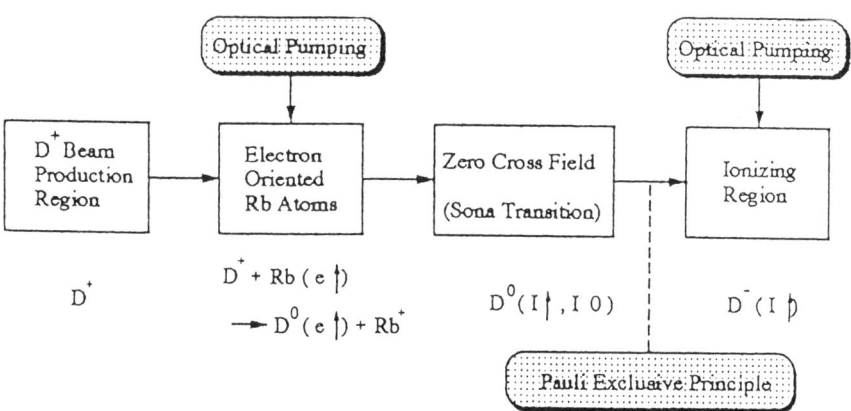

Fig.2 Block diagram of the dual-optically-pumped polarized ion source.

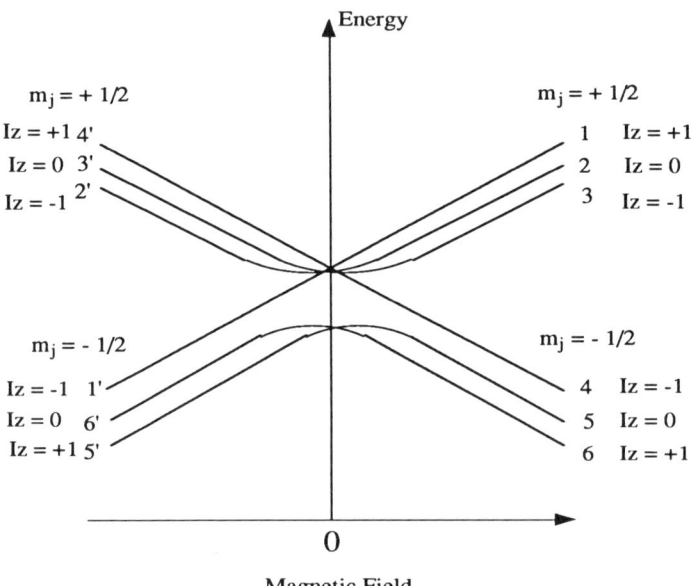

Fig.3 Hyper-fine sublevels of deuterium atom in Sona transition.

physics experiments with polarized deuteron beams in the KEK-PS are under discussion.

An ordinary deuteron beam was successfully accelerated to an energy of 11.2GeV (5.6GeV/u) in the KEK-PS, the limiting energy of the ring, in 1992. [10] The beam intensity of the accelerated deuterons reached more than $2 \times 10^{12}$ ppp which was almost the same as that for protons. Acceleration of polarized deuterons in the synchrotron is rather easier than that of polarized protons. Since the anomalous magnetic moment (G) of the deuteron is -0.1426, which is 10 times less than that of the proton, the number of depolarization resonances that are caused by the betatron motions of the beam and imperfection of the ring are quite few during the acceleration of polarized deuterons in the synchrotron compared with polarized protons.

As for polarized ion source, an optically pumped polarized ion source(OPPIS) has been used for generating nuclear-spin polarized negative hydrogen ions so far. It has been believed that this type of polarized ion source is not useful to produce a highly nuclear-spin polarized(vector and tensor) deuterium ions. In this paper, we report the preliminary experimental results which showed that the highly vetor polarized negative deuterium ions could be produced by OPPIS with dual-optically-pumped technique.

In Fig.2,  a block diagram of the dual-optically-pumped polarized ion source is shown schematically.   An energetic electron-spin polarized hydrogen beam is generated via pickup of a polarized electron by a proton beam of a few keV in an optically pumped alkali vapor.   Then the atomic polarization of the hydrogen beam is transformed into nuclear polarization by a diabatic transition between hyperfine sub-levels (Sona transition).   Finally, the nuclear-spin polarized hydrogen beam is ionized.

Although OPPIS is very useful to generate polarized protons, it has been thought that OPPIS is inadequate for making highly polarized deuterons.   In deuterium atoms, because the nuclear spin, I=1, three hyperfine sub-levels ($I_z$=+1,0,-1) exist.   High polarization can not be expected if only a Sona transition is used because of the $I_z$=0 state.   The theoretical maximum polarizations, in this case, are +-2/3 for vector polarization ($P_z$) and -1/3 for tensor polarization ($P_{zz}$).

To achieve a high polarization, a new scheme which  selects a pure nuclear-spin state is  necessary.  In 1988, Schneider and Clegg[11] proposed a new nuclear-

PRINCIPLE OF DUAL-OPTICALLY PUMPED
POLARIZED NEGATIVE DEUTERIUM ION SOURCE

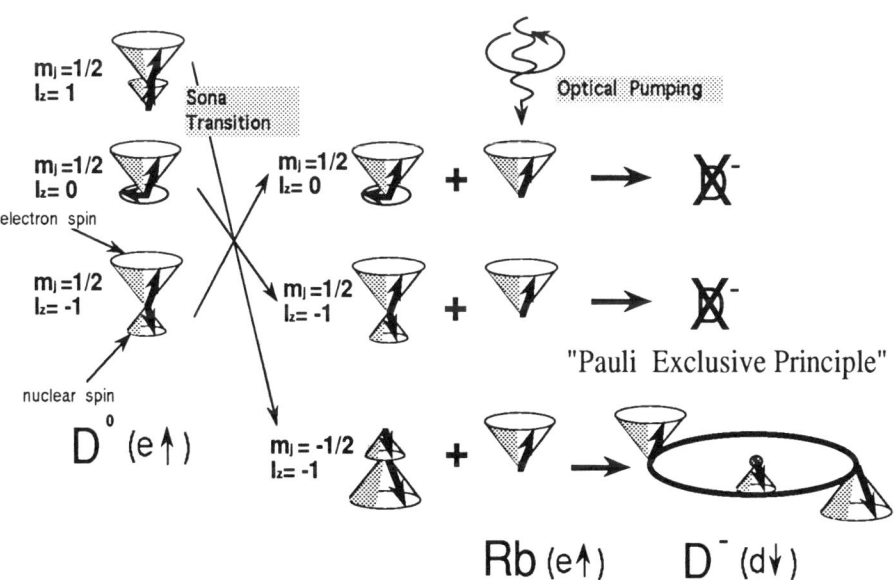

Fig. 4  Principle of OPPIS for polarized negative deuterium ions with dual optical pumping.

spin state selection scheme.    Their idea is as follows: After picking up the polarized electrons from optically pumped alkali atoms, deuterium atoms are electron-spin polarized, for example, in the state of $m_j = +1/2$ as shown in Fig.3.   These electron-spin polarized deuterium atoms equally populate three hyperfine sub-levels $I_z = +1, 0$, and -1 in a high magnetic field, which are labeled the states 1, 2, and 3, respectively in Fig.3.   Using the Sona transition, the state 1 ($m_j = +1/2$, $I_z = +1$) goes to the state 1' ($m_j = -1/2$, $I_z = -1$), the state 2 ($m_j = +1/2$, $I_z = 0$) goes to the state 2' ($m_j = +1/2$, $I_z = -1$), and the state 3 goes to the state 3' ($m_j = +1/2$, $I_z = 0$), respectively as shown in Fig.2.   Therefore, the deuterium atoms with only the hyperfine level of $I_z = -1$ (state 1' in Fig.3) has an opposite electron-spin state, $m_j = -1/2$, of the other two sub-levels (2' and 3') after Sona transition.   When the alkali atoms in the ionizer are also optically pumped and their electrons are to be spin polarized in the $m_j +1/2$ state, only deuterium atoms with the electron-spin state of $m_j = -1/2$ ( state 1' ) can form negative ions because of the Pauli exclusion principle.    This process is shown in Fig.4 schematically.   The nuclear-spin state of the negative deuterium ions in this case is $I_z = -1$, the nuclear vector polarization becomes -1.   The nuclear tensor polarization is, in this case, -1.

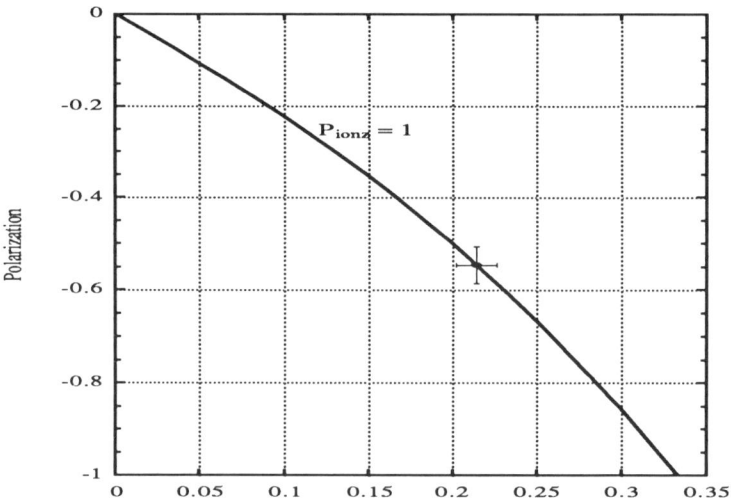

Fig.5 Relation between $P_D$ and $\varepsilon$ when $P_{IONZ} = 1$.    The closed circle in the figure shows the experimental result.

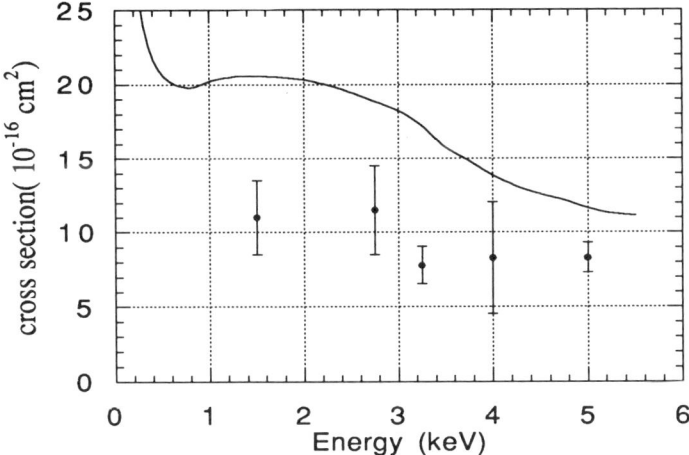

Fig. 6 Spin-exchange cross sections of H-Na system as a function of atomic hydrogen beam energy. The solid line shows the theoretically calculated one and the closed circles presents the experimental results.

Using a proper rf transition simultaneously, a pure nuclear tensor polarization of -2 may become possible.

In spite of this possibility of making a highly polarized deuteron beam by optical pumping, they concluded eventually in their paper that this dual optical pumping scheme might be not practical because efficient optical pumping of the thick target in the ionizer is difficult due to radiation trapping. Radiation trapping is a re-absorption process of flourescense photons in optical pumping and it limits the maximum polarization of the pumped atoms. However, their conclusion was qualitative and they did not estimate quantitatively the   effect of radiation trapping. Recently, we have re-examined the dual-pumped scheme in detail and found that radiation trapping was not a serious problem and highly polarized deuterons could be obtained with the dual-pumped scheme.[12]

Recently, a preliminary experiment for proving the principle of the dual-pumped scheme has been carried out at KEK.   The result of the experiment is shown in Fig. 5.   The vertical axis in the figure presents the nuclear vector polarization of negative deuterium ions.   The horizontal axis shows the relative change of the beam intensity of the negative deuterium ions by switching the optical pumping of the alkali atoms in the ionizer on and off.    Deuterium atoms in only one hyperfine sub-level can become negative deuterium ions by picking up polarized electrons from the optically pumped alkali atoms in the ionizer.    Thus, the beam intensity of negative deuterium ions depends upon the population of deuterium atoms in each hyperfine sub-level after the Sona transition.    This means that the deuteron vector polarization ($P_D$) and the electron-spin polarization of optically pumped alkali atoms in the ionizer($P_{IONZ}$) affect the beam intensity of negative deuterium ions.    These values are related each other as expressed in the following equation.

$$P_D = -2\varepsilon/P_{IONZ}(1-\varepsilon).$$  (1)

Here, $\varepsilon = (I_{off}-I_{on})/I_{off}$, where $I_{off}$ and $I_{on}$ are the beam intensities of negative deuterium ions when the optical pumping of the alkali atoms in the ionizer is turned off and on, respectively.    The solid line in Fig.4 presents the relation between $P_D$ and $\varepsilon$ when $P_{IONZ} = 1$.    The closed circle in the figure shows the experimental result.    The electron-spin polarization of alkali atoms in the ionizer ($P_{IONZ}$) was measured using a Faraday rotation method.    The errors shown in the figure present the fluctuations of the data taken at different times.

In the preliminary experiment, we obtained $P_D = -0.55 +- 0.04$.    In the present apparatus where the magnetic field strength at the neutralizer is 1.1T, the theoretical maximum polarization is limited to less than 80%.    This is because the spin-orbit coupling in neutral deuterium atoms, created by picking up polarized electrons from the optically pumped alkali atoms in the neutralizer, reduces the electron-spin polarization at this  magnetic field strength.    Thus, the obtained deuteron-spin polarization was almost 70 % of the maximum limiting value.    This is very encouraging and it may be said that the dual-optically pumped scheme for producing highly polarized negative deuterium ions has worked in principle.    Our results may open up a new possibilities for the optically pumped polarized ion sources.    There seems to be no fundamental problem for polarized deuteron acceleration in the KEK-PS.

## 4. Future of OPPIS with spin-exchange scheme

The principle of the spin-exchange optically pumped polarized hydrogen or deuterium ion source is based on the electron-spin polarization transfer of alkali

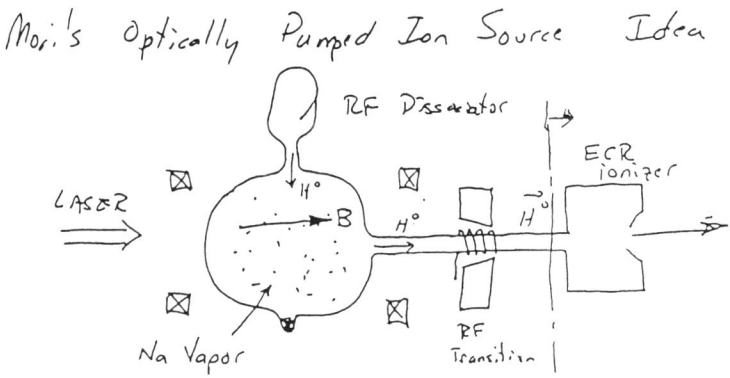

Mori's Optically Pumped Ion Source Idea

(a) Spin-exchange for thermal hydorgen atoms.

Spin-Exchange :  fast H⁰    B→2kG                    ←1.5kG

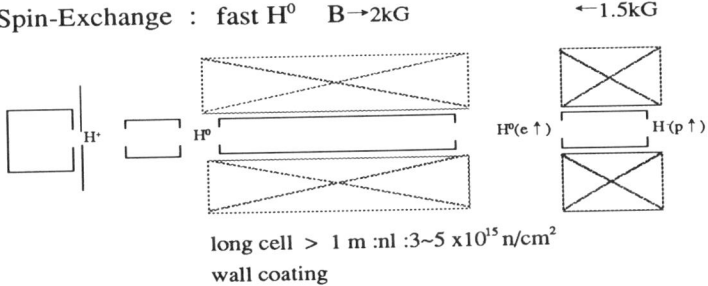

long cell > 1 m :nl :3~5 x10$^{15}$ n/cm$^2$
wall coating

* acceptance : 0.18 mm.mrad(100% normalized)
*multiple sactterign

(b) Spin-exchange for fast hydrogen atomic beam.

Fig. 7 OPPIS with spin-exchange scheme.

atoms to the hydrogen or deuterium atoms by spin exchange collision. The way of making nuclear polarization of hydrogen or deuterium atoms by the spin-exchanged collision is as follows;(1) Atomic hydrogen or deuterium beam either which is thermal or energetic(~keV) is generated. (2) Electron-spin polarized alkali atoms are produced by optical pumping. (3) The electron-spin polarization is transferred to hydrogen or deuterium atoms by spin exchange collisions. (4) The electron-spin polarization of hydrogen or deuterium atoms are transferred to the nuclear-spin polarization by hyperfine interaction. (5) The nuclear-spin polarized hydrogen or deuterium atoms are ionized.

In the mutual atomic collision process between hydrogen/deuterium atoms and alkali atoms, the spin-exchange differential cross section can be written by,

$$\frac{d\sigma_{ex}}{d\theta_{cm}} = \frac{|f_t(\theta_{cm}) - f_s(\theta_{cm})|^2}{4} \qquad (2)$$

$$f_s(\theta_{cm}) = \frac{1}{2ik} \sum_{l=1}^{\infty} (2l+1)[\exp\{2i\eta_l^s(v_{rel})\} - 1]P_l(\cos\theta_{cm})$$

$$f_t(\theta_{cm}) = \frac{1}{2ik} \sum_{l=1}^{\infty} (2l+1)[\exp\{2i\eta_l^t(v_{rel})\} - 1]P_l(\cos\theta_{cm})$$

Here, $f_t$ and $f_s$ are the scattering amplitudes for the singlet and triplet states of the two atomic molecular system, respectively and each of them can be  estimated from the phase shift which is calculated from an electronic potential curve corresponding to each state.

The calculated total spin-exchange cross sections are summarized for thermal H-Na, H-K, H-Rb and H-Cs systems in Table 2. [14]

Table 2    Total spin-exchange cross section for thermal H-Na, H-K, H-Rb and H-Cs systems.

| | |
|---|---|
| H-Na: | $2.2 \times 10^{-15}$ cm$^2$ |
| H-K : | $2.7 \times 10^{-15}$ cm$^2$ |
| H-Rb: | $2.9 \times 10^{-15}$ cm$^2$ |
| H-Cs : | $3.4 \times 10^{-15}$ cm$^2$ |

The spin-exchange cross sections between energetic hydrogen or deuterium atoms of several keV and thermal alkali atoms are calculated by Swenson et al. The calculated cross section for H-Na system is shown in Fig.6 as a function of the incident energy of hydrogen atomic beam.   The measurement of the spin-exchange cross section for H-Na system was carried out at KEK with modifying their OPPIS. The results are also shown in Fig. 6.   The measured values are somewhat smaller than those expected from the calculation, however, still larger than $1 \times 10^{-15}$ cm$^2$ .

The atomic polarization of thermal hydrogen atoms induced by spin-exchange collisions with optical pumping polarized alkali atoms can be expressed in the following form.

$$P_H = \frac{\gamma_{SE} \, P_A}{\gamma_{SE} + \gamma_r} \, [1 - \exp\{-(\gamma_{SE} + \gamma_r) \, t\}] \qquad (3)$$

Here, $P_A$, $\gamma_{se}$ and $\gamma_r$ are the atomic polarization, the spin-exchange rate($=<\sigma_{ex}v>n_A$) and the polarization relaxation time of hydrogen atoms, respectively. Apparently, when $\gamma_{se} > \gamma_r$, eq. (3) becomes,

$$P_H \sim P_A. \qquad (4)$$

In order to get a high polarization of hydrogen atoms, the density of sodium atoms $n_A$ should be more than $10^{13} n/cm^3$ and the relaxation rate $\gamma_r$ should be less than $10^3$ $s^{-1}$.

In optical pumping at a simple three level system, the polarization becomes,

$$P \approx [ \, 1 - \frac{2\gamma}{\Phi + 2\gamma} \, ] \qquad (5)$$

Here, $\Phi = I / h\nu$, where I is a laser power, and $\gamma$ is the relaxation rate, respectively. When I = 1W, $\Phi$ becomes more than $10^7$ photons/s. Even if $\gamma$ is $10^4$ $s^{-1}$, the polarization is still close to 100%. However, when alkali atom density exceeds more than $10^{12}$ n/ cm$^3$, an emitting photon in spontaneous decaying of the excited state is re-absorbed by another atom and eventually the polarization should be destroyed. This multiple process is named radiation trapping.

The polarization decrease due to the radiation trapping was calculated theoretically by Tupa and Anderson[15] and their calculations have been clarified experimentally.[16] The polarization can not reach 100% because of radiation trapping when the relaxation time is small. In order to achieve high polarization of more than 90% at the sodium density of about $10^{13}$ n/cm$^3$, the relaxation time should be longer than at least 100$\mu$sec.

Another important effect induced by radiation trapping is that the polarization of optically pumped alkali atoms are very much affected by the external magnetic field strength. In a weak magnetic field, the hyperfine sublevels for ground and excited states of alkali atom are not well resolved and the absorption line widths for those sublevels are overlapped each other by a Doppler broadening. Therefore, re-absorption probability of emitted photons from the excited state becomes large and the polarization results in decreasing.   On the other hand, in a strong magnetic field, the hyperfine sublevels are well resolved and the polarization decrease caused by radiation tapping can be reduced.   Of course, electron-spin polarization of hydrogen atoms is realized in a strong magnetic field by spin-exchange collision, therefore RF transition  is essential to make nuclear-spin polarization.

Recently, the group led by Holt in Argonne National Laboratory has been carrying on the development works of the spin-exchange type of polarized deuterium gas source for electron storage rings.[17]   The obtained polarized deuterium flux was $2.1 \times 10^{17}$ n/s and the electron-spin polarization of deuterium atoms of 73+-3% was achieved in a strong magnetic field of 2.2kG.

A schematic layout of the OPPIS with spin-exchange is shown in Fig. 7. The expected beam intensity reaches more than 10 mA if 30 % of the ionization efficiency in the ECR ionizer, which is a very reasonable value for ordinary ECR ion source,  is assumed.

## 5. Conclusion

Present status of OPPIS for polarized negative hydrogen ions are breifely summarized.   And a new dual-optically-pumped scheme to obtain a high deuteron-spin polarization in an optically pumped polarized ion source has been examined. The results of the preliminary experiment are very encouraging and it is shown that the new scheme, in principle, has worked.

The spin-exchange technique looks very  interesting for future OPPIS, in which the beam intensity becomes comparable with that of ordinary unpolarized ion source .   We may conclude that OPPIS has a large potential.

The authors would like to appreciate Profs. M.Kihara, Y.Kimura, and H. Sugawara for their continuous encouragement.   They are indebted to Mrs. M/ Kinsho, K.Ikegami, and A. Takagi for their helps in the experiments.   They   are also grateful to Drs. A. Zelenskii and C.D.P.Levy for their valuable suggestions.

REFERENCES
[1] W.L.Anderson: Nucl. Instr. and Meth. 167,363(1979).

[2] Y.Mori, K.Ikegami, Z.Igarashi, A.Takagi,and S.Fukumoto: AIP Proc. 117,New York(1983)123.

[3] R.L.York, O.B. Van Dyck, D.R.Swenson and D.Tupa: Proc. of the Int. Workshop on Poarized Ion Sources and Polarized Gas Jets, KEK Report 90-15(1990),page 142.

[4] L.Buchmann,C.D.P.Levy, M.McDonald, R.Ruegg, and P.W.Schmor: ibid page 161.

[5] A. Zelenskii, S.A.Kokhanovskii, V.G.Polushkin and K.N. Vishnevskii:ibid page 154.

[6] Y.Mori and M.Kinsho;Proc. of High Energy Spin Physics Conference, Nagoya,1992.(in press)

[7] M.Kinsho, Y.Mori; Proc. of Particle Accelerator Conference, Washington D.C.,1993.(in press)

[8] E.A.Hinds, W.D.Cornelius and R.L.York; Nucl. Instr. Meth.,189,599(1981).

[9]. Y.Mori, A.Takagi,S.Fukumoto; Bull. Am. Phys. Soc. 26,129(1981).

[10]Y.Mori ; Proc. of Particle Accelerator Conference, Washington D.C.,1993.(in press)

[11] M.B.Schneider and T.B.Clegg: Nucl. Instr. and Meth.,A254,630(1987).

[12]Y.Mori ; submitted to Nucl. Instr. Meth.

[13] M.kinsho, K.Ikegami, A.Takagi, and Y.Mori; contribution in this conference.

[14] A.Ueno,Y.Mori, K.Ohata, Y.Wakuta, I.Kumabe; Nucl. Instr. Meth.,A271 ,343(1988).

[15]D.Tupa and L.W.Anderson; Phys.Rev.,A31, 3722(1985).

[16]Y.Mori et al.; Nucl. Instr. Meth.,A268,270(1988).

[17] A. Zghiche et al.; Proc. of the Workhop on Polarized Gas Targets for Storage Rings,Heidelberg, 1991, page 103.

# A STUDY OF SPIN-EXCHANGE POLARIZATION TRANSFER IN HYDROGEN-RUBIDIUM COLLISIONS

A.N. Zelenski and S.A. Kokhanovski

Institute for Nuclear Research, Russian Academy of Sciences, 117312 Moscow

and

TRIUMF, 4004 Wesbrook Mall, Vancouver, B.C. V6T 2A3

## ABSTRACT

The results of experimental studies of basic limitations on optical pumping of high density rubidium vapor are presented. A maximum 40% proton polarization was obtained in a spin exchange optically pumped polarized H$^-$ ion source using a 30 cm long dry-film coated cell. About 50% polarization was measured with a 90 cm long copper uncoated cell at a Rb thickness of $8 \times 10^{14}$ atoms/cm$^2$. A pulsed Ti:sapphire laser was used for optical pumping of Rb with up to 1 kW pulsed power, 200 $\mu$sec pulse duration and 10-15 GHz linewidth.

## INTRODUCTION

The recent progress made in the development of optically pumped polarized H$^-$ ion sources is very impressive. The polarization in excess of 80%, which was obtained from the dc TRIUMF source, is actually higher then was expected taking into consideration all possible depolarization processes. The record current of polarized H$^-$ – over 20 $\mu$A in dc mode (TRIUMF)[1] and above 400 $\mu$A in pulsed operation (INR, Moscow)[2] – was also obtained from optically pumped sources. Furthermore, the full capabilities of optically pumped sources have not been realized as yet, due to the very low efficiency with which the laser power is used. A 5 W Ti:sapphire laser used for optical pumping of alkali-metal vapor yields an electron polarization production rate of up to 10$^{19}$ atoms/sec and a theoretical limit on equivalent polarized proton current of up to 1 A. This may be a way of producing multiampere polarized beams for fusion applications. The much smaller operational currents mentioned above are due to short polarization relaxation times and limits on production of low divergence high current proton beams. It is clear that the INR-type OPPIS in pulsed, low duty factor operation is capable of producing at least 1-2 mA of polarized H$^-$ within a normalized emittance of 1.5 $\pi$ mm-mrad, suitable for high energy accelerators.[3] The possibility of further improvement will be limited by the space-charge effect of Rb ions on the proton beam, which is apparently already significant in the dc OPPIS.

A possible way to avoid the above limitation and better realize the capabilities of optical pumping is to use the spin-exchange technique of polarization.[4] Polarization transfer through spin-exchange collisions at thermal energies is well known. The thermal cross sections for alkali-alkali spin-exchange are above 10$^{-14}$ cm$^2$. Recently, hydrogen and deuterium atoms were successfully polarized in collisions with optically pumped potassium. An important calculation was done by Swenson *et al.*,[5] it showed that the spin-exchange cross section of hydrogen-alkali collisions is $(1 - 3 \times 10^{-15} \text{cm}^2)$ at hydrogen beam energies of 1-10 keV. The conclusion was that spin-exchange should slightly increase the polarization in OPPIS and is also important for the "collisional pumping" technique of polarization. "Collisional pumping" of proton polarization due to the hyperfine electron-proton interaction can occur only in a low magnetic field.[6] The maximum density of optically-pumped alkali vapor is strongly limited by radiation trapping, as was shown by Tupa *et al.*,[7] and to polarize a $10^{15}$ atoms/cm$^2$ thickness we need a 10 m long cell at low field. But if we use spin-exchange only for electron polarization in a high magnetic field and then use the conventional Sona method of polarization transfer, the technique looks feasible, since a $5 \times 10^{12}$ atoms/cm$^3$ polarized density has already been achieved.

A layout of the spin-exchange optically-pumped polarized ion source is presented in Fig. 1. In the original INR source, a neutral hydrogen beam was injected from an external source into a solenoid, where a helium ionizer cell and an optically pumped cell are situated.[8] The only difference between the INR charge-exchange source and spin-exchange source is the absence of the helium ionizer cell in the latter. The hydrogen beam is neutral and we avoid any possibility of emittance degradation caused by Rb ion space-charge. Spin-exchange occurs between the ground states of the hydrogen and alkali atoms, and therefore only a 5-8 kG magnetic field is required for optical pumping of high density vapour, in contrast with the charge-exchange OPPIS, where over 20 kG is required to prevent depolarization due to the spin-orbital interaction in excited states.

The results of calculations of H-Rb and H-Cs spin-exchange polarizations at a hydrogen beam energy of 1-2 keV are presented in Fig. 2. The calculations show that production of 90% electron polarization of hydrogen atoms (and consequently about 85% proton polarization) requires at least a $10^{15}$ atoms/cm$^2$ thickness of optically pumped Rb vapour with polarization in excess of 95%. The ability to optically pump high density alkali vapour is crucial for the feasibility of spin-exchange sources. In contrast, the charge-exchange source can produce high polarization even at low Rb thickness, if background is kept low by reducing neutralization on residual gas, or by energy separation. For example, at the TRIUMF OPPIS, polarization drops only at a Rb cell thickness less than $2 \times 10^{13}$ atoms/cm$^2$. The experimental limit on optical pumping of high density vapour is about $5 \times 10^{12}$ atoms/cm$^3$ and on thickness about $10^{14}$ atoms/cm$^2$ (with a 20 cm long sodium cell in pulsed mode of operation at INR, Moscow[8] and KEK[9]). Increasing the cell length is a simple way to get more thickness at limited density (supposing laser power is unlimited, as in the case of pulsed operation), but in practice it will limit the polarized current, because losses of the hydrogen beam in a 200 cm cell will be too high. (The divergence of the hydrogen beam from the best sources is about 20 mrad at 1-2 keV beam energy). So, to realize the advantages of spin-exchange polarization it is necessary to increase the density of optically pumped vapor to at least $(1-2) \times 10^{13}$ atoms/cm$^3$. The first experimental results on spin-exchange polarization in hydrogen-sodium collisions were presented at the last workshop at KEK.[4] Recent progress has been made due to the switch to optical pumping of Rb and pulsed Ti:sapphire lasers.

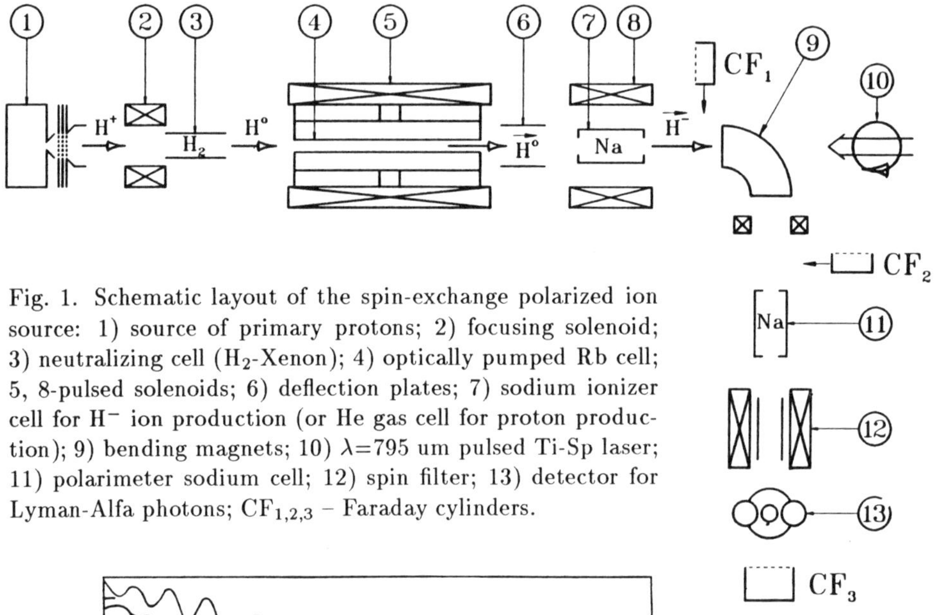

Fig. 1. Schematic layout of the spin-exchange polarized ion source: 1) source of primary protons; 2) focusing solenoid; 3) neutralizing cell ($H_2$-Xenon); 4) optically pumped Rb cell; 5, 8-pulsed solenoids; 6) deflection plates; 7) sodium ionizer cell for $H^-$ ion production (or He gas cell for proton production); 9) bending magnets; 10) $\lambda$=795 um pulsed Ti-Sp laser; 11) polarimeter sodium cell; 12) spin filter; 13) detector for Lyman-Alfa photons; $CF_{1,2,3}$ – Faraday cylinders.

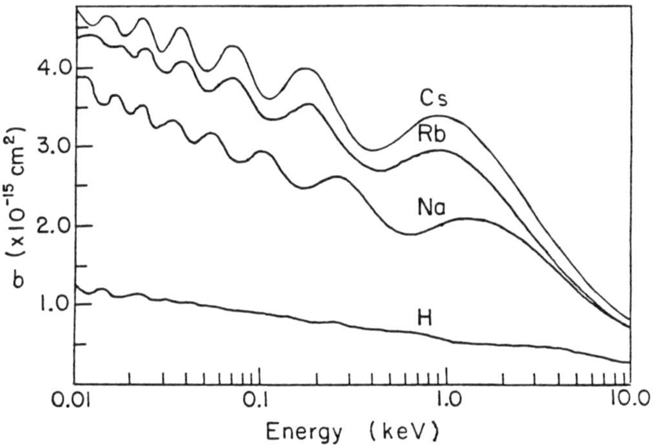

Fig. 2. Calculated spin exchange cross sections as a function of the energy for fast $H^0$ atoms incident on H, Na, Rb and Cs. (from Ref.[5])

By using a dry-film wall coating the polarization relaxation time was increased to up to 600 $\mu$sec. From the calculation of radiation trapping it was expected that the maximum vapor density in the optically pumped cell, given a 600 $\mu$sec relaxation time, would be at least $5 \times 10^{13}$ atoms/cm$^3$ (see Ref.[10]). The experimental results are apparently in contradiction with this calculation, and probably there is another polarization loss process, which will be discussed later.

## EXPERIMENTAL SETUP FOR SPIN-EXCHANGE POLARIZATION STUDIES

The experiments were done at the test-bench of the INR pulsed OPPIS. Downstream of the optically pumped spin-exchange cell everything has the same configuration as in the charge-exchange source. Electron polarization is transferred to the proton by a Sona transition and then H$^-$ ions are produced by second electron capture in the alkali-metal vapor ionizer cell. It seems that the optimal energy of the primary neutral hydrogen beam for spin-exchange sources will be about 1-2 keV. At this energy the spin-exchange cross sections for H-Rb and H-Cs collisions are up to $3 \times 10^{-15}$ cm$^2$. Also at this energy very intense neutral hydrogen sources are available, which were developed for plasma diagnostics in fusion research.[11]

In experiments on spin-exchange polarization we used a neutral hydrogen beam injector (the prototype was developed at Budker INP, Novosibirsk),[11] which yielded at 7 keV up to 30 mA of equivalent neutral current through the optically pumped cell and helium ionizer. The source of primary protons was optimized for an extraction voltage of 7-10 keV and produced substantially lower current at 2-2.5 keV beam energy, where we did most of the experiments on spin-exchange polarization. We had about 200-400 $\mu$A of polarized proton beam, which was enough for polarization measurements even at low duty factor. The source operates at a 1 Hz repetition rate, limited by the pulsed solenoid cooling system and 100 $\mu$sec pulse duration. For proton polarization measurements we used our conventional technique, based on analysis of metastable hydrogen atoms which are produced by neutralization of polarized protons in a polarimeter sodium cell.[8]

Most of the experiments on spin-exchange polarization were done with a 35 cm long and 1.4 cm diam Rb cell. The cell was made from stainless steel with a copper inner tube (liner). The copper liner was mechanically and electrochemically polished. We used a procedure of dry-film coating described in detail by Swenson et al.[12] Rubidium vapor from a 30 cm long reservoir enters the cell through a 0.1 cm wide, 25 cm long slit. In the first experiments Rb was loaded directly into the reservoir. It was very inconvenient and unsafe handling the long cell, loaded with very reactive Rb. When exposed to air Rb is coated by an oxide layer, and to break through this layer it is necessary to raise the cell temperature much higher than operational. The Rb density then increases suddenly, sometimes causing problems for further cell operation, in particular, often destroying the dry-film wall coating. We later invented "self-opening" Rb containers. Such a container consists of two pieces of stainless steel tube, each closed at one end by a copper plug. The two tubes fit closely together to reduce contact of Rb with air, and a compressed spring is installed between them. After loading with Rb in a glove box, the tubes are fastened together by two strips of copper foil soldered on with low melting point alloy. Once the containers are loaded, we can handle them easily, since they are almost hermetically sealed. Inside the reservoir, the solder melts at 70-90° C and the spring forces the tubes apart, well below the operational temperature of 150° C. For a 35 cm cell we used 4 containers to provide a homogeneous distribution of Rb density along the cell. This technique was even more important for experiments with a 90 cm cell.

We used a Faraday rotation technique for measuring the Rb vapor thickness. A small movable mirror, installed on the source axis between the Rb cell and the proton neutralizer cell, directed a laser beam to an analyzing polarizer after the beam had passed through the Rb cell. We used the pumping laser as a probe for thickness measurements, and therefore could not use the Faraday rotation technique for polarization measurements. The rubidium and hence proton polarization were varied from pulse to pulse by scanning the laser wavelength with a tilting intracavity etalon. The laser wavelength was measured every pulse by a wavemeter, based on analysis of the interference pattern from a Fizeau interferometer. The laser wavelength was tuned by measuring the laser power absorption curve.

## EXPERIMENTAL RESULTS OF THE POLARIZATION STUDIES

We studied the polarization relaxation time by measuring the polarization as a function of delay time between laser pulse and measuring gate. The relaxation time strongly depends on the cell wall surface conditions. The 600 $\mu$sec relaxation time has been measured with a fresh dry-film coated cell and 200 $\mu$sec for a fresh uncoated copper liner (see Fig. 3).

It follows from calculations of radiation trapping in optically pumped high density vapor, that even at infinite laser power there is a maximum polarization limit which depends on the polarization relaxation time.[10] For a 600 $\mu$sec relaxation time, radiation trapping limits polarization above a Rb density of about $(3-5) \times 10^{13}$ atoms/cm$^3$. The results of proton polarization measurements as a function of Rb thickness are presented in Fig. 4. In contrast with calculations, polarization drops at a thickness above $5 \times 10^{14}$ atoms/cm$^2$, corresponding to a density of $1.5 \times 10^{13}$ atoms/cm$^3$. Apparently, there are other depolarization processes besides radiation trapping. For example, at high density and high pumping power, Rb atoms in excited states become important. We discovered very efficient ionization in collisions between sodium atoms in excited states.[13] We can roughly estimate the density of excited states from measurements of

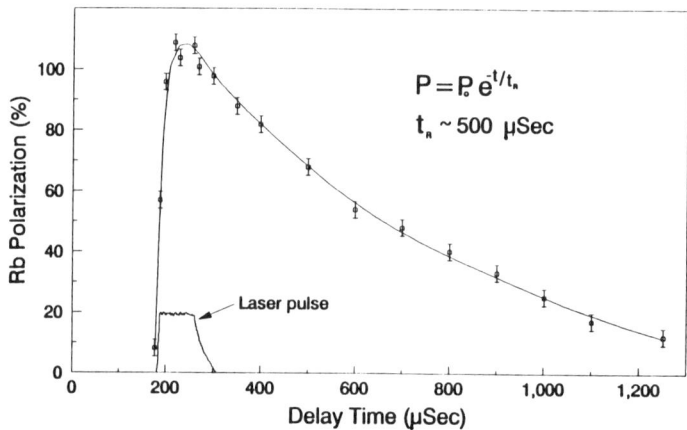

Fig. 3. Rb polarization relaxation time for copper liner with fresh dry-film coating. Measuring gate is 20 $\mu$sec.

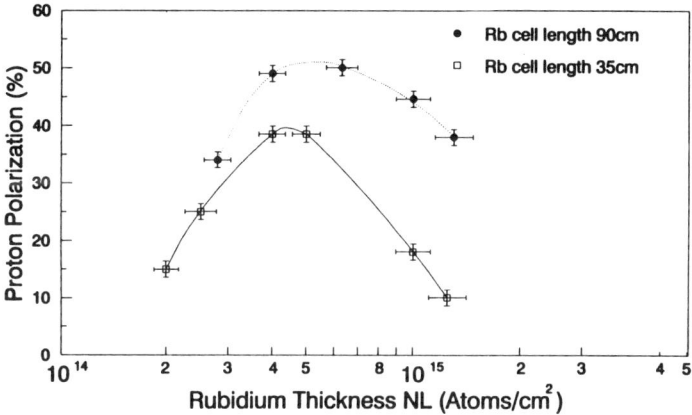

Fig. 4. Proton spin-exchange polarization dependence on optically pumped Rb vapour thicknesses. Short 35 cm cell with dry-film coating. Long 90 cm cell uncoated. Magnetic field - 8 kG.

absorbed laser power and the time of Rb 5p-state spontaneous decay. The result is an excited state density of $10^{12}$ atoms/cm$^3$ at a Rb density of $10^{13}$ atoms/cm$^3$. The process of excitation exchange might be the major depolarizing process at this density, since the cross section of the excitation exchange process:

$$Rb\left(5P_{\frac{1}{2}}\right) + Rb_g \rightarrow Rb\left(5P_{-\frac{1}{2}}\right)$$

is about $10^{-11}$ - $10^{-12}$ cm$^2$ (see Ref.[14]). After exchange, the polarized atom can lose polarization by spontaneous decay. An increase in relaxation time reduces both the pumping power and the rate of exited states production. But excitation exchange limits the optically pumped cell density. Apparently, the density limit of $(1-2) \times 10^{13}$ atoms/cm$^3$ is lower than we expected from radiation trapping calculations alone.

To complete our study of optical pumping of monatomic vapor, we built a 90 cm long cell. The coppers liner's inner diameter was 1.6 cm. The cell reservoir was loaded with 8 "self-opening" Rb containers. The cell temperature was controlled by circulation of glycerin at up to 180° C. The results of polarization measurements with the long cell are presented in Fig. 4. We obtained up to 50% proton polarization at a Rb vapor thickness of $6 \times 10^{14}$ atoms/cm$^2$. This result is in good agreement with extrapolation of data from the 35 cm uncoated cell. Of course, the real advantage of the longer cell is only realized with a dry-film wall coating. Unfortunately, we have not yet the equipment for wall-coating the long cell.

## OPTICAL PUMPING OF ALKALI-METAL VAPOR MIXTURES

As mentioned above, polarization transfer in spin-exchange collisions is a more efficient process at thermal energy, due to the higher cross section. Therefore when one species in a mixture of two alkali-metal vapors is optically pumped, the other also becomes polarized due to spin-exchange collisions. Initial discussions of using a mixture of alkali-metal vapors in the optically pumped cell of a polarized ion source only took into account the buffering effects of the second species.[15] It was then realized that optical pumping of mixtures reduces radiation trapping depolarization, since we can keep the density of optically pumped vapor below $10^{13}$ atoms/cm$^3$ and nevertheless get high total density ($5 \times 10^{13}$ atoms/cm$^3$) due to a high buffer vapor density. The use of a mixture also avoids depolarization due to excitation exchange, since the cross section of excitation exchange between different alkalis should be a few orders less than the resonant cross section between identical atoms. In our first experiments on optical pumping of potassium-sodium mixtures, we obtained $7 \times 10^{14}$ atoms/cm$^2$ total mixture thickness with 70% polarization.[16] Apparently the best choice of alkalis is the rubidium-cesium mixture. Optical pumping of Rb vapor by Ti:sapphire lasers is very efficient and well studied. The spin-exchange cross section of Rb-Cs collisions is very high ($2 \times 10^{-14}$ cm$^2$). Also, the spin-exchange cross sections are highest for H-Rb and H-Cs collisions at 1-2 keV hydrogen beam energy. The use of a Rb-Cs mixture allows low temperature operation of the optically pumped cell and the use of a dry-film wall coating for increasing the polarization relaxation time.

The cell for studying Rb-Cs mixture optical pumping is shown in Fig. 5. Rb and Cs are loaded in two reservoirs having independent temperature control, allowing separate control of the Rb and Cs fluxes and hence of the density ratio in the cell. The dry-film coated copper liner is 30-50° C hotter than the reservoirs to prevent any condensation on the wall coating. We are planning to study optical pumping of Rb-Cs mixtures in collaboration with TRIUMF.

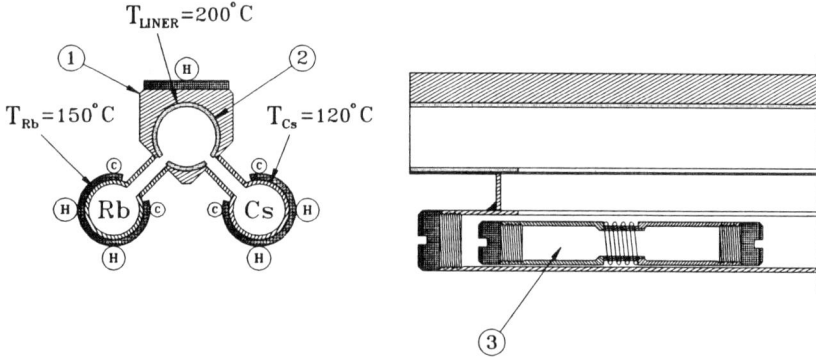

Fig. 5. Cell for experiments on optical pumping of Rb-Cs mixture. 1) stainless steel cell body; 2) copper liner with dry-film coating; 3) "self-opening" container; H) heaters; C) cooling tubes.

## ESTIMATION OF CURRENT FROM SPIN-EXCHANGE SOURCES

The hydrogen beam undergoing polarization is neutral in spin-exchange sources and therefore current estimation is quite simple and directly follows from experimental results on neutral injector development and the source geometry. We will make all estimates for production of polarized H$^-$ ions, since that is the best choice for high energy accelerators, but as is well known, the OPPIS is capable of producing ten times higher polarized proton beam by using ionization in a helium cell.

The geometry of the ionizer is determined by the specified polarized beam emittance, since the divergence of the final polarized beam is produced mostly when $H^-$, created by charge exchange inside the ionizer magnetic field, leaves the solenoidal field.[17] The initial neutral beam emittance is negligible. At typical ionizer parameters – B=1.5 kG ionizer solenoid magnetic field and 2R=1.5 cm diam of ionizer cell – the emittance increase is equal to 1.3 $\pi$ mm-mrad, close to the maximum emittance acceptable for high energy accelerators. The typical length of the ionizer cell is about 30 cm, which could be reduced to 15-20 cm by using a jet ionizer. The distance between the ionizer and optically pumped cell varies from 25 cm (INR OPPIS with small 4.2 cm bore of pulsed solenoid) to 50 cm (LAMPF, TRIUMF OPPIS with large 15 cm bore of superconducting solenoid). The magnetic field of 5-8 kG required for the spin-exchange source might be produced by a conventional solenoid, allowing a more compact design than a superconducting solenoid. We assume a 30 cm distance between the optically pumped cell and the ionizer for the spin-exchange source. The length of the optically pumped cell should be in the range of 30-100 cm. A short cell would be used in the case of using alkali mixtures, the long cell with monatomic Rb vapor. The diameter of the optically pumped cell is limited by the available laser power in dc operation. A reasonable choice is 1.0 cm. In pulsed operation the laser power is not a limitation and especially for the case of a long cell and converging primary proton beam it is reasonable to use a larger (2-3 cm) diam cell. The absence of a strong limitation on the diameter of the optically-pumped cell is another important advantage of spin-exchange polarization, since in the OPPIS with ECR source, increasing the cell entrance diameter raises gas loading and neutralization on residual gas, consequently reducing the polarization.

The high intensity convergent atomic hydrogen beams are produced by neutralization of a high-brightness proton beam just after a focusing solenoidal lens.[11] (A convergent proton beam might also be produced by using a spherical extraction system). In such a way, up to 1 $A/cm^2$ of equivalent neutral hydrogen current was obtained at a distance of 100 cm from the source at 1 keV beam energy in pulsed operation at 10 Hz repetition rate and 200 $\mu$sec pulse duration.[18] As discussed above, the full length of the optically pumped cell and ionizer assembly should be 110-160 cm. The beam focus should be at the exit of the optically pumped cell or closer to the Sona transition region. A small beam diameter at the Sona region ensures that no polarization is lost there. At least 100 mA of equivalent neutral current will pass through a 1.5 cm diam ionizer cell. Rb vapor appears to be the best choice for the ionizer cell. The $H^-$ yield from a Rb ionizer should be about 20% and the resulting polarized $H^-$ current will be 20 mA. That is close to unpolarized source currents, and development of such a source will provide great opportunities for high energy spin-physics experiments.

## CONCLUSION

Optically pumped polarized H⁻ ion sources now produce higher current than the best atomic beam sources in both pulsed and dc operation, with the promise of much higher capabilities to come. We believe that the development of the spin-exchange polarization technique will lead to polarized H⁻ currents in excess of 10 mA. Detailed study of the processes involved in optical pumping of high density alkali-metal vapor and of hydrogen beam interaction with such media must be accomplished to fully realize the power of the optical pumping technique.

## ACKNOWLEDGEMENTS

We would like to thank V.M. Lobashev, G. Dutto, L.W. Anderson, P.W. Schmor, Y. Mori, C.D.P. Levy, D.R. Swenson for helpful discussions and encouragement in carrying out this work.

## REFERENCES

1. C.D.P. Levy, *et al.*, "Status of the TRIUMF OPPIS", this workshop.
2. A. Zelenski, *et al.*, Proc. of 7$^{th}$ Int. Symp.on High Energy Spin Physics, Protvino, USSR, 1986, pp. 154-167.
3. G. Dutto, *et al.*, "Proposal for a pulsed H⁻ OPPIS for high energy accelerators", to be published in Proc. 1993 IEEE PAC Conf.
4. A. Zelenski *et al.*, Proc. of Int. Workshop on Polarized Ion Sources and Jets KEK Report, No. 90-15, 310-320, (1989).
5. D.R. Swenson, D. Tupa, L.W. Anderson, J. Phys. B, **18**, 4433 (1985).
6. L.W. Anderson *et al.*, Phys. Rev. Lett., **52**, 609, (1984).
7. D. Tupa, L.W. Anderson, Phys. Rev. A., **36**, 2142, (1987).
8. A. Zelenski *et al.*, Nucl. Instrum. and Methods, **A245**, 223, (1986).
9. Y. Mori *et al.*, Nucl. Instrum. and Methods, **A268**, 270 (1988).
10. D. Tupa, L.W. Anderson, Phys. Rev. A, **33**, 1045, (1986).
11. V.I. Davydenko *et al.*, Dokl. Acad. Nauk USSR, **271**, 1380, (1983).
12. D.R. Swenson, L.W. Anderson, Nucl. Instrum. and Methods **B29**, 627, (1988).
13. A. Zelenski, *et al.*, Pis'ma Zh. Eksp. Teor. Fiz., **44**, 21, (1986).
14. P. Scalinski and L. Krause, Phys. Rev. A, **26**, 3338 (1982).
15. W. Cornelius, Y. Mori, Phys. Rev. A, **31**, 3718, (1985).
16. A. Zelenski, *et al.*, Proc. 8$^{th}$ Int. Symp. on High Energy Spin Physics, AIP No. 187, 1208, (1988).
17. G.G. Ohlsen, *et al.*, Nucl. Instrum and Methods, **73**, 45, (1969).
18. Yu.I. Belchenko, *et al.*, "Sources at the Novosibirsk Institute of Nuclear Physics", Rev. Sci. Instr., **61**, 378, (1990).

# Proposal for a Pulsed Optically Pumped Polarized H⁻ Ion Source For High Energy Accelerators

A.N. Zelenski

Institute for Nuclear Research, Russian Academy of Sciences,
117312 Moscow, Russia

C.D.P. Levy, P.W. Schmor W.T.H. van Oers, and G. Dutto
TRIUMF, 4004 Wesbrook Mall, Vancouver, B.C., Canada V6T 2A3

Y. Mori
KEK, Oho 1-1, Tsukuba-shi, Ibaraki-ken 305, Japan

## ABSTRACT

The acceleration of polarized protons in multi-GeV machines is a great challenge for accelerator physicists. An essential part of such development is the primary source of polarized H⁻ ions. Recent studies at TRIUMF and INR, Moscow, showed that pulsed Optically Pumped Polarized Ion Source (OPPIS) could produce up to 1.2 mA H⁻ ion beam with polarization in excess of 80% within 1.5 $\pi$ mm-mrad normalized emittance. This current is ten times higher than the best currently available from atomic beam sources. The pulsed OPPIS is quite inexpensive in comparison with atomic beam source, and is ideally suited for high energy accelerator applications.

## INTRODUCTION

The importance of spin effects in the multi-GeV range of energies has been realized in recent years and as a result a number of experiments employing accelerated polarized proton beams have been proposed for KAON,[1] FNAL,[2] BNL,[3] and SSC.[4] Accelerator physicists have met this challenge by implementation of the "Siberian snake" technique of preserving the polarization during acceleration.[5] Another important part of such facilities is the primary source of polarized H⁻. Conventional atomic beam sources are gradually being improved, and recent progress on a D⁻ plasma ionizer will probably lead to increased polarized H⁻ currents of up to 100 $\mu$A in 1.5 $\pi$ mm-mrad emittance, even though H⁻ is not a favored ion for atomic beam sources.[6] Typical currents of unpolarized H⁻ ion injectors are about 50 mA.[7] A big difference in the polarized and unpolarized current will substantially restrict the possibilities of studying polarization phenomena at the low duty-factor high energy accelerators.

## PROPOSAL FOR A PULSED HIGH CURRENT OPPIS

An alternative to the atomic beam source is the comparatively new technique of optically pumped polarized ion sources, which is particularly suitable for H⁻ production. In the last few years optically pumped polarized H⁻ sources have been put into routine operation at KEK,[8] LAMPF[9] and TRIUMF.[10] This has improved substantially the facilities for polarization studies at these laboratories.

Two basic OPPIS configurations are in use. At KEK, LAMPF and TRIUMF, ECR sources are used for producing the primary proton beam. Very similar results have been obtained with this arrangement for both pulsed and cw modes of operation, and presently the H⁻ current is limited to less than 200 $\mu$A in a 1.0 $\pi$ mm-mrad normalized emittance. This limitation is to occur as a result of the ECR plasma temperature (higher than 2 eV). The KEK pulsed OPPIS, for example, nearly achieved this current limit.[8] The influence of rubidium ion space charge on emittance degradation is also very important and is not yet well understood. Another approach, implemented at INR, Moscow overcomes these problems[11] (see Fig. 1.). In this technique a high intensity neutral atomic hydrogen beam is injected into a strong longitudinal magnetic field, where it is ionized in a pulsed gas helium (or neon) cell. The resulting proton beam is then injected into an optically pumped Na (or Rb) cell, which is situated in the same solenoidal field as the ionizer. In effect, the ionizer cell acts as a proton source in a high magnetic field. The proton yield from He at hydrogen beam energies of 5-8 keV is about 70% and about 40% for neon. The He cell is isolated and biased at -1 kV, allowing energy separation of the protons from the primary neutral hydrogen beam. A conventional electromagnetic pulsed gas valve cannot be used in a high magnetic field and a piezoceramic or pneumatic valve must be used. A very bright neutral injector (the prototype was developed at the Budker INP, Novosibirsk) produces up to 30 mA equivalent transmitted atomic beam current through the ionizer cell and the optically pumped cell. It is based on production of a low divergence proton beam, which is extracted by a four-electrode multiwire system from an expanded plasma.[12] The proton beam is focused by a solenoidal magnetic lens and neutralized in a pulsed hydrogen or alkali vapour cell. The geometry of the source extraction system and focusing lens has to be chosen carefully to provide the conditions for space-charge compensation during beam formation, in order to avoid increasing the beam divergence.

The resulting polarized current is very close to estimates obtained from the measured initial beam and the efficiency of ionization and charge-exchange processes in helium and sodium. The polarized current doesn't depend on the magnitude of the magnetic field in the optically pumped cell, but there is a loss of about 30%, because some additional divergency is introduced during deceleration of the proton beam at the exit of the He ionizer. The polarized H⁻ current obtained at INR for a maximum polarization of 65% is 400 $\mu$A in an emittance of

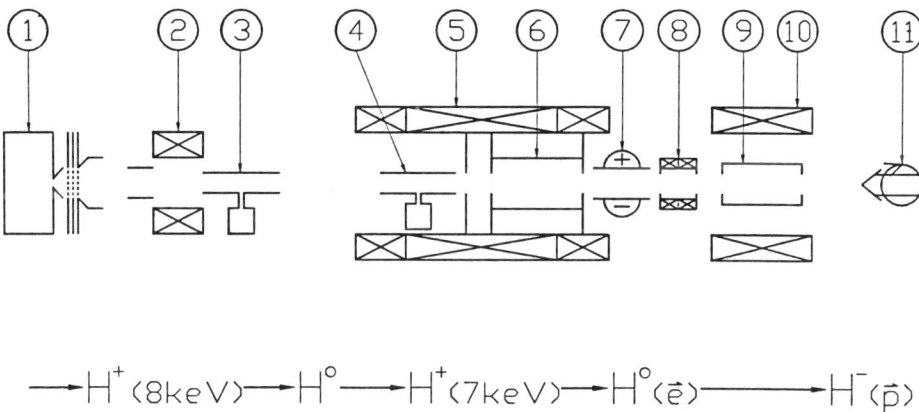

Fig. 1. Schematic layout of the pulsed optically pumped polarized H⁻ ion source: 1-source of primary protons; 2-focusing lens; 3-neutralizer cell; 4-pulsed helium ionizer cell; 5-superconducting solenoid 25-30 kG; 6-optically pumped Rb cell; 7-deflecting plates; 8-Sona transition magnetic shield and trim coil; 9-sodium ionizer cell; 10-ionizer solenoid 1.5 kG; 11-pulsed Ti:Sapphire laser.

$1 \pi$ mm-mrad. At higher Na thicknesses in the optically pumped cell, the current increases to 600 $\mu$A but the polarization drops to 45%, because of radiation trapping within the sodium vapour.

Recent progress in OPPIS development has been the result of switching from dye lasers and optical pumping of sodium to solid state Ti:sapphire lasers and optical pumping of rubidium[13] and potassium.[9] The advantages of Rb as a medium for optical pumping and the high power of Ti:sapphire lasers have greatly improved the OPPIS parameters. At TRIUMF even in a cw mode of operation, up to $8 \times 10^{13}$ Rb atoms/cm² with an electronic polarization of over 95% was produced. An efficiency of 50% for capture of polarized electrons by the incident proton beam was obtained. A high power pulsed Ti:sapphire laser was tested in the INR source, where a 90% neutralization efficiency at the highest polarization was achieved, nearly doubling the polarized H⁻ current compared with using sodium. In such a way, it should be possible to produce at least 800 $\mu$A polarized H⁻ beam in a 1.0 $\pi$ mm-mrad emittance. This current is expected to scale with emittance and, if 1.5 $\pi$ mm-mrad emittance is acceptable – such emittance is specified in the FNAL proposal on acceleration of polarized protons[2] -1200 $\mu$A could be available from a pulsed OPPIS. As for possible improvements, the current scales with the neutral beam intensity and development in that area is definitely not exhausted. For example, the above current of 400 $\mu$A was measured at only half the neutral hydrogen beam intensity, which has been obtained from the Budker Institute prototype source.

This technique is particularly suitable for high energy accelerators having low repetition rates of 10-15 Hz (FNAL, SSC). At such low rates, the ionizer helium consumption is only $3 \times 10^{17}$ atoms/sec and vacuum pumping of the He ionizer cell is accomplished easily by one 1000 l/sec turbomolecular pump. A very simple, inexpensive laser system based on a pulsed Ti:sapphire laser could be used to produce high polarization of the Rb vapor. The INR pulsed Ti:sapphire laser produces up to 1 kW power in a pulse duration of 200 $\mu$sec, with a linewidth of 10-12 GHz and a repetition rate up to 25 Hz. A longer pulse duration could be realized by using two such lasers, or an alexandrite laser.

Very important results have been obtained recently at TRIUMF.[14] Proton polarization of over 80% was obtained in a cw mode of operation, at a high magnetic field of 25 kG in the optically pumped cell. In the INR-type source, energy separation of the protons produced by ionization of the primary neutrals provides better background conditions and the polarization should be even higher.

Combining the INR results of highest current production and the experience with pulsed Ti:sapphire lasers and the TRIUMF results of highest polarization, we propose the development of a pulsed polarized ion source which will produce high current polarized H$^-$ beam having the specifications shown in Table 1.

Table 1. Pulsed OPPIS parameters.

| | |
|---|---|
| Repetition rate | 10-15 Hz |
| Pulse duration | 100 $\mu$sec |
| Pulsed polarized H$^-$ current | > 1.0 mA |
| Pulsed polarized H$^+$ current | 10 mA |
| Polarization | 80-85% |
| Normalized emittance | 1.5 $\pi$ mm-mrad |

Such a pulsed OPPIS will produce at least a factor of ten times higher polarized H$^-$ current than the best atomic beam source currently available. It's construction is less expensive than that of an atomic beam source and we believe it is ideally suited to be used at high energy accelerators. An important feature of the proposed OPPIS is the capability of further development by using a spin-exchange technique of polarization.[15] In that technique there is no space-charge current limitation, since polarization takes place in collisions between neutral hydrogen and alkali-metal atoms. Future spin-exchange optically-pumped sources will likely produce polarized H$^-$ ion currents in excess of 10 mA, and may finally solve the problem of a polarized injector for high energy accelerators.

## CONCLUSION

There is a great deal of interest in high energy spin physics experiments at fixed target, collider and storage ring setups. The development of a high performance pulsed optically pumped polarized H⁻ ion source should be considered for the most efficient use of these facilities with polarized beams.

Anderson (Univ. of Wisconsin USA) and Mori (KEK National laboratory, Japan) were awarded the 1993 IEEE Particle Conference Technology Award for their invention and development of the optically pumped polarized negative ion source and in recognition of successes of the first generation OPPIS. We believe the optically pumped polarized H⁻ ion sources of the next generation, which have been discussed in this paper, will produce polarized H⁻ ion currents of 1-10 mA, i.e. close to the currents of unpolarized ion sources.

## REFERENCES

1. W.T.H. vanOers, "Spin-physics at KAON", in Proc. $9^{th}$ Intern. Symp. on High Energy Spin Physics edited by K.H. Althoff (Springer Verlag, 1991) p. 335.

2. Report on Acceleration of Polarized Protons to 120 and 150 GeV in the Fermilab Main Injector, SPIN collaboration, Michigan, Indiana, Fermilab, N. Carolina/TUNL, Protvino, Moscow, KEK, March 1992.

3. S.Y. Lee, Particle and Field Series 42, AIP Conf. Proc. No. 223, 30, (1990).

4. Proc. 1985 Ann Arbor Workshop on Polarized Beam at the SSC, AIP Conf. Proc. 145 (AIP New York 1986).

5. Ya.S. Derbenev, A.M. Kondratenko, Proc. $10^{th}$ Intern. Conf. on High Energy Accelerators, IHEA, Serpukhov (Protvino, USSR 1977), Vol. 2, 70.

6. A. Belov *et al.*, "A source of polarized H⁻ ions with deuterium plasma ionizer" (to be published in Proc. $10^{th}$ Intern. Symp. on High Energy Spin Physics, (Nagoya, Japan 1992).

7. J.G. Alessi *et al.*, Rev. Sci. Instr., **61**, 625, (1990).

8. Y. Mori, Proc. $8^{th}$ Intern. Symp. on High Energy Spin Physics, (Minneapolis 1988), AIP Conf. Proc. **187**, 1200, (1989).

9. R.L. York *et al.*, Operation of the OPPIS at LAMPF, Proc. 1991 IEEE PAC, p. 1928.

10. C.D.P. Levy *et al.*, "Status of the TRIUMF OPPIS", in Proc. $13^{th}$ Intern. Conf. on Cyclotrons and Their Applications, (World Scientific, Singapore, 1992) p. 322.

11. A. Zelenski *et al.*, Nucl. Instrum. and Methods **A245**, 223, (1986).

12. V.I. Davydenko *et al.*, Dokl. Akad. Nauk, USSR, **271**, 1380, (1983).

13. C.D.P. Levy *et al.*, "A dc OPPIS based on optically pumped rubidium", in Proc. 4$^{th}$ Intern. Conf. on Ion Sources, ed. B.H. Wolf (Bensheim 1991) Rev. Sci. Instr. **63**, 2625, (1992).

14. A. Zelenski *et al.*, "Optimization studies of proton polarization in the TRIUMF OPPIS", (to be published in Nucl. Instrum. and Methods).

15. A. Zelenski *et al.*, Proc. Intern. Workshop on Polarized Ion Sources and Jets KEK Report, No. 90-15, 310, (1989).

# STATUS OF THE TRIUMF OPTICALLY PUMPED POLARIZED H⁻ ION SOURCE

C.D.P. Levy, A.N. Zelenski *, K. Jayamanna, M. McDonald, R. Ruegg, P.W. Schmor and G. Wight

TRIUMF, 4004 Wesbrook Mall, Vancouver, B.C., Canada V6T 2A3

## ABSTRACT

The optically pumped polarized H⁻ ion source at TRIUMF reliably produces 20 $\mu$A dc of 80% polarized beam within an emittance of 1.0 $\pi$ mm mrad. The source is now being prepared for an upcoming experiment at TRIUMF (E497) that will measure parity violation in $pp$ scattering at 230 MeV. The laser system upgrade for fast spin flip up to 200 s⁻¹ and simultaneous monitoring of Rb thickness and polarization is described.

## 1. INTRODUCTION

Figure 1 is a schematic diagram of the optically pumped dc source at TRIUMF (OPPIS). A 28 GHz, 800 W electron-cyclotron-resonance (ECR) proton source is situated within a longitudinal magnetic field. Protons extracted at an energy of 2 - 3 keV enter a Rb vapour oven where the field strength is 2.5 T. The Rb vapour is optically pumped by 8 - 9 W of circularly polarized light from two titanium:sapphire (TiS) lasers tuned to the $D_1$ transition of Rb at 795 nm.

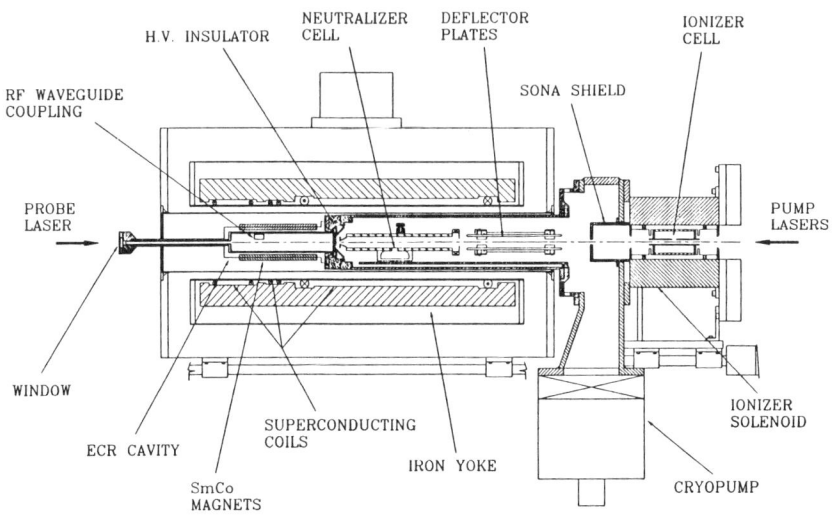

Fig. 1. Schematic diagram of the TRIUMF OPPIS.

*Visitor from INR Moscow, Russian Academy of Sciences, Moscow, Russia

Typically, the Rb electron polarization is 95 - 100%. The protons pick up polarized electrons and become electron polarized neutral hydrogen. Deflector plates downstream of the neutralizer oven sweep out any charged species, and the beam of neutrals passes through a region in which the magnetic field reverses. The electron polarization is passed to the nucleus by a hyperfine process known as a Sona transition, and the resulting nuclear polarized atoms then pass through a Na vapour cell where approximately 10% of the neutrals are negatively ionized to form H$^-$ ions. They are then accelerated to 300 keV and transported to the TRIUMF cyclotron.

The TiS laser bandwidth is narrowed to 3 GHz by a 0.5 mm thick uncoated intracavity etalon. Coarse tuning is by a 3-plate birefringent filter and fine tuning is with the etalon. The TiS laser beams naturally diverge, are brought to a focus near the source axis 8 m from the Rb cell, pass through a circular polarizer, and expand to a width of 6-7 mm (FWHM) in the optically pumped Rb cell. The polarization and Rb thickness are determined by measuring the Faraday rotation of counter-propagating probe light at 12808.0 cm$^{-1}$. Typically the Rb thickness is about $5 \times 10^{13}$ atoms cm$^{-2}$, and the Rb polarization is highly saturated. Dropping the laser power in half reduces both the Rb polarization and the final beam polarization by about 5%. We use two pump lasers for the extra polarization and as insurance against laser failure during operations.

The recent performance and reliability of OPPIS are reported below, as well as upgrades to the laser system to meet the requirements of an experiment (E497) at TRIUMF[1] to measure the parity-violating analyzing power in $pp$ scattering at 230 MeV. High spin flip rates (up to 200 sec$^{-1}$) are specified in E497, which required modification of the laser optical pumping systems and all diagnostics. In addition, very stringent limits on spin-correlated beam intensity, position and energy fluctuations, will require precise control of laser power, frequency and polarization. Measurements of such spin-correlated changes are reported in a companion contribution to this conference.[2]

## 2.   SOURCE PERFORMANCE

OPPIS has been described in detail previously.[3] Since then, the performance of the source has been improved, so that 20 $\mu$A of 80% polarized current are reliably obtained within a normalized emittance of 1.0 $\pi$ mm mrad. The main reason for the improvement has been the change to optical pumping of Rb vapour by TiS laser light, rather than pumping Na vapour with dye laser light at 590 nm.[4] The total laser power is higher, the photon flux per watt is greater, and the heavier, cooler Rb atom is slower and suffers fewer depolarizing wall collisions. In addition, Rb has a 50% greater absorption bandwidth than Na, reducing radiation trapping, and is a better match to the laser bandwidth.

The construction of two low energy polarimeters has freed us from relying on cyclotron beam time to develop the source. One polarimeter operates at the source energy of a few keV and is based on the detection of Lyman-alpha radiation from metastable hydrogen.[5,6] It provides high relative accuracy quickly enough to use it for tuning the source. The other polarimeter is based on the $^6$Li(p,$^3$He)$\alpha$ reaction at 286 keV.[7] It is much slower than the Lyman-$\alpha$ polarimeter but is useful for checking absolute accuracy.

Figure 2 shows the beam polarization at 369 MeV, over a three week run of the charge symmetry breaking experiment at TRIUMF (E369). The TRIUMF cyclotron was run in a phase restricted tune[8] with pulse widths reduced to 0.8 ns from the usual 4 ns, with typical current at a liquid deuterium target of close to 2.5 $\mu$A. This was the maximum current which could be handled by the target cooling system. The polarization of the secondary neutron beam was up to 70%. The source polarization was very stable, and such fluctuations as were seen are well understood. Between runs 930 - 950, the ionizer magnetic field was reduced to increase the beam current. After run 1000 a vacuum leak in the ECR chamber was fixed and an old quartz liner was replaced, thus increasing the current and allowing us to raise the source emittance by increasing the ionizer field, which in turn preserved more polarization during charge exchange. Poor cyclotron tuning reduced the polarization at run 1040. By run 1054 the cyclotron tune had been optimized and at run 1062 the laser system was retuned after running three days unattended. Losses in the TRIUMF cyclotron due to a depolarizing resonance are dependent on the cyclotron tune and have never been observed to be less than 3%. From this we conclude that the source polarization was at least 80%, and measurements at 300 keV have shown polarization of up to 82%.

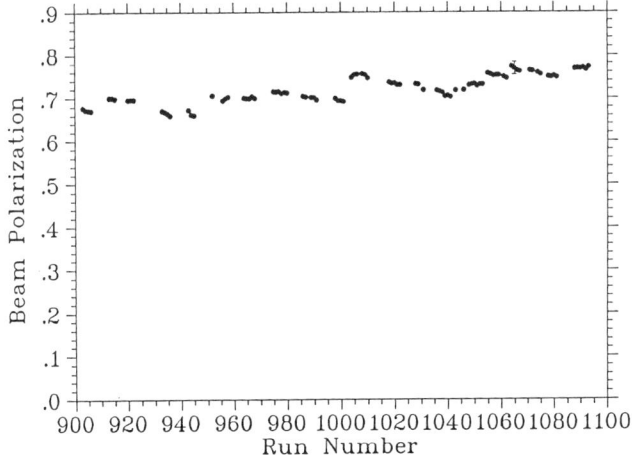

Fig. 2. Beam polarization at 369 MeV over a three week period.

## 3.   RELIABILITY IMPROVEMENTS

Previous reliabilty of the source had been compromised by erosion and sparking of the extraction electrodes, clogging of the ionizer cell apertures with Na, and vacuum leaks around the quartz liner of the ECR proton source.

Most of the Rb deposited on the extraction electrodes came from $Rb^+$ created in the optically pumped cell by charge exchange with the proton beam. A bias of $+(20–50)$ V applied to the last extraction electrode eliminated $Rb^+$ deposition. Further, the minimum spacing between the 1 mm apertures in the multi-aperture extraction grids was increased from 100 to 200 $\mu$m. The two changes increased the lifetime of the extraction electrodes from 1 week to 6 weeks.

The ionizer solenoid was increased in length to 30 cm, producing a much flatter magnetic field profile. This reduced the peak field in the centre of the solenoid that was required to preserve polarization at the edges of the Na vapour distribution, and therefore reduced the emittance of the source. A small further gain in field uniformity was realized with separate control of the centre and end coils. The previous ionizer cell had an active length of approximately 7 cm, and because of its high vacuum conductance it was necessary to recycle the Na, which led to problems of clogging and Na deposition on other source elements. The new cell has an active length of 15 cm and Na vapour leaving the cell is trapped in room temperature chambers at the ends. One 20 gm charge of Na typically lasts 600 hours before being exhausted. Typically, a 10 gm charge of Rb lasts about 800 hours in the optically pumped cell. Of course, larger charges can be used.

The vacuum sealing arrangement for the ECR proton source was redesigned. A quartz tube forms the vacuum chamber, and it was found that indium O-rings eliminated vacuum leaks around the quartz, which otherwise ruin the source performance. Extra care was also taken to protect the quartz from electron beam heating and inadvertant contact with ECR zones.

## 4.   LASER SYSTEM UPGRADE FOR FAST SPIN REVERSAL

From the beginning, OPPIS was designed with fast spin flip in mind, at rates up to at least 100 sec⁻¹. Spin flipping is accomplished by changing the laser light frequency (by about 94 GHz at a magnetic field of 2.5 T) and reversing the direction of circular polarization. The tilt angle of the intracavity etalon determines the laser frequency. A laser frequency stabilization system based on scanning interferometers has been previously described.[9] Until recently we were limited by the VAX control system to a maximum spin flip rate of 1 s⁻¹. We are now upgrading the laser control system electronics. A schematic of the circuit used for fast frequency switching is presented in Fig. 3. Current

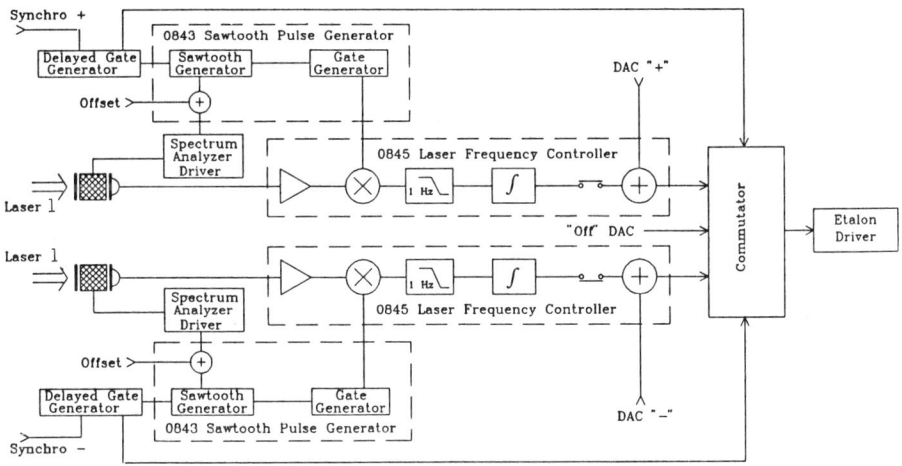

Fig. 3. Block diagram of laser frequency control for fast spin flip, for one laser.

signals driving each galvanometer-mounted etalon are produced separately for spin states "up" and "down" in two feedback circuits, for a total of four feedback circuits. A gated fast electronic switch or commutator switches between the appropriate channels and applies the required "up" or "down" signal to the etalon galvanometer, thus avoiding computer and CAMAC system limitations to the spin flip rate. No stabilization signal is required for the "off" state since the laser frequency is not critical in that case. The spin flip rate is now limited by the response time of the galvanometer. Initially, we used G-102 (General Scanning Inc.) galvanometers and obtained a frequency switching time of 1.5 - 2.0 ms. Recently, G-120 galvanometers were installed, with position readback and feedback for precise control of angular position and damping of transients. As a result, the polarization switching time was reduced to less than 0.5 ms (see Fig. 4, where Faraday rotation-versus-time signals during spin flip are shown). The new galvanometer permits spin flip rates up to 200 s$^{-1}$ with less than 10% deadtime.

Laser power stabilization is based on minimizing the power difference between opposite helicity states, rather than keeping the absolute power in each state constant. Analog power signals from a photodetector are converted to a frequency, then gated with spin flip and integrated in CAMAC scalars for about 10 s of integration time. The difference is used to drive the birefringent filter position, which affects laser power. This system operates very reliably and reduces the variations in laser power between spin states up and down to less than 0.2%, while preventing laser mode hops.

The Faraday rotation system has been upgraded to give on-line measurements of Rb thickness and polarization, while spin flipping. The previous sys-

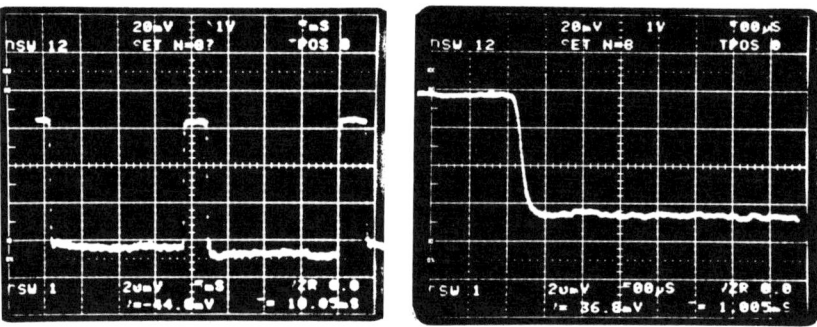

Fig. 4. Faraday rotation signals during spin flip: a) two sequential pulses "up" and "down", 18 ms pulse duration; b) 0.5 ms/div, transition time less than 0.5 ms.

tem relied on a rotating analysing polarizer and photodetector to generate a transmission-versus-angle curve of the probe light. A least squares fit to the data was used to compute the minimum transmission angle to an accuracy of about 0.05°. This process took about 1 min and the polarization state of the Rb was required to remain constant for that time. In the new system, after passing through the source, the probe light is split into two orthogonally polarized beams by a calcite crystal. The two emerging beams are parallel and separated by 4 mm, and impinge on a single silicon photodetector. A small shutter, mounted on a G-124 galvanometer, is placed between the calcite and the photodetector and alternately blocks each beam, at twice the spin flip rate. The output of the photodetector is gated with the shutter, and measurement of the intensities of the orthogonal components allows the input polarization angle to be calculated. The advantage of using one photodetector and a shutter, rather than the older method of using a polarizing beamsplitter and two photodetectors,[10] is that drift between the detectors is eliminated. Bench tests of the system produced an accuracy of 0.01° every second, while driving the shutter with a 100 Hz square wave. In actual use, the accuracy is comparable with the rotating analyzer method, but no disruption of spin flip is required. If spin "off" states are included in the spin flip sequence, then the Rb polarization and thickness can be accurately monitored on-line, and the possibility is open to provide correction signals to keep the absolute polarization and thickness constant.

## 5.   ACKNOWLEDGEMENTS

We wish to thank J. Welz and the ion source group for their assistance in maintaining OPPIS, and P. Bennett, S. Kadantsev and M. Mouat for their help in developing the laser control system. The polarization data at 369 MeV was supplied by J. Zhao.

## REFERENCES

1. J. Birchall *et al.*, Nucl. Phys. **A553**, 823c (1993).

2. C.D.P. Levy *et al.*, "Spin-correlated current modulation in OPPIS" (these proceedings).

3. L. Buchmann *et al.*, Nucl. Instrum. Meth. **A306**, 413 (1991).

4. C.D.P. Levy *et al.*, Rev. Sci. Instrum. **63**, 2625 (1992).

5. A.N. Zelenskii *et al.*, Nucl. Instrum. Meth. **A245**, 223 (1986).

6. A.N. Zelenskii *et al.*, Nucl. Instrum. Meth. (in press).

7. L. Buchmann, Nucl. Instrum. Meth. **A301**, 383 (1991).

8. W.D. Ramsey *et al.*, Nucl. Instrum. Meth. **A327**, 265 (1993).

9. C.D.P. Levy *et al.*, Proc. Int. Workshop on Polarized Ion Sources and Polarized Gas Jets, KEK Report 90-15, p. 180 (1990).

10. F. Strumia, Nuovo Cimento, **XLIV B**, 387 (1966).

# SPIN CORRELATED CURRENT MODULATION IN OPPIS

C.D.P. Levy, K. Jayamanna, W.D. Ramsey and P.W. Schmor
TRIUMF, 4004 Wesbrook Mall, Vancouver, B.C., Canada V6T 2A3

and

A.N. Zelenski
INR Russian Academy of Sciences, 117312 Moscow, Russia

## ABSTRACT

The modulation of polarized current correlated with spin-reversal has been measured for the first time in an Optically Pumped Polarized Ion Source (OPPIS). The modulation amplitude depends on the density of the optically pumped Rb vapour and the energy of the primary proton beam in OPPIS. Several physical processes contribute to the current modulation. By optimizing some of the source parameters, the current modulation was reduced to less than $10^{-5}$, as measured at 220 MeV in a transverse electric field ionization chamber, part of a parity violation measurement detector system. This is close to the stated requirements of the approved parity violation experiment. A technique of laser control for fast (up to 200 times per second) spin-flip has been developed which should provide for precision control of optical pumping and maintain the spin correlated current modulation below the $10^{-5}$ level.

## INTRODUCTION

The measurement of the parity-violating (PV) analyzing power $A_z$ in proton-proton scattering to an accuracy of $2 \times 10^{-8}$ is a challenge for the polarized ion source community, because of the extremely high beam stability requirements, especially with respect to any current or beam energy modulation correlated with spin-reversal.

At Los Alamos a Lamb-shift source produced a low current (less than 1 $\mu$A at the source) polarized H$^-$ beam for experiments at 15 MeV and 800 MeV. The current, emittance and beam position fluctuations due to the alternating magnetic and electric fields (required for fast spin reversal) were the major factors in the resulting experimental errors – $A_z = (2.4 \pm 1.1) \times 10^{-7}$ (see Ref.[1]).

Atomic beam polarized proton sources were used for $A_z$ measurements at 13.6 MeV at Bonn University and at 45 MeV at PSI. The use of different rf transitions for spin-reversal gives rise to differences in the proton polarization. An asymmetry in ionization efficiency by polarized electrons leads to some current modulation. The difference between polarized and unpolarized current could be about 1% and is proportional to polarization.[2] Hence, the current modulation is proportional to the difference in polarization for opposite spin states. As a result a current modulation correlated with spin flip of magnitude $(2-8) \times 10^{-5}$ was observed at PSI.[3] The electron beam, or ECR ionizers used in the atomic beam source produce unstable ionization efficiency and beam position. The low beam velocity and long relaxation time of the electron polarization in the ionizer are basic limitations, which restrict the spin reversal rate to less than 100 times per second. Nevertheless, after a careful, detailed study of the errors the measurements of $A_z$ yielded a most impressive level of accuracy with $A_z = -(1.5 \pm 0.22) \times 10^{-7}$ at 45 MeV at PSI[3] and $A_z = -(1.5 \pm 0.5) \times 10^{-7}$ at 13.4 MeV at Bonn.[4]

The aim of the TRIUMF experiment is to measure the $^3P_2 - {}^1D_2$ partial wave contribution to the PV longitudinal analyzing power in proton-proton scattering at 222 MeV, where the $^1S_0 - {}^3P_0$ amplitude crosses zero. The absolute value of the PV amplitude at 222 MeV is expected to be only $(4 - 7) \times 10^{-8}$, less than at 45 or 800 MeV, and a corresponding accuracy of at least $\pm 2 \times 10^{-8}$ must be achieved in the experiment.[5] In particular, a new approach in polarized beam production is required to meet PV experiment specifications.

In recent years, optically pumped sources of polarized $H^-$ ions have been developed which overcome many of the above problems. The dc OPPIS at TRIUMF reliably produces 20 $\mu A$ of $H^-$ ion current with a proton polarization in excess of 80%.[6] This has substantially enhanced TRIUMF's facilities for polarization experiments. Data taking has recently been completed in a second generation experiment measuring charge-symmetry breaking in neutron-proton elastic scattering at 350 MeV. A polarized proton beam of about 2.5 $\mu A$ was delivered to a liquid deuterium target in a phase-restricted mode of cyclotron operation, which allowed a reduction in the time-spread of the incident proton beam to less then 0.8 ns (FWHM). A maximum current of about 10 $\mu A$ could be extracted, with slightly lower polarization. In pulsed operation, currents of 10 to 20 $\mu A$ were obtained at LAMPF,[7] 60 $\mu A$ at KEK[8] and 400 $\mu A$ at INR, Moscow,[9] with decreasing duty factors respectively, at polarizations of 65 to 70%. The significance of optically pumped sources to PV experiments was realized from the beginning of their development. Spin reversal in OPPIS is achieved by reversing the circular polarization of the pump laser light and adjusting the laser frequency to allow for the Zeeman shift of about 94 GHz at 2.5 T in the alkali-metal vapour charge exchange target. The high field is required to minimize spin-orbital depolarization as excited state hydrogen atoms decay to the ground state. It would appear this should have no effect on the beam current. In addition, the high beam velocity and short polarization rise time of 3 $\mu s$ allows spin reversal rates up to 1 kHz. Synchronous detection techniques can be used to cancel drifts of the polarized beam, target and detector parameters.

Problems of laser power and frequency stabilization do exist particularly if the highest spin reversal rates are required. It is of course possible to use different lasers for each spin state, although this does not eliminate the need for producing identical power, linewidth and alignment. However, a change in the absolute magnitude of the polarization is not so important in itself, but only insofar as it produces a spin-reversal Correlated Current Modulation (CCM).

## BASIC PROCESSES AFFECTING CURRENT MODULATION

Contrary to naive expectations, we have found optical pumping affects the OPPIS current, due to a variety of atomic collision processes. Consequently, a difference in optical pumping conditions for opposite polarization directions will give rise to CCM. The sensitivity of these processes to optical pumping is a function of many OPPIS parameters, such as: optically pumped vapour density, energy of the primary proton beam, laser power, polarization relaxation time, proton neutralization on residual gas, etc. As a result, a complicated dependence of CCM on the OPPIS parameters is expected. There are a number of atomic collision processes in the optically pumped vapour which contribute to CCM.

H$^-$ production by double charge exchange with ground state Rb: The best understood process is the CCM caused by asymmetry in H$^-$ production in polarized and unpolarized alkali-metal vapour. In an unpolarized vapour, after neutralization, some of the hydrogen atoms can undergo a second charge exchange reaction to become H$^-$. Since H$^-$ has a single bound state $^1S_0$ , it cannot be formed by a polarized hydrogen atom picking up an electron from Rb polarized in the same direction. The H$^-$ yield in the optically pumped Rb is given by[10]:

$$I(\text{H}^-) \sim I_0 C \sigma_{o-} N L (1 - TP^2) \tag{1}$$

where $I_o$ is the atomic hydrogen yield in the Rb, $\sigma_{0-} = 3 \times 10^{-16}$ cm$^2$ is the cross section for H$^-$ production, $NL$ is the Rb cell thickness, $P$ is the Rb polarization, and $T = 0.94$ is the polarization transfer ratio between Rb and hydrogen at a magnetic field of 2.5 T. $C$ is a correction factor accounting for integration of H$^-$ production along the length of the cell. The H$^-$ ions which are produced in the Rb cell are swept out of the beam by deflection plates between the Rb cell and the sodium negative ionizer cell (see Fig. 1), giving rise to a dependence of the final polarized H$^-$ ion current on Rb polarization. For Rb polarization Pup and Pdown ($P_u$ and $P_d$), and corresponding polarized H$^-$ currents I$_u$, I$_d$, the following is true:

$$CCM_d^u = \frac{I_u - I_d}{I_u + I_d} \sim \frac{1}{2} C T \sigma_{0-} N L (P_u^2 - P_d^2) \sim C T \sigma_{o-} N L (P_u - P_d) \tag{2}$$

since $Pu^2 - Pd^2 \sim 2(P_u - P_d)$, $P_u \sim P_d \sim 1$. For useful PV measurements, in order to exclude the nonlinear response of the ionization chamber detector to the current modulation, the CCM must be of the order of $10^{-5}$ or less. The above estimate implies that an upper limit on Rb polarization asymmetry is about 0.5%. It can be realized for a thin Rb target with highly saturated polarization, by precision control of laser frequency, power and circular polarization. Later, we will discuss a technique which has been developed for this control in the fast spin-flip mode of OPPIS operation.

Proton neutralization on Rb atoms in exited states: During optical pumping, there is some equilibrium density of Rb atoms in excited states N*=N(Rb*), where Rb*=Rb(5P$_{\frac{1}{2}}$). This density depends on Rb vapour density, laser power and polarization relaxation time. At a typical Rb cell density of about $2 \times 10^{12}$ atoms/cm$^3$, radiation trapping gives rise to an increase of the effective spontaneous decay time to more than 10 times the natural 28 ns decay time. We can estimate the density of Rb* from measurements of absorbed laser power and the effective spontaneous decay time. It is about 1-3% of the full Rb vapour density.

The cross section of electron capture from Rb atoms in excited states is less than the neutralization cross section in collision with Rb atoms in the ground state $- \sigma_{+0}$ (H$^+$ + Rb*) $\sim 0.6\ \sigma_{+0}$(H$^+$ + Rbg) at a hydrogen beam energies of 1-5 keV.[11] Therefore, the CCM caused by this effect has an opposite sign to the "double polarized electron capture CCM" discussed above.

H$^-$ production on Rb atoms in excited states: Part of the primary proton beam is neutralized by capture of electrons from residual gas (mostly hydrogen). As a result, unpolarized hydrogen atoms are produced giving rise to a loss of polarization. This unpolarized current depends on the ECR operation, gas efficiency, cell geometry and vacuum pumping. It can be estimated from the H$^-$ ion polarization dependence on optically pumped Rb cell vapour thickness, in paricular from polarization drop at low vapour thicknesses. This current is about 3-5% for the TRIUMF OPPIS. For these unpolarized atoms, the capture of a second polarized electron is not forbidden. When Rb vapour is optically pumped, there is some density of Rb atoms in excited states. The yield of H$^-$ ions for H$^0$ - Rb* collisions is much higher than in collisions with Rb in the ground state.[12] This effect gives rise to CCM of opposite sign to the above considered CCM due to double polarized electron capture. This is because more H$^-$ ions are produced in collisions with excited Rb atoms and then swept out of the beam by the electric field of the deflection plates. This effect will dominate at low Rb thicknesses, where the background current becomes significant and the "double capture CCM" is small. These two effects should cancel each other at some Rb density which apparently will be optimal for OPPIS operation with PV experiments.

Associative ionization of Rb atoms: Another process producing coherent current modulation is the ionization of Rb vapour by resonant D1 line laser light. Current modulation due to this effect was first seen in sodium at INR, Moscow.[13] A reduction in current of close to 100% was observed when the sodium density was more than $5 \times 10^{13}$ atoms/cm$^3$ and high pulsed laser power was used. The ionization is caused by a process involving collisions between excited state atoms, and is highly non-linear with respect to both laser power and vapour density. Since the ionization process decreases the effective vapour density and hence the polarized current, this modulation is also opposite in sign to the double charge exchange process. It is difficult to calculate the magnitude of the modulation directly, although, from the experimental results at INR, it is estimated to be at the level of 1-2% for 9 W of laser power and a Rb density greater than $5 \times 10^{12}$ atoms/cm$^3$. The modulation will tend to zero as the density is increased to very high values, given a finite laser power.

## EXPERIMENTAL RESULTS

A detailed description of the laser system upgrade for fast spin reversal is presented in Ref.[6]

The current modulation inside the source at H$^-$ beam energies of 2-3 keV, at 300 keV in the cyclotron injection beam line and at 220 MeV in an ionization chamber (part of the PV detector system), has been studied.

CCM measurements at 2-3 keV: Most of the experiments were done at low energy inside OPPIS. Downstream of the ionizer are electrostatic bender plates which can deflect either $H^-$ or $H^+$ ions into a low energy polarimeter. Current is measured with a Faraday cup, which intercepts the beam just after the bender plates. The CCM between polarized $I_{on}$ and unpolarized $I_{off}$ currents is defined as follows:

$$CCM_{off}^{on} = \frac{I_{on} - I_{off}}{2I_{off}} \tag{3}$$

In the first experiments the unpolarized current was measured with a blocked laser. A rotating laser beam chopper (Princeton Applied Research M197) modulated a pump laser at a rate of 50-200 times per second. The $H^-$ current from the Faraday cups or ionization chamber was converted to a frequency in a linear current digitizer (with a 4 MHz clock). The frequency signal was gated with laser states "on" and "off", or if fast spin-flip was being used, with "up" and "down" spin states. The gate lengths were identical for both laser states and shorter than the laser modulation times, so that the Rb polarization had time to reach equilibrium. The gated signals were integrated in two CAMAC scalers. The clock frequency was gated by using the same gates and integrated in another two scalers and used to compensate current measurements for the difference in integration time for opposite spin states. Two Ti:sapphire lasers, each pumped by a 20 W Coherent argon laser produced about 9 W of light at 795 nm for optical pumping of Rb. The resulting proton polarization was high and fairly constant over a wide range of Rb vapour thickness. The highest proton polarizations of 75 - 80% were attained at Rb thicknesses of $(2-6) \times 10^{13}$ atoms/cm$^2$.

The results of current modulation measurements are shown in Fig. 1. Over the thickness range of $(2 - 8) \times 10^{13}$ atoms/cm$^2$, the current asymmetry rises linearly as expected due to the double charge exchange process and shows quantitative agreement with the above estimates. At still higher thicknesses, modulation changes sign. At a thickness of $9 \times 10^{13}$ atoms/cm$^2$, the CCM has a maximum negative value of about 1%, within expectations. The sudden change from one dominant process to the other at high Rb density ($N$ about $3 \times 10^{12}$ atoms/cm$^2$) may be evidence of strong nonlinearities of radiation trapping and other processes discussed above. Increasing the density of Rb* states produces larger negative CCM, due to the lower neutralization cross section in collisions with Rb*, and the decrease in the Rb atom density due to associative ionization in Rb*–Rb* collisions. At higher Rb densities, the absolute value of CCM decreases because of rapid laser power attenuation in the vapour. At low Rb density CCM decreases and becomes negative due to the lower neutralization cross section for protons on Rb* and the higher cross section for $H^-$ production from collisions of unpolarized hydrogen atoms with Rb atoms in excited states. In Fig. 1 we can see some indications an even more complicated behaviour of CCM at Rb thicknesses of $1 - 2 \times 10^{13}$ atoms/cm$^2$. Fine CCM structure was resolved using improved temperature control of the Rb cell, as shown in Fig. 2. The temperature of the Rb cell was accurately controlled in this case with circulating hot ethylene glycol solution regulated by a thermostat. Such behaviour of

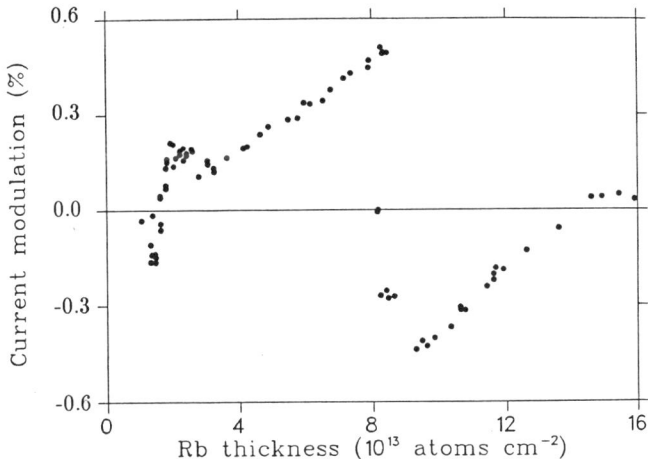

Fig. 1. Modulation of H⁻ source current correlated with the laser state "on" and "off", as measured at 2.8 keV beam energy in FC inside the OPPIS.

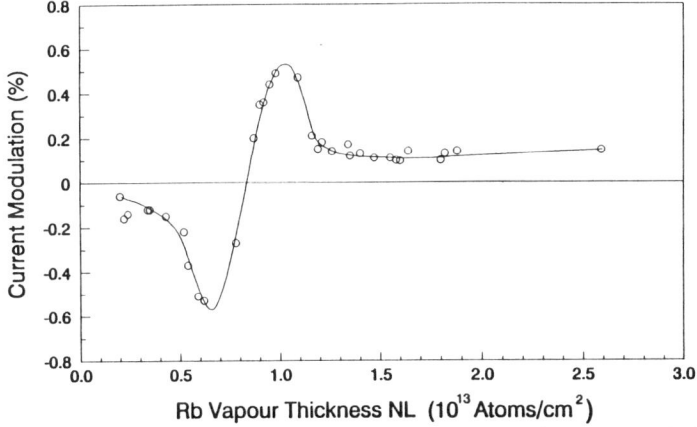

Fig. 2. Spin-correlated current modulation $CCM_{off}^{on}$ dependence on optically pumped Rb vapour density with improved cell temperature control. Energy of primary proton beam 2.8 keV.

CCM at low Rb density was very surprising so we investigated other processes which could simulate CCM. For example, in CCM measurements at low energy, the real current modulation could not be separated from the spin-reversal correlated beam energy modulations. Our current measuring system, consisting of an electrostatic bender (which allowed the polarized beam to be moved off the laser axis) and Faraday cup, has energy resolution. Therefore, coherent energy modulation could be detected as a current modulation. The CCM measurements at 300 keV are much less sensitive to the beam energy modulations since the first beam bend is at 300 keV and the relative energy modulation is 100 times smaller. A comparison of results at low and 300 keV energies is another way of distiguishing CCM from coherent energy modulations.

Coherent current modulation measurements at 300 keV. Results of CCM measurements at 300 keV are presented in Fig. 3. For these experiments, the fast spin-flip technique was used instead of the rotating-wheel laser beam chopper. In general, the CCM dependence on Rb vapour thickness at 300 keV is similar to that as measured at 2.8 keV, but strong oscillations at low thickness disappeared. A possible explanation of this difference is the energy modulation influence at low energy. More experiments are planned to answer this question. At low Rb density we can calculate the CCM by combining the effects of: double polarized electron capture; neutralization on Rb in excited states; and H⁻ production from unpolarized hydrogen beam, as follows:

$$CCM_{\text{off}}^{\text{on}} \sim -\frac{N*}{N}\left(1 - \frac{\sigma_{+0}^*}{\sigma_{+0}} + B\frac{\sigma_{0-}^*}{\sigma_{0-}}\right) + CT\sigma_{0-}NL\,P^2 \qquad (4)$$

where $\sigma_{0-}^*/\sigma_{0-}$ is the ratio of the H⁻ production cross sections in excited and ground states; $B = I_0/I^+$ is the fraction of the primary proton beam $I^+$ which is neutralized by the residual gas. This equation contains all the above discussed qualitative features of CCM. For example, at low Rb density, the first negative term determines the negative sign of the CCM. However, for a detailed quantitative analysis there are too many parameters: $N^*$, $B$, $\sigma_{0-}^*$, which are not known to the required degree of accuracy.

A CCM dependence on primary proton beam energy could be also explained by using the equation (4) and cross section dependence on beam energy. Experimental results are presented in Fig. 4. Measurements were taken inside the source. At low energy and low Rb thickness the term $B(\sigma_{0-}^*/\sigma_{0-})$ dominates, since the $\sigma_{0-}^*$ cross section maximum occurs at a lower energy than for $\sigma_{0-}$. At higher energy, the "double polarized electron capture" term is responsible for the positive maximum of CCM, and then CCM is decreased due to the drop. This behaviour strongly depends on Rb density. At higher density, the ratio $N^*/N$ increases due to radiation trapping. Therefore, at higher beam energy, where $\sigma_{0-}$ is smaller, the first negative term might change the CCM sign again to negative. From the expression (4), the CCM(up-down) may be represented as follows:

$$CCM_d^u \sim C_1\left(\frac{N_u^*}{N} - \frac{N_d^*}{N}\right) + C_2(P_u - P_d) \qquad (5)$$

Ionization effects have not been included. They could be represented as a quadratic term of $(N^*/N)$ and therefore is small at low Rb density. There is no simple relation between the Rb atom polarization and the density of Rb atoms in excited states. Both depend on the Rb density, laser power and polarization relaxation time. To minimize the current modulation at spin-reversal, both terms should be minimized, since cancellation at some point will be inherently unstable. For example, only increasing the laser power for better saturation of Rb polarization (and hence reducing the second term in (5)) is not enough for CCM reduction, since at higher laser power, the ratio $N^*/N$ could also increase, resulting in a larger CCM. At low Rb density, the CCM caused by double polarized electron capture is smaller, but $N^*/N$ is not, in the absence of radiation trapping. Therefore it is necessary to reduce the factor $B(\sigma_{0-}^*/\sigma_{0-})$.

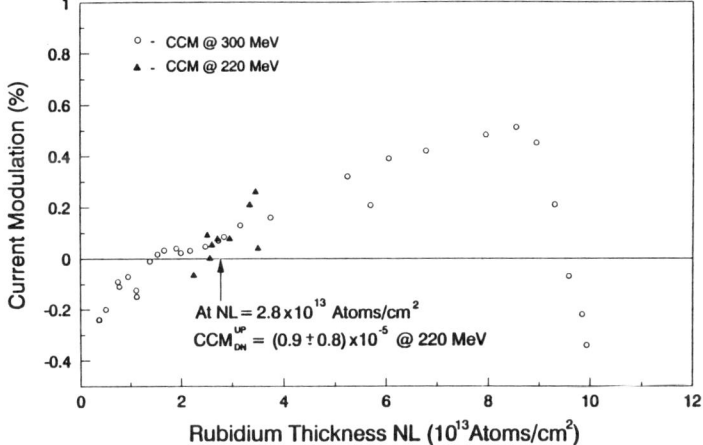

Fig. 3. $CCM_{off}^{on}$ dependence on optically pumped Rb vapour thickness.

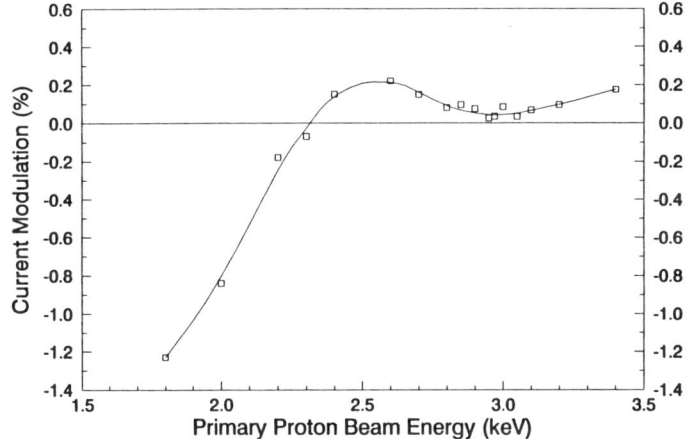

Fig. 4. Spin-correlated current modulation $CCM_{off}^{on}$ dependence on energy of primary proton beam.

CCM measurements at 220 MeV. Results of $CCM_{off}^{on}$ measurements are presented in Fig. 3. They are similar to the results of CCM measurements at 300 keV. At 220 MeV the $CCM_d^u$ was studied for optimal OPPIS parameters, where $CCM_{off}^{on}$ is minimal. During injection, acceleration and extraction of the polarized beam, some additional fluctuations might be introduced. The synchronous detection technique should, in principle, cancel these fluctuations if their frequency is not correlated with the spin-flip rate. A major fluctuation frequency is ac line frequency of 60 Hz. To reduce the influence of this frequency on the CCM measurements, one spin state duration was 18 ms and the length of the measuring gate equal to the period of the ac line (about 16 ms). The spin direction was reversed 25 times per second. Under these conditions, any modulations caused by the ac line must average to zero in a first approximation. As a result at optimal

OPPIS parameters: proton beam energy of 2.8 keV and Rb vapour thickness of $2.5 \times 10^{13}$ atoms/cm$^2$ – the following CCM have been measured:

$$CCM_{\text{off}}^{\text{on}} = (45 \pm 8) \times 10^{-5}$$
$$CCM_d^u = (0.9 \pm 0.8) \times 10^{-5}$$

The possibility of reducing CCM to the level specified for PV experiments has thus been experimentally proven. But the problem of OPPIS stabilization and on-line control of CCM at the $10^{-5}$ level is not completely solved. A much more detailed study of all the basic physical processes must be undertaken to determine the sensitivity of CCM to changing OPPIS parameters. An appropriate system of OPPIS parameter controls and stabilization will be developed to meet these requirements.

## CONCLUSIONS

In the first measurements of correlated current modulation in the optically pumped polarized ion source, a complex dependence of CCM on the polarized target density and beam energy has been observed. This dependence may be described as a combined effect of at least four different atomic collision processes. An estimate was made of the degree of optical pumping stability needed to meet the requirements of PV experiments. For optimal OPPIS parameters, CCM was reduced to less than $10^{-5}$. CCM measurements at low (2-3 keV) beam energy indicate that modulation of the beam energy during spin flip also occurs. We are now planning a direct measurement of such energy modulation, since a strong limitation on this modulation (less than 3 eV) is specified by the PV experiments.

## ACKNOWLEDGEMENTS

We would like to thank W.T.H. vanOers, L.W. Anderson, W. Haeberli, Y. Mori, S.A. Page, D.R. Swenson, D. Tupa, V. Dudnikov for helpful discussions. We wish to acknowledge the help of the engineering and technical staff of the Ion Source Injector (ISIS) group for assistance in OPPIS maintenance, and especially the contributions of M. McDonald and R. Ruegg. As well, we would like thank P. Bennett for the development hardware for fast spin flip control, S. Kadantsev and M. Mouat for development software for the laser control, G. Wight for assistance in OPPIS development and in the preparation of this paper. Finally we would like to thank G. Dutto for the strong support and encouragement he has provided in the carrying out of this work.

## REFERENCES

1. V. Yuan, *et al.*, Phys. Rev. Lett., **57**, 1680 (1986).

2. S. Jaccard, AIP Conf. Proc. **80**, 95 (1982).

3. W. Haeberli, Can. J. Phys. **66**, 485 (1988).

4. P.D. Eversheim, *et al.*, Phys. Rev. B. **256** 11 (1991). Private communication "Final analysis gave the result $A_z = -(0.72 \pm 0.28) \times 10^{-7}$.

5. TRIUMF Experimental Proposal E497, spokespersons, J. Birshall, S.A. Page, and W.T.H. vanOers (1987).

6. C.D.P. Levy, *et al.*, "Status of TRIUMF OPPIS" (these proceedings).

7. R.L. York, *et al.*, Proc. 1991 IEEE Particle Accelerator Conference, **3**, 1928 (1991).

8. Y. Mori, AIP Conf. Proc., **187**, 1200 (1989).

9. A.N. Zelenski, *et al.*, Nucl. Instrum. & Methods, **A245**, 223 (1986).

10. W.D. Cornelius, ANL-84-50, 385 (1984).

11. M. Kimura, *et al.*, Phys. Rev. **A26**, 3113 (1982).

12. B.M. Smirnov, Sov. Fiz. Dokl., **166**, 922 (1965).

13. A.N. Zelenski, *et al.*, Pis'ma Zh. Eksp. Teor. Fiz., **44**, 21 (1986).

# PERFORMANCE OF THE LAMPF OPTICALLY PUMPED POLARIZED ION SOURCE

R.L. York, D. Tupa, D.R. Swenson, M.W. McNaughton,
and O.B. vanDyck
Los Alamos National Laboratory, Los Alamos, NM 87545

## ABSTRACT

In 1992, the LAMPF optically pumped polarized ion source (OPPIS) was used in experiments demanding a wide range of currents and polarizations. OPPIS was operated in different configurations to meet the differing current and polarization requirements for each experiment. We describe methods used to increase beam polarization at the expense of current for experiments that were count rate limited. OPPIS can be operated at 50 μA, giving 56% polarization, 25 μA with 65% polarization, or 2 μA with 77% polarization. The source reliability in 1992 was excellent, easily exceeding 95%. Contributions to experimental systematic errors made by the source were measured in 1992. We speculate about further improvements that can be made to OPPIS.

A more complete description of OPPIS can be found elsewhere.[1] There are many factors that affect its performance. Some of these, such as the design of the ECR source, ECR extraction optics, or thickness of the alkali vapor in the ionizer cell, primarily affect the beam intensity, I. Others, such as the magnetic field at the polarizer cell or design of the Sona region, affect the H⁻ beam polarization, P. There are many variables, however, that affect I and P simultaneously - usually in such a way that there is a tradeoff between the two. Understanding the variables and how to compromise between I and P has allowed us to meet the needs of different experimenters during the LAMPF 1992 run cycle.

One way of trading P for I is by varying the thickness of the polarized K vapor target. The H⁻ beam intensity increases linearly with K vapor thickness over a wide range of target thicknesses. However, the vapor thickness that can be pumped to maximum polarization is restricted by the limited laser power and radiation trapping. This relationship has been reported by many.[1,2,3,4]

Another factor that determines the relationship between I and P is the geometry of the extraction optics of the ECR source. The plasma lens has a hexagonal close-packed array of 1.0 mm diameter holes with 0.25 mm spacing between holes. The H⁺ current extracted from the

ECR (like the final H⁻ beam) is proportional to the total extraction area and therefore the number of holes in the array; arrays of 7, 19, and 37 holes were used in 1992. The array size determines the diameter of the H⁺ beam in the K cell. The K is more highly polarized along the axis of the polarizer cell because of depolarizing wall collisions at the walls and a higher laser intensity on the axis. Thus, the polarization is higher when more $H^0$ beam is formed near the axis of the cell.

The range of I and P is also affected by aperturing the beam in the 750 kV transport line. This result indicates a spatial dependence in the beam polarization. With a smaller aperture, current is lost, but the polarization is increased.

Figure 1 demonstrates how different source operating parameters can produce different values of I and P. Data taken with three different H⁺ beam diameters show how P is increased at the cost of I. For each extraction geometry, the thickness of the K in the polarizer cell can be varied to give a range of source performance. The figure also shows how the polarization can be increased by inserting a small aperture at 750 kV.

In 1992, we examined contributions that OPPIS might make to systematic errors in precise experiments.[5] These measurements, descibed in the next two paragraphs, were made while the source was in normal operation with the 19-hole ECR extraction array and with no aperture in the 750 kV beam line.

Systematic errors will result if the position, polarization, or current of the beam fluctuates in a way correlated to its spin state. We

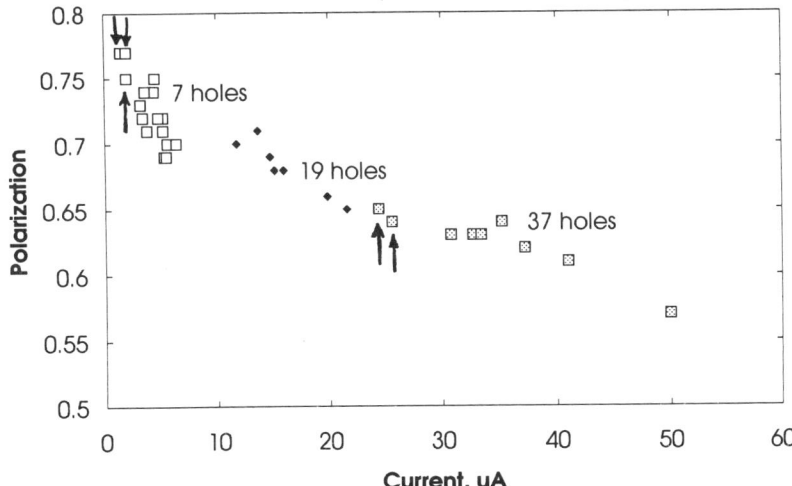

Fig. 1 The I and P of OPPIS can be varied by using ECR extraction arrays with different numbers of holes, as shown by the different symbols. The different data points in each series are from placing an aperture in the 750 kV beam line (denoted with arrows) or from changing the thickness of the K vapor.

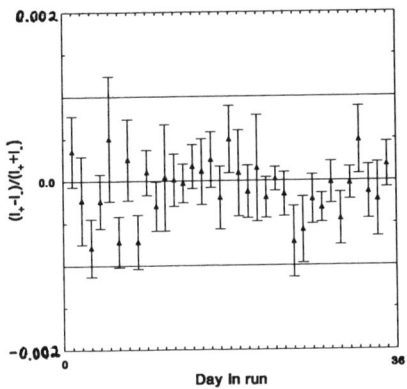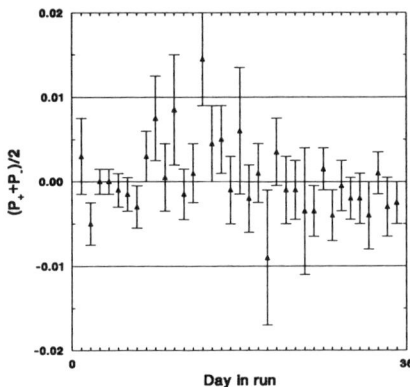

Fig. 2  (a) Intensity asymmetry i and (b) polarization asymmetry p vs. time. The time interval was from June 15 to July 25, 1992, during normal OPPIS beam production. In some cases, data from several days have been combined, when the beam was off for a long time.

measured the motion of the 5 mm diameter beam to be 0 +/- 0.01 mm. The asymmetry in the intensity, $i = (I_+ - I_-)/(I_+ - I_-)$, averaged $(-12 +/- 5) \times 10^{-5}$ over a month long period. The asymmetry in the polarization, $p = (P_+ - P_-)/2$, averaged $(-5 +/- 5) \times 10^{-4}$. Figure 2 shows i and p measured daily over a month.

There are other characteristics of the source that an experimenter may need to know in order to avoid systematic errors. Depending upon the linac tune, the center of the full energy 800 MeV beam had relative polarizations of up to 3% higher than the outer edge. The beam pulses are produced at 120 Hz, each about 850 μs long. Under some conditions, the relative polarization at the beginning of the pulse was up to 2% higher than at the end.

Table I outlines the performance of OPPIS since 1989. There are still some modifications that should be explored to improve the performance of OPPIS. Measurements made at TRIUMF[6] confirm calculations[7] showing that higher polarizations are obtained if the magnetic field at the polarizer cell is increased. Well over 10% relative improvement in polarization with no loss in current can be expected if a magnet of higher field is installed. Studies show that the $H^+$ beam intensity from the ECR is still limited by the amount of available rf power. The overall source performance is limited by the laser power available, and data indicate that the K vapor is less polarized near the walls than in the center of the target. Therefore, more laser power would improve the current and polarization of the source.

In conclusion, OPPIS has proven to be a source of remarkable versatility. The range of its current and polarization can be balanced to accommodate a wide range of demands from experiments. OPPIS reliability was well over 95% in 1992. The optically pumped source is well suited for high precision experiments, such as charge symmetry breaking or parity violation. There are many avenues that promise even better performance for OPPIS.

| Table I. A history of the current and polarization produced by OPPIS since its inception. | | | |
|---|---|---|---|
| | Current | Polarization | $P^2I$ |
| Lamb-Shift Source | 0.8 μA | 75% | 0.45 μA |
| 1989 OPPIS | 4.0 μA | 45% | 0.81 μA |
| 1990 OPPIS | 25 μA | 55% | 7.6 μA |
| | 15 μA | 62% | 5.8 μA |
| 1991 OPPIS | 38 μA | 60% | 14 μA |
| | 20 μA | 64% | 8.2 μA |
| 1992 OPPIS | 50 μA | 56% | 16 μA |
| | 25 μA | 65% | 11 μA |
| | 2 μA | 77% | 1.2 μA |

# REFERENCES

1. R.L. York et al., in Proceedings of the International Workshop on Polarized Ion Sources and Polarized Gas Jets, KEK-90-15, 170 (1990).
2. Y. Mori et al., "Report on optical pumping of sodium vapor with a pulsed dye laser," KEK Report 86-2 (1986).
3. L. Buchmann et al., in Proceedings of the International Workshop on Polarized Ion Sources and Polarized Gas Jets, KEK-90-15, 161 (1990), also ref. 6.
4. A.N. Zelenski et al., in Proceedings of the International Workshop on Polarized Ion Sources and Polarized Gas Jets, KEK-90-15, 154 (1990).
5. D. Tupa et al., "Systematic errors in an optically pumped polarized ion source," accepted for publication in Nucl. Instrum. Methods.
6. C.D.P. Levy et al., "Optimization studies of proton polarization in the TRIUMF optically-pumped polarized ion source," accepted for publication in Nucl. Instrum. Methods.
7. E.A. Hinds et al., Nucl Instrum Methods 189, 599 (1981).

# POLARIZATION DIAGNOSTICS AND OPTICAL PUMPING DEVELOPMENT FOR OPPIS AT LAMPF

**D.R. Swenson, D. Tupa, R.L. York, M. Dulick, and O.B. van Dyck**
**Los Alamos National Laboratory, MP-5, H838, Los Alamos, NM 87544**

## ABSTRACT

We report improvement of the polarization diagnostics and their use in the development of the Optically Pumped Polarized Ion Source (OPPIS).

## INTRODUCTION

We have developed an improved low-energy polarimeter (LEPO) based on the reaction $^6$Li(p,$^3$He)$^4$He at 750 keV[1,2]. The unique features of the polarimeter are the use of permanent magnet momentum analysis[3] to separate the reaction products from the elastically scattered protons, and the use of both $^3$He and $^4$He data to determine the polarization. The polarimeter has been used in source optimization studies. We have continued development of Faraday rotation diagnostics for measuring the alkali vapor density and polarization. The improvements include new calculations and improved technique. We have studied the effect of the laser spectral distribution on the beam polarization.

## LOW ENERGY $^6$Li POLARIMETER

The LEPO polarimeter allows the nuclear polarization of the beam to be measured without using scarce and expensive accelerator time. LEPO operation requires resolving the $^3$He and/or $^4$He reaction products in the presence of a much larger background of elastically scattered protons. Past attempts using silicon surface barrier detectors with energy resolution or coincidence techniques have suffered because the electronics were not fast enough to handle the background rate. We have solved this problem by adding a permanent magnet momentum filter to the detection system as shown in Fig. 1. The momentum filter consists of a pair of 2.54-cm-long SmCo magnets with a peak field of 6.7 kG and a gap spacing of 3.0 mm. The integrated field along the particle trajectory (measured to be 15.7 kG·cm) deflects the 750-keV protons by 4.7 mm while the 2.4 MeV $^4$He and 3.0 MeV $^3$He (which have nearly the same momentum) are deflected by 2.6 mm. Two collimating slits (1.0 x 2.4 mm) reduce the spot diameter to 4.3 mm at the detector and define the scattering angle. A third slit transmits the He while blocking the protons.

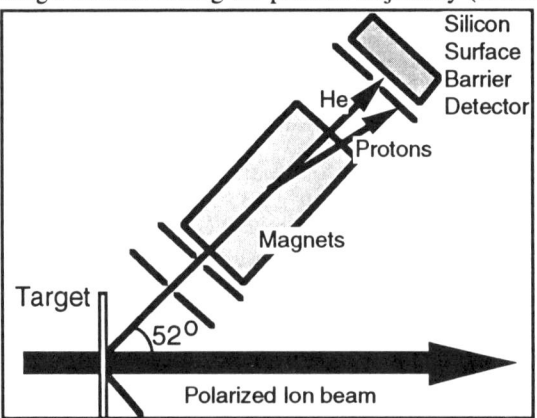

Figure 2 shows energy spectra with and without momentum analysis. In the latter case, both the $^4$He and $^3$He peaks are clearly resolved and can be used to calculate the polarization.

Fig. 1. The momentum analyzed detector system for the LEPO polarimeter.

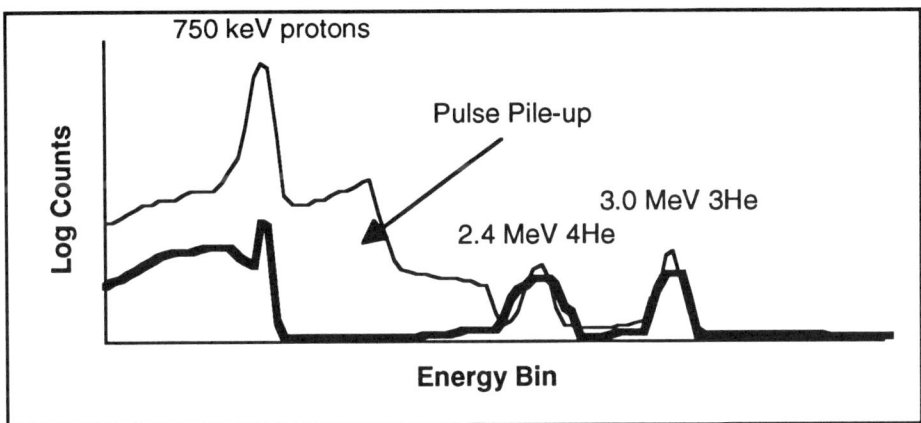

Fig. 2. Typical Multi-channel Analyzer (MCA) data measured with (thick line) and without (thin line) momentum analysis.

Indeed, it is possible to measure the polarization using only one detector and the ratio of the two peak areas. For our experiments, two symmetric detectors mounted at 52 degrees above and below the beam were used to detect forward scattered particles. The polarization can be determined using the conventional "ratio method" using spin reversal with the two detectors and either the $^3$He or $^4$He data[4]. If the polarization is the same in both spin states, the ratio of the analyzing powers for the two reactions can be determined[5]. Another way of determining the polarization uses data from both peaks on both detectors, the "four peak method". The four peaks can be used to form the ratio:

$$R = \frac{N_{u3}N_{d4}}{N_{u4}N_{d3}} = \frac{\Omega_{u3}\Omega_{d4}}{\Omega_{u4}\Omega_{d3}}\frac{(1+\alpha_3 P)(1-\alpha_4 P)}{(1+\alpha_4 P)(1-\alpha_3 P)} = R_0 \frac{(1+\alpha_3 P)(1-\alpha_4 P)}{(1+\alpha_4 P)(1-\alpha_3 P)}, \tag{1}$$

where N is number of counts, $\Omega$ is the detector acceptance solid angle, $\alpha$ is the analyzing power, P is the beam polarization, and $R_0$ is R measured with unpolarized beam. The subscripts distinguish between the $^3$He or $^4$He peaks on the "up" or "down" detectors. The quadratic equation can be solved for P as a function of $R/R_0$ and the known analyzing powers. This method allows the polarization for each spin state to be determined independently, and it is less sensitive to changes in the false asymmetry or beam steering. The different methods provide cross checks to limit systematic errors. Experimental results using these methods are discussed below.

## LEPO RESULTS

Results with the momentum analyzed version of LEPO are encouraging, although improvements are needed. The most serious problem was the limited lifetime of the targets. The targets consisted of 120 $\mu$g/cm$^2$ of $^6$LiF evaporated onto a 40 $\mu$g/cm$^2$ carbon foil. To achieve a lifetime of a few days, it was necessary to limit the average beam current to 0.5 $\mu$A. Count rates for each peak were typically 1 count per second per $\mu$A. The proton background rate was 5 times larger. Count rates were sensitive to the beam steering. The beam had to be positioned precisely to get equal rates in the two arms, which indicates that the alignment was not ideal. One would expect the ratio of the $^3$He and $^4$He peaks measured with one detector for unpolarized beam would be constant because the momentums are nearly the same and hence the $\Omega$'s should be the same. If this were so, with proper alignment, $R_0$ would equal 1 and would not need to be measured. This was not always the case, indicating possible misalignment of the detectors or problems with the target. Measurements taken during one-day indicated that $R_0$ did not drift. For example, during a one day run with unpolarized beam, five measurements of P yielded a standard deviation of 0.9%, in agreement with the counting statistics. However, from day to day, measurements of $R_0$ ranged from 0.81 to 1.02. More study of the systematic errors is needed. Using the measured $R_0$, the "four-peak method" gave consistent polarization measurements.

LEPO was used in source optimization studies. The measurements showed better polarization with a 7-hole extraction lens (3.3-mm-diameter ion beam) than with a 19 hole extraction lens (4.3-mm-diameter ion beam). The polarizations were 69±2% versus 63±1%, which agreed well with later measurements at 800 MeV. A tubular μ-metal shield was installed around the zero-crossing region of the Sona transition. The polarization without the shield was 62.6±1.7%; with the shield it was 63.7±1.1%, thus showing little improvement. Measurements varying the laser size by 200% indicated little dependence on laser size.

## FARADAY ROTATION DIAGNOSTICS

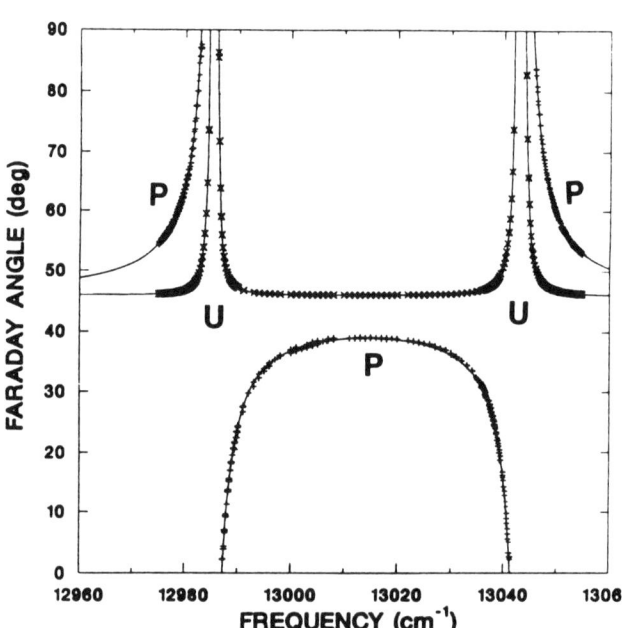

Faraday rotation is a useful diagnostic for measuring the thickness and electron spin polarization of an alkali vapor. Extracting useful information from the measurements requires accurate calculations of the Verdet constant for unpolarized and polarized vapor. M. Dulick has written a computer program[6] that calculates these constants for each of the alkalis, for all probe laser frequencies, and for all magnetic fields. A copy may be obtained from the authors. Measurements with Na and K vapor at various frequencies and magnetic fields are consistent with the calculations. Figure 3 is an example of our data at 3 kG.

Fig. 3. Measured and calculated Faraday rotation angles for unpolarized (U) and polarized (P) K vapor. Thickness is $5.0 \times 10^{13}$ cm$^{-2}$ and polarization 72%.

At the last conference[7], we described several techniques for measuring Faraday rotation. The measurements of Fig. 3 were made using the "Two-detector method". We have further developed the "Rotating λ/2 plate method" and have found it to be the method of choice. The "Rotating λ/2 plate method" gives accurate, stable, real-time measurements and does not require frequent calibration. It was used to test the accuracy and stability of the OPPIS automated spin flip over several days .

## LASER FREQUENCY DISTRIBUTION

The spectral output of our Ti:Sapphire laser is narrowed by two uncoated etalons (200 GHz and 20 GHz free spectral range) to match the Doppler-broadened absorption line of K vapor (1 GHz FWHM). It consists of four or more discrete longitudinal cavity modes spaced 225 MHz apart, as shown in Fig 4a. The discrete nature of the distribution may limit the optical pumping efficiency because the mode spacing is large compared to the natural line width. To test the effect of a more uniform laser spectrum, we have obtained a prototype vibrating laser cavity mirror[8]. The mirror is mounted on a high-frequency acoustic horn, which is excited by a piezo-electric modulator. This allows the laser cavity length, and hence the frequency of the cavity modes, to be modulated at 140 kHz, which is faster than the average wall collision rate (45 kHz) of the K vapor. Figure 4b shows the laser frequency distribution with the modulator on. In limited tests conducted during accelerator operations, the modulator did not affect the polarization of the beam as measured by the 800-MeV polarimeter. Further tests are needed. A larger improvement is expected for Rb vapor because the linewidth is greater.

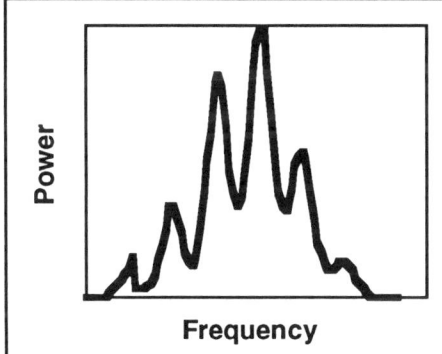

Fig. 4a.    Laser Spectrum with modulator off. Laser power 3.8 W.

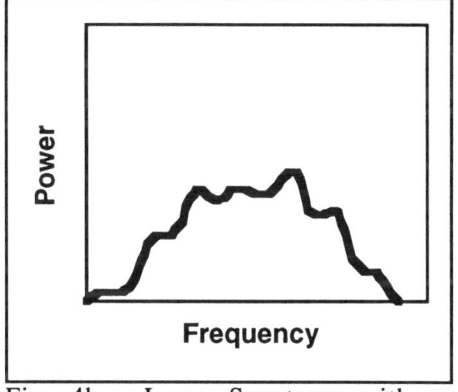

Fig. 4b.    Laser Spectrum with modulator on. Laser power 3.5 W.

[1]Louis Brown, and Claude Petitjean, Nucl. Phys. A117 343 (1968)

[2]L. Buchmann, Nucl. Inst. and Meth.A301 383 (1991)

[3]We acknowledge L.J. Rybarcyk for suggesting the idea, and helpful discussions, H.E. Williams for engineering, and K.W. Jones, J. D. Wieting, W.P. Potter, T. J. Wehner, J.D. Paul, and M. McNaughton.

[4]W. Haeberli, Ann. Rev. Sci. 17 373 (1967)

[5]We measured $\alpha_4/\alpha_3$ to be 1.4. The value obtained from reference 1 was 1.6.

[6]M. Dulick, Program "Faraday.for", VAX FORTRAN 77 Version, LA-UR 91-1577, LANL (1991)

[7]D.R. Swenson et al., KEK Report 90-15 A 187 (1990)

[8]Gaylen Erbert, LLNL, L-463, PO Box 808, Livermore, CA 94551, U.S. patent 5132979

# PRODUCTION OF POLARIZED HEAVY IONS FOR INTERMEDIATE-ENERGY NUCLEAR PHYSICS

Masayoshi TANAKA[+], Takashi OHSHIMA[++], Kenji KATORI*, Mamoru FUJIWARA,
Hiroshi OGATA, Noriyuki SHIMAKURA**, and Michiya KONDO
  Research Center for Nuclear Physics, Osaka University, Mihogaoka 10-1, Ibaraki, Osaka 567, Japan
+ Kobe Tokiwa Jr. College, Ohtani-cho 2-6-2, Nagata, Kobe 653, Japan
++The Institute of Physical and Chemical Research, 2-1 Hirosawa, Wako-shi, Saitama 350-01, Japan
* Laboratory of Nuclear Study, Osaka University, Machikaneyama-cho 1-1, Toyonaka, Osaka 565, Japan
**Department of General Education, Niigata University, Igarashi Ninomachi 8050, Niigata 950-21, Japan

## ABSTRACT

The present status and future prospect of the polarized heavy ion source based on OPPIS are presented emphasizing on the production of the polarized $^3$He beam. It was ensured by means of the beam foil spectroscopy that the polarized $^3$He beam with the polarization degree better than 4-5% was produced. Toward the next phase of our project the whole source composition was remounted on a high voltage platform so that the polarized $^3$He$^+$ ion might be accelerated and stripped to form the $^3$He$^{2+}$ ion for injecting it inot the injector cyclotron. In order to increase the beam intensity a 2.45GHz ECR ion source used for the initial phase was replaced by a new ECR ion source, Neomafios-10GHz.

## 1. Introduction

The central topics of a new ring cyclotron facility at RCNP are oriented to nuclear physics at an intermediate energy typified by 100 MeV/A. After finishing the construction of the ring cyclotron and related experimental equipments a variety of experiments have started with polarized proton and unpolarized light ion beams. Recent budget allowed us to construct a new polarized proton and deuteron ion source and an ECR ion source for the production of heavy ions in addition to the improvement of the vertical injection line of the injector cyclotron.

As one of the future physics program, intermediate-energy nuclear physics with polarized heavy ion beams was several years ago offered and a polarized heavy ion source has been developed at RCNP because we believe our plan would be one of new frontiers in view of the fact that no other institute had ever attempted at such an incident energy region except for the Saturne, France, where 750 MeV/A $^6$Li beams are now available[1]. Since all of the polarized heavy ion sources practically available so far have been an atomic beam type, the polarized ion species have been limited only to alkali atoms. On the other hand, an OPPIS (optical pumping polarized ion source) is a universal ion source which can polarize not only proton but any nuclei irrespectively of ion species. Guided by this aspect, we have developed an OPPIS applied to $^3$He as the first step because of technical easiness relative to other heavy ion candidates and physical importance. The importance of polarized $^3$He beams is exemplified as follows; first of all a polarized $^3$He ion is approximated by a polarized neutron which is not easily obtainable by other methods and secondly the $^3$He beam energy achieved at RCNP, i.e., 180 MeV/A, is quite suitable for the study on the excitation mechanism of the spin isospin mode in nuclei. In particular, in combination with the high resolution spectrograph, Grand Raiden, the high resolution study on the spin structure of the Gamow Teller giant resonace will be cleared through the ($^3$He,t) reaction with $^3$He polarized beams.

## 2. Present Status of OPPIS Applied to $^3$He Polarization

Our OPPIS is applied for the polarization of $^3$He, in which we use spin and charge exchange collisions between incident $^3$He$^{2+}$ ions and sodium atoms polarized by means of the laser optical pumping. An 2.45GHz ECR ion source produces $^3$He$^{2+}$ beams and they are transported to a sodium vapor cell located inside a 3kG solenoidal coil through a magnetic analyzer and focusing elements. Meanwhile, alkali atoms are polarized by a pumping laser provided by a single mode ring-dye laser and the degree of the atomic polarization and sodium vapor thickness are monitored by observing a Faraday rotation angle of another prove laser whose wavelengths are tuned between D$_1$ and D$_2$ lines of a sodium atom. The electron polarization of $^3$He$^+$ ions formed by spin and charge exchange collisions is then transferred to the $^3$He nucleus through the hyperfine interactions when ions emerge from the solenoidal coil. The nuclear polarization of $^3$He$^+$ ions is finally measured by a polarimeter based on, so called, a beam foil spectroscopy as described somewhat in detail in the following. The polarized $^3$He$^+$ ion is introduced to a thin carbon foil after being separated from other components by a 90 deg. electrostatic analyzer keeping the polarization direction unchanged. $^3$He$^+$ ions are almost neutralized, i.e., $^3$He I, but in the excited states after they pass through the foil. During photon decay processes of $^3$He I atoms, a certain amount of the nuclear polarization is transferred to the atomic electron of $^3$He I atoms. Emitted photons, thereafter, are circularly polarized periodically in accordance with hyperfine coupling frequencies of the $^3$He I. This means that the $^3$He nuclear polarization can be determined from the measurement of the circular polarization degree of the emitted photons.

Through above studies, it was pointed out that one of the serious nuclear depolarizations specific to heavy ions was induced by an insufficient LS decoupling during spin and charge exchange collisions because of a weak magnetic field applied to the sodium cell. A simple calculation of atomic physics tells us that the magnetic field enough strong to decouple the LS couplings is proportional to $Z^4/n^3$[2], where Z is an atomic number of an ion to be polarized and n is a principal quantum number of an ionic state to which spin and charge exchanged electron transfers. Only 1 or 2 Tesla is enough for the proton case, while an extremely large value of the magnetic field, i.e., more than 16 Tesla is needed even for the $^3$He case. This strongly suggests that the full decoupling is unpractical for heavier ions than proton. It is, however, expected that even under the absence of the magnetic field a sizable amount of the nuclear polarizations should survive as demonstrated for the proton case[3]. This is also true for the polarization of other heavy ions. For instance, the survival value of the nuclear polarization for the $^3$He$^{2+}$ + Na system is approximated[4] by

$$P_{Nucl}(3He) \cong \frac{1}{2} P_{elect}(Na)\{0.301 + 0.201 A_0^{col}(L = 2)\},$$

(1)

assuming that the 3s electron of the sodium atom is predominantly transferred to the 3d state of the $^3$He$^+$ ion[5], where sodium polarization, where

$P_{Nucl}(^3He)$ and $P_{elect}(Na)$ are, respectively, the produced nuclear polarization of the $^3He^+$ ion and the $A_0^{col}$ (L=2) is an alignment factor of the $^3He^+$ state initially formed as defined by

$$A_0^{col}(L) = \frac{\sum_{m=-L}^{L}[3m^2 - L(L+1)]\sigma(m)}{L(L+1)\sum_{m=-L}^{L}\sigma(m)}.$$

(2)

Here, $\sigma(m)$ is a cross section of spin and charge exchange collisions going to a magnetic substate m of an orbital angular momentum L. In our case, L=2 (d sate) is assumed. From eq. (1), it is suggested that the amount of $P_{Nucl}(^3He)$ would depend on the $^3He^{2+}$ incident energy, since $A_0^{col}$ (L=2) depends on the collision mechanism. Therefore, the measurement of the incident energy dependence of $P_{Nucl}(^3He)$ is indispensable to search for a suitable incident energy where the nuclear polarization becomes maximum.

### 3. Experimental Results and Discussion

Photon spectra emitted from ions penetrating the carbon foil used for the beam foil spectroscopy are taken by a monochrometer. Among various lines one can see a prominent peak at 389 nm corresponding to the transition from the $3^3P_J$ (J=0,1,2) to $2^3S_1$ state in $^3He$ I and this photon peak is used for determining the nuclear polarization[6]. In order to obtain the optimized condition, under which the resulting nuclear polarization is maximum, the measurement of $P_{Nucl}(^3He)$ is carried out by varying the $^3He^{2+}$ incident energy from 20 to 28 keV. The result shows that $P_{Nucl}(^3He)$ is 4-5% irrespectively of the incident energy. This suggests that the alignment factor, $A_0^{col}$ (L=2) defined by eq. (2) is also constant throughout this energy range. To see it more clearly, $A_0^{col}$ (L=2) is evaluated from the observed $P_{Nucl}(^3He)$ and $P_{elect}(Na)$ by using eq. (1) and plotted in Fig. 1. Though the error bars are too large to say definitely, each data of $A_0^{col}$ (L=2) deviate constantly to the negative direction. The theoretical approach based on a semi-classical impact parameter method[7] is performed to see whether such an energy dependence is reproduced or not. The calculations involving transitions to 3s, 3p, and 3d states of a $^3He^+$ ion are done at incident $^3He^{2+}$ energies of 16, 22, and 28 keV. The sum of the cross sections

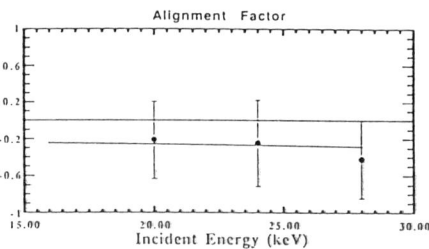

Fig. 1 Observed and theoretical alignment factors plotted as a function of the $^3He^{2+}$ incident energy.

to these states is 1.5, 1.4, and 1.2 in the unit of $10^{-14}$ cm$^2$ , which is in rough agreement with the observed values[5]. In addition it is found that the main contributing cross section is due to the transition to the 3d state is at each incident energy. The interpolated result of the calculation for $A_0^{col}$ (L=2) for the transition to the 3d state is shown by a solid line in Fig. 1. It is interesting that the observed trend is qualitatively reproduced by the theory.

## 4. Conclusion and Future Plan

Through the present work, it was demonstrated that the $^3$He nuclear polarization was 4-5% in an incident energy ranging from 20 to 28 keV. The beam intensity of the polarized beam was only about 40 enA. This is mainly due to a low production rate of $^3$He$^{2+}$ ions from the 2.45GHz ECR ion source and an insufficient beam optics. Aiming at the second phase of our project, we remounted the whole system of the polarized ion source on a high voltage platform so that the polarized $^3$He$^+$ ions might be accelerated and stripped to form $^3$He$^{2+}$ ions for injection into the injector cyclotron. The 2.45GHz ECR ion source was replaced with a new ECR ion source, i.e., Neomafios-10GHz[9] which will be used also for the production of unpolarized heavy ions. Note that the $^3$He$^{2+}$ beam intensity more than 200 e$\mu$A was extracted with ease. To be startled, this value is two orders of magnitude larger than the performance of our old 2.45GHz ECR ion source. In addition, we are now preparing following improvements; a replacement of sodium optical pumping with Rb one by using a Ti:Sapphire laser pumped by powerful argon ion lasers, and a use of Sona field[8] with higher magnetic field. With these improvements we hope that our polarized ion source will be improved as summarized in the following;

1. $^3$He nuclear polarization will increase to about 30%,
2. the beam intensity of the polarized $^3$He$^{2+}$ ions after stripping will amount to 20 e$\mu$A.

In the final phase of our project, we should consider the polarization of other heavier ions by means of OPPIS. Various light heavy ions up to, say, $^{19}$F can be polarized with the polarization degree of 10-20% but an amount of the achievable polarization seems to be decreased according as an atomic number. This is due to the fact that the depolarization caused by LS couplings becomes more significant for heavier elements. Nevertheless, the acceleration of these polarized ions is hopeful.

## References

1. Chamouard, P.A., et al., 1991, Coll. de Phys. **C6**, 565.
2. Bransden, B.H., and Joachain, C.J., *Physics of Atoms and Molecules*, 1983, Longman Scientific & Technical, New York.
3. Hinds E.H. et al., 1981, Nucl. Instr. Meth. **189**, 599.
4. Ohshima, T. et al., to be published in Hyperfine Interactions.
5. Dubois R. D. et al., 1985, Phys. Rev. **A31**, 3603.
6. Ohshima, T. et al., 1992, Phys. Let., **B274**, 163.
7. Shimakura, N. et al, 1993, to be submitted to Phys. Rev A.
8. Sona, P.G., 1967, Energia Nucl., **14**, 295.
9. Geller, R., 1990, Ann. Rev. Nucl. Part. Sci., **40**, 15.

# TARGET FOR PRODUCING POLARIZED $^{21}$Na BY OPTICAL PUMPING

P. A. Voytas, J. E. Schewe, P. A. Quin, and L. W. Anderson
Physics Department, University of Wisconsin-Madison.
Madison, Wisconsin 53706

R. E. Miers
Physics Department, Purdue University, Fort Wayne, Indiana 46805

## ABSTRACT

The development of a target for producing polarized $^{21}$Na by optical pumping is detailed. Maximum nuclear polarizations of 0.63 have been obtained with this target. Results of investigations of the effects of the passage of a charged particle beam on the optical pumping process are also presented.

## I. INTRODUCTION

Many tests of the weak interaction are possible by studying observables in nuclear beta decay. Nuclear polarization dependent observables can be particularly sensitive in tests for physics beyond the standard model of the weak interaction. The development of the target described here was motivated by an interest in performing two such tests. One is the measurement of the polarization of the betas emitted in the decay of a polarized parent nucleus[1]. The other test is the measurement of the ground state beta asymmetry parameter in a mirror-nucleus decay[2]. The sensitivity for both experiments increases rapidly with the polarization of the parent nucleus.

$^{21}$Na has properties which make it a good candidate for such experiments if adequate nuclear polarization is obtained. It is a mirror-nucleus decay which reduces the influence of nuclear structure effects, and the decay has a branch to an excited state of $^{21}$Ne which is important in measuring the ground state beta asymmetry parameter using the method of reference (2). We describe here the target we have developed for producing and polarizing $^{21}$Na by optical pumping to be used in such tests.

## II. OPTICAL PUMPING

### 1. METHOD

We use optical pumping as the method of producing nuclear polarization. This process, originally developed for use with natural sodium, relies on the transfer of angular momentum to the alkali atom from circularly polarized light at the $D_1$ transition wavelength[3]. Since $^{21}$Na has the same nuclear spin (3/2) as natural sodium, the hyperfine structure is similar. Furthermore, the isotope shift of the $D_1$ transition is 1.6 GHz[4] which is comparable to the Doppler broadening of sodium at room temperature. Thus, results of optical pumping studies on natural sodium are directly applicable to $^{21}$Na.

To reduce the frequency of wall collisions, a buffer gas is usually introduced into the cell containing the sodium, increasing the time it takes an atom to reach the wall. Gases used in this capacity should not react chemically with sodium and should have a small cross section for depolarizing sodium atoms in

© 1994 American Institute of Physics

sodium-buffer-gas collisions. Helium and neon are both good choices in these respects[5]. We chose neon because $^{21}$Na is easily produced in the $^{20}$Ne(d,n)$^{21}$Na reaction. Thus, neon serves as both the target nucleus and the buffer gas for the optical pumping process.

## 2. LASER SYSTEM

The laser system consists of a transverse pumped tunable dye laser driven by a pulsed Cu vapor laser. Operating at 7 kHz, the Cu vapor laser produces $\approx$2 W average power in each of two Cu lines (at 510.6 nm and 578.2 nm). The Cu laser pulse is $\approx$50 ns FWHM, giving a peak power of 6 kW in each line. The Cu laser output beam is focused with a cylindrical lens to a line 0.16 mm wide by 25 mm long on the dye cell through which is circulating Rhodamine chloride dye in ethanol. The dye absorbs only the 510.6 nm line of the Cu laser. The optical cavity of the dye laser is defined on one end by a diffraction grating with which the wavelength is selected and on the other end by a 10° wedge which provides optical feedback from one surface only. Between the dye cell and the grating is a telescopic beam expander of magnification 7.

The average dye laser output power is about 200 mW with a measured bandwidth of 0.06 nm which is matched to the collisionally broadened resonance transition of sodium in our targets. Circular polarization of the dye laser output is achieved with a linear polarizing beamsplitter followed by a $\lambda/4$ plate for the 589.6 nm $D_1$ line of sodium. Following this is a movable $\lambda/2$ plate for changing the handedness of the outgoing light.

The laser system was tested off-line to determine its effectiveness in performing the optical pumping. In particular we wanted to know that we could achieve substantial nuclear polarization with our pulsed laser system. Results of off-line tests on natural sodium with neon buffer gas in glass cells, using the techniques of reference (5), gave electronic polarizations of about 0.75 and nuclear polarizations of about 0.80.

## III. ON-LINE OPTICAL PUMPING

### 1. GENERAL

Having demonstrated the feasibility of optical pumping with our laser system, we now needed a target cell to produce the $^{21}$Na for on-line pumping. The on-line tests proceeded in two phases. First a small, proof of principle cell was constructed and tested. This prototype cell was designed to demonstrate our ability to produce $^{21}$Na in an environment clean enough that very little $^{21}$Na would react chemically with impurities. We also used the test cell to investigate the optical pumping process in the presence of a charged particle beam. In the second phase, we used the results from the prototype cell to design a production cell for beta decay experiments.

### 2. PROTOTYPE TARGET

The prototype target was an all metal and glass target constructed around a six-way 3.38 cm Conflat-style cube. Metal foils (0.0025 cm thick nickel or HAVAR) were used as entrance and exit foils for the deuteron beam and as the exit foil for the betas. All foils were sealed by sandwiching them between two polished copper gaskets. A pyrex window opposite the beta exit foil allowed the laser light in. A reservoir of natural sodium on the bottom port provided natural sodium to do optical tests of the interaction of the charged beam with the optical pumping process as well as acting as a chemical getter for any contaminants. A second pyrex window on the top port allowed optical monitoring of the interaction of the particle beam and the optical pumping process. Access for gas handling and vacuum access was obtained through modified double sided flanges mounted between the cube and the top/bottom flanges. The cell was cleaned thoroughly, installed on the beamline and baked under vacuum at 400 °C for several days.

We were able to optically pump the natural sodium in the test cell. The cell was heated to 150°C to get a large enough number density of sodium to observe the fluorescence with a photomultiplier tube (PMT). The cell was then filled with natural neon to a pressure of $10^5$ Pa. The change in fluorescence when the magnetic field defining the quantization axis is turned off and on gives a rough indication of the polarization. For the test cell the nuclear polarization was in the 0.7-0.8 range.

Two tests were performed to study the effects of the passage of the deuteron beam through the cell. For both tests 8.5 MeV deuterons passed through the cell for 1.0 second and the PMT output was monitored as a function of time. Results for beam currents of 30 nA and 200 nA were similar for both tests. For one test, the laser was left on during the entire measurement and the results are shown in Fig. 1(a). The vapor starts out polarized then the deuteron beam disturbs the polarization, giving an increase in fluorescence. As the beam ionizes the neon and sodium, the number of sodium atoms (as opposed to ions) available to fluoresce decreases, leading to an equilibrium level of fluorescence with the deuteron beam on.

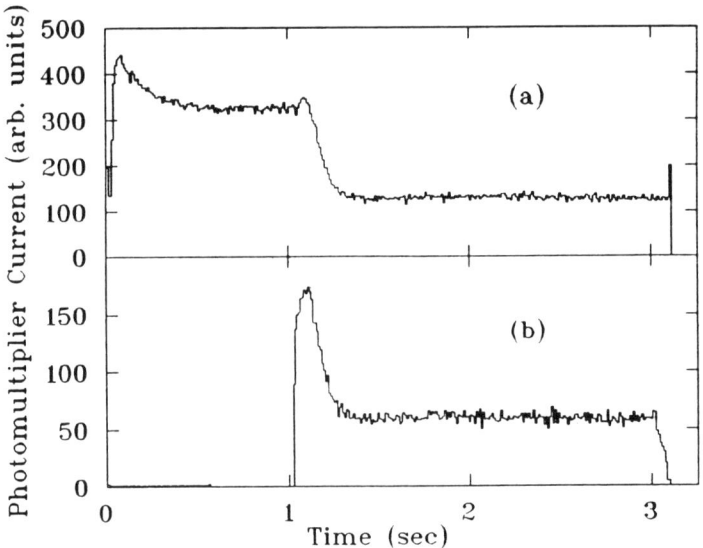

Figure 1. See text for details

When the deuteron beam is removed, the sodium and neon ions neutralize, the sodium atoms become polarized again, and the fluorescence returns to its initial level. For the second test, the laser beam was shuttered off during the time the deuterons were passing through the cell and then turned on as the deuteron beam was removed. Results of this test are shown in Fig. 1(b). Here the sample completely depolarizes during the period when the deuteron beam is on. When the deuteron beam is taken away and the laser turned on there is a large rise in fluorescence as the sample re-polarizes.

From these two experiments, we infer that the deuteron beam effectively destroys the optical pumping process. However, the system recovers on a fairly short time scale of about 200 ms after removal of the deuteron beam.

## 3. NUCLEAR TESTS

To test the on-line optical pumping of $^{21}$Na we abandoned optical techniques of determining the polarization because the deuteron beam passing through the neon gas in the cell produces too few atoms to detect their fluorescence. We obtain a good estimate of the nuclear polarization by taking advantage of the parity violating beta-asymmetry in the decay of the $^{21}$Na we produce. The angular distribution of betas from the decay of a polarized nucleus is

$$W(\theta) \propto (1 + \beta P A cos(\theta)) \tag{1}$$

where P is the nuclear polarization, $\theta$ is the angle between the beta velocity, v, and the nuclear polarization direction, and $\beta$ is v/c for the betas. "A", the beta asymmetry parameter, depends on weak interaction coupling constants and nuclear matrix elements. $^{21}$Na has two beta decay branches of significant strength. The branch (94.9%) to the ground state of $^{21}$Ne has $A_{g.s.} = 0.86$ and the branch (5.1% ) to an excited state of $^{21}$Ne has $A_{ex} = -0.6$. We use this information to determine the nuclear polarization as follows. A detector placed at the beta exit foil measures the beta rate after activation of the target by the deuteron beam. This process is performed repeatedly for both $\sigma^+$ and $\sigma^-$ light. The rates for the two different circular polarization directions (and hence nuclear polarization states) are used to determine the asymmetry

$$\mathcal{A} \equiv \frac{W(0°) - W(180°)}{W(0°) + W(180°)} = P[(BR_{g.s.})A_{g.s.} < \beta_{g.s.} > +(BR_{ex})A_{ex} < \beta_{ex} >] \tag{2}$$

where the velocity average is over those betas whose asymmetry we measure, and the BR's are the respective branching ratios. We assume that the polarization is the same magnitude for both $\sigma^+$ and $\sigma^-$ pumping. From the measured asymmetry, $\mathcal{A}$, we obtain a good estimate of the nuclear polarization. We measured the asymmetry using the prototype target with natural neon as the target and obtained results which indicate a maximum nuclear polarization of 0.69(7).

## 3. PRODUCTION TARGET

Having demonstrated optical pumping of $^{21}$Na, we designed a production cell incorporating all the features necessary for the experiments described in the Introduction. We needed a cell large enough and with a high enough pressure to give a diffusion time to the walls comparable to the $^{21}$Na half-life of 22.5 s. We still require an all metal and glass system to ensure cleanliness and lack of contaminants. The final cell design is shown in Fig. 2. A re-entrant fused quartz window lets the laser light in. The deuteron beam enters through a 0.001 cm thick by 1.59 cm diameter HAVAR foil and exits through a 0.0025 cm by 3.81 cm HAVAR foil. The betas exit through a 0.0025 cm by 3.81 cm HAVAR foil as well. All gas handling and vacuum access is done through one port on one side, leaving the other side of the main body of the cell clear of obstructions so the excited state gamma rays are more easily and cleanly detected. At $3\times10^5$ Pa, the diffusion time to the walls is calculated to be 17 seconds which improves our duty cycle by a factor of 15 over the prototype cell.

Nuclear tests as described in Sec. III.3 were performed and the results are shown in Fig. 3. For these tests we used separated isotope $^{20}$Ne for the target gas. The maximum asymmetry corresponds to a nuclear polarization of 0.63 and note that the asymmetry decays on a roughly 17 second time scale.

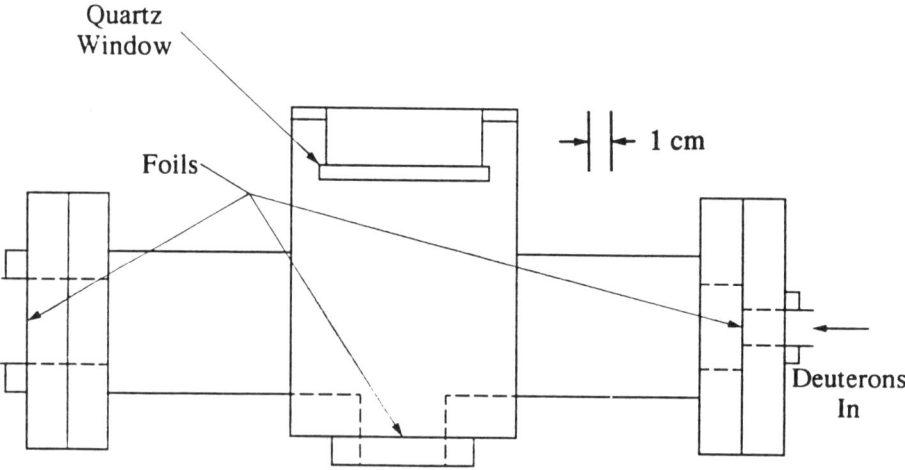

Figure 2. Schematic Diagram of Production Target

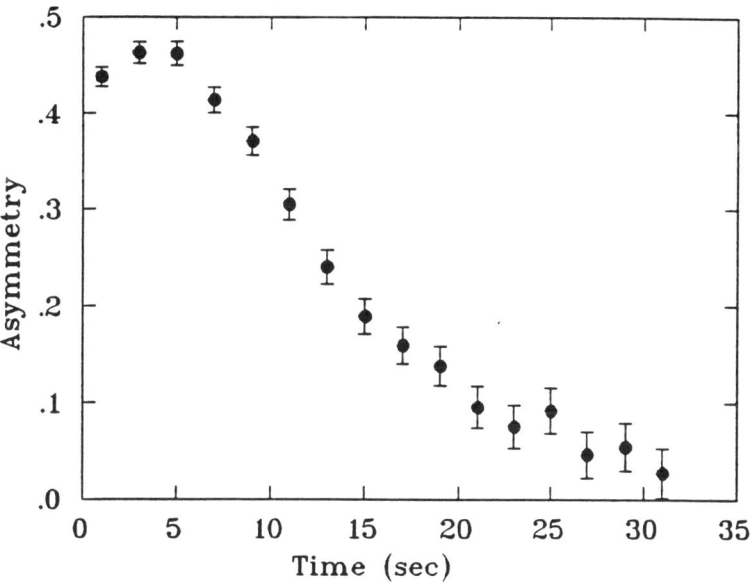

Figure 3. Asymmetry vs. Time after activation

## IV. CONCLUSIONS

We have successfully produced $^{21}$Na and polarized it by optical pumping, achieving maximum nuclear polarizations of 0.63(3). This degree of polarization and the large activity of the target make it a promising tool in studying the weak interaction in nuclear beta decay.

**Bibliography**

1. P. A. Quin, T. A. Girard, Phys. Lett. B **229**, 29 (1989).

2. J. D. Garnett, et al. Phys. Rev. Lett. **60**, 499 (1988); G. S. Masson, P. A. Quin, Phys. Rev. C **42**, 1110 (1990); A. Converse, et al., Phys. Lett. B **304**, 62 (1993).

3. W. Franzen, A. G. Emslie, Phys. Rev. **108**, 1453 (1957).

4. G. Huber, et al., Phys. Rev. C **18**, 2342 (1978).

5. L. W. Anderson, A. T. Ramsey, Phys. Rev. **132**, 712 (1963).

# OPTICALLY PUMPED POLARIZED LITHIUM SOURCE AT FSU*

E.G. Myers, A.J. Mendez, B.G. Schmidt, K.W. Kemper,
P.L. Kerr and E.L. Reber
Department of Physics
Florida State University
Tallahassee, FL 32306-3016, USA

## ABSTRACT

The dye-laser optically pumped polarized $^{6,7}$Li$^-$ ion source of the FSU tandem-linac heavy-ion accelerator has completed $1\frac{1}{2}$ years of operation. Highly polarized beams of $^6$Li have been accelerated with on target currents typically greater than 50 particle nA. The source uses electro-optically modulated circularly polarized light to optically pump an atomic beam into a single magnetic substate. Adiabatic rf transitions to other substates enable the polarization to be changed. The polarization of the atomic beam is measured by laser induced fluorescence using a second electro-optic to modulate the beam from the same laser, and Zeeman tuning. The polarized atomic beam is surface ionized to Li$^+$ and undergoes charge exchange to Li$^-$. The tandem-linac can accelerate the beam up to 60 MeV.

## INTRODUCTION

In 1988 it was decided to undertake a major upgrade of an existing polarized $^6$Li$^-$ source at FSU. The aim was to obtain sufficient current and polarization of $^{6,7}$Li to enable nuclear reaction studies[1], with good statistical quality, to be obtained with the recently completed tandem-superconducting linac. The developments, which followed those made at the polarized alkali source at MPI Heidelberg[2-5], consisted of:

1) using laser optical pumping, without a Stern-Gerlach type magnet to nuclear polarize the atomic beam,

2) using laser induced fluorescence for measuring the polarization of the atomic beam,

3) improving the beam optics by using an ionizer geometry with extraction at 90° to the holding magnetic field, and

4) remounting the whole source on a high voltage platform, with computer control, enabling an injection energy into the tandem up to 105 keV.

*Supported in part by the National Science Foundation.

Optical pumping to a single magnetic substate, together with a single rf transition unit, enables in principle, all possible beam polarizations to be achieved at their maximal values. For $^6$Li one ideally achieves $P_z = \pm 1$, $P_{zz} = +1, -2$, while with the conventional source, with two rf transitions, the limits were $P_z = \pm\frac{2}{3}$, $P_{zz} = \pm 1$. Further, because with sufficient laser power, a larger solid angle of atomic beam from the oven can be polarized than can be accepted by a Stern-Gerlach magnet, the source produces more current for a given lithium consumption.

The cost of the items purchased for the upgrade was \$215,000, of which \$150,000, provided by the State of Florida, was used for the laser system, optics and rf components. The remainder was used for vacuum equipment, high voltage power supplies and transformers, a computer control system and material. With the exception of the ionizer magnet, all the ion beam forming and ion beam optical components of the source were constructed in-house.

## PRODUCTION OF POLARIZED ATOMIC BEAM

A schematic of the atomic beam portion of the ion source is shown in Fig. 1. Some construction details are shown in Fig. 2.

### 1. Atomic beam formation

The lithium atomic beam is obtained from a lithium boiler, operated at 750° - 850°C, contained within a vacuum insulated oven[6]. The boiler is loaded with a 25 gram load of isotopically enriched $^6$Li or $^7$Li. The beam is formed by a separately heated Laval nozzle with an opening of 0.5mm, and collimated at a distance of 5mm by a heated skimmer (1mm opening), and at a distance of 35mm by a room temperature collimator of diameter 1mm. Thirty collimators are mounted on a wheel and new collimator is inserted every day of operation to prevent blockage. The oven power consumption is about 350W and the housing is water cooled.

In an offline setup, using $^7$Li, the axial velocity distribution of the atomic beam from the oven was measured by observing the Doppler shifted and broadened laser induced fluorescence line shape obtained by scanning the frequency of a collinear laser beam. The observed distribution, at an oven temperature of 800°C, had a most probable velocity of $(2.2\pm0.1)$Kms$^{-1}$ and a FWHM of $(2.0\pm0.1)$Kms$^{-1}$. The corresponding values for an effusive velocity distribution at this temperature are 1.96 Kms$^{-1}$ and 2.03 Kms$^{-1}$.

### 2. State selection by optical pumping

The optical pumping cell is mounted on the oven exit flange. The cell consists of a stainless-steel cuboid 25mm × 25mm × 37.5mm through which three 12.5mm diameter holes have been bored, along the axes of symmetry. The atomic beam travels the length of the cuboid while the other four faces are fitted with anti-reflection coated windows. The optical pumping laser beam enters and exits

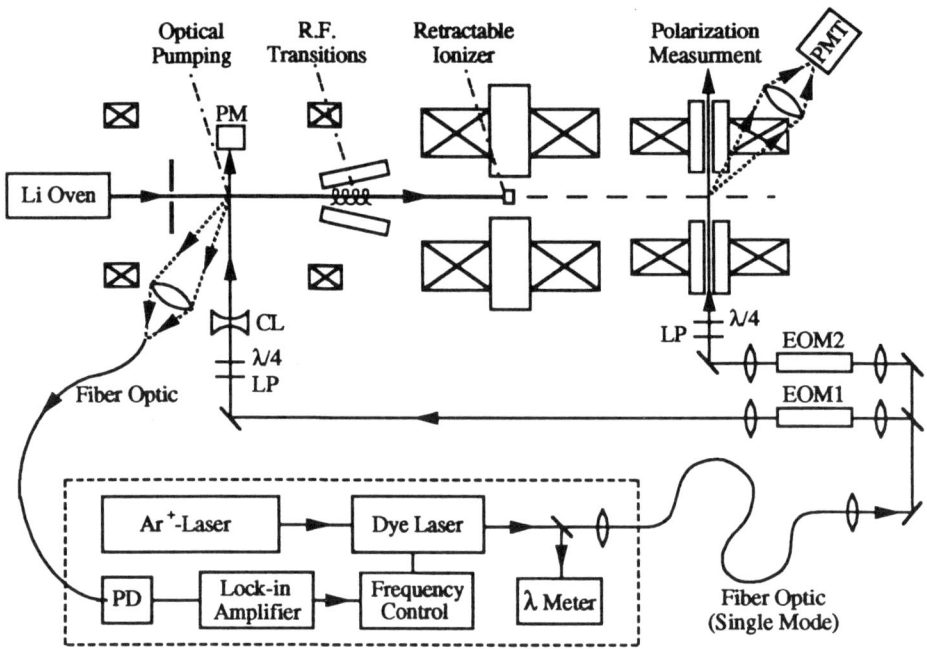

Fig. 1: Schematic of the laser system and atomic beam part of the polarized source. LP = linear polarizer, $\lambda/4$ = quarter wave plate, PMT = photomultiplier tube, CL = cylindrical lens, PD = photodiode, EOM = electro-optic modulator.

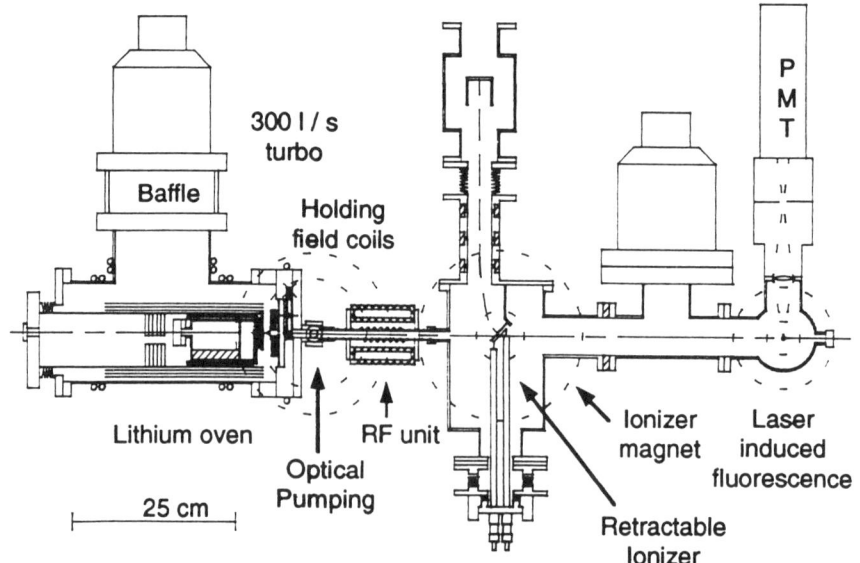

Fig. 2: Sectional drawing of the atomic beam part of the source.

horizontally through two of these and irradiates the atomic beam transversely at a distance of 7cm from the oven nozzle. The upper window permits visual observation of fluorescence from the interaction region, while fluorescent light passing through the lower window is focused by a lens into a fiber optic and is transported to a photodiode connected to the laser frequency control electronics. A "holding" magnetic field of 10 gauss, directed (anti-) parallel to the laser beam is provided by a pair of 22cm O.D. coils plus two smaller pairs of trim coils.

The optical pumping light, which is circularly polarized, is obtained from a single frequency dye laser which is tuned midway between the ground-state hyperfine components of the lithium D1 resonance line at 670.8nm, see Fig. 3. The light, however, is electro-optically modulated at half the $2S_{\frac{1}{2}}$ hyperfine splitting (114 MHz for $^6$Li, 402 MHz for $^7$Li) so that the first order sidebands coincide in frequency with the two $2S_{\frac{1}{2}}, F \rightarrow 2P_{\frac{1}{2}}$ ground state D1 hyperfine components, (the HFS of the $2P_{\frac{1}{2}}$ state is not resolved). If the light has $\sigma^+$ polarization absorption occurs with the selection rule $\Delta M_F = +1$, while the spontaneous decay ($2P_{\frac{1}{2}}$ mean lifetime = 27ns) occurs with the selection rule $\Delta M_F = 0, \pm 1$. Hence in the few microseconds the atoms take to cross the laser beam they can undergo many cycles of excitation and decay with the result that the population accumulates in the stretched state $M_F = I + \frac{1}{2}$, designated state (1) in the Breit-Rabi diagram.

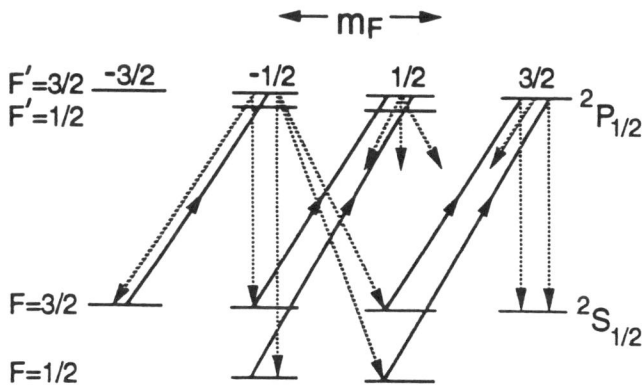

Fig. 3: Optical pumping of $^6$Li using two-frequency laser excitation. The heavy lines indicate the laser induced transitions, and the dotted lines indicate the spontaneous decays, $\Delta M_F = 0, \pm 1$. For clarity, the spontaneous decays have been completed only for the $M_F = -\frac{1}{2}$ state. Also not shown are the laser induced transitions to the $F' = \frac{1}{2}$ level, which occur because the hyperfine splitting of the $2P_{\frac{1}{2}}$ level is not resolved.

Analysis of laser induced fluorescence spectra[7] show that when optically pump-ing $^6$Li to state 1 we typically obtain nuclear polarizations of $P_z = 0.94$, $P_{zz} = 0.84$. With optical pumping to state 1 followed by an rf transition[8] to state (2) we obtain $P_z = 0.07$, $P_{zz} = -1.67$. With optical pumping to state 4 we obtain $P_z = -0.93$, $P_{zz} = 0.81$. The efficiency of the optical pumping is consistent with estimates based on numerical solution of the relevant rate equations[9].

## PRODUCTION OF Li$^-$ ION BEAM

The ion beam components of the source are based on the design of the Hei-delberg MP-tandem source[3]. With careful optimization we have obtained, for 8.2 $\mu$A extraction, 6.1 $\mu$A of focused Li$^+$ and -0.225 $\mu$A of Li$^-$, giving an overall transmission of 2.75%. More typically the ratio of Li$^-$ current to extraction is 2%, falling to 1.6% at the highest extraction currents. As regards transmission through the tandem, we have obtained 150nA of analyzed Li$^{3+}$ from a Li$^-$ source output of -200nA.

The performance of the lithium oven is variable, and the output, at a given temperature, depends on the heating history and remaining lithium charge. Stable operation for a week at 750°-800°C with 10-20 $\mu$A extraction is typical. Higher extraction currents, 30-40 $\mu$A can be obtained for several days at temperatures up to 850°C. The highest extraction current observed to date has been 57 $\mu$A with -0.9 $\mu$A of Li$^-$. The average consumption of lithium from the boiler has been found to be 3 mg per $\mu$A-hour of extraction current. This is consistent with an effusive type angular distribution from the oven nozzle.

## MEASUREMENT OF ON TARGET POLARIZATION[10]

Using the reaction $^{12}$C($^6\vec{\text{Li}},\alpha$)$^{14}$N*$(0^-,4.92$ MeV), which should have an ana-lyzing power $A_{yy}=1$ at all angles and energies, an absolute measurement of the on target tensor polarization of the $^6\vec{\text{Li}}$ beam can be obtained. For a 33 MeV $^6\vec{\text{Li}}$ beam produced by optical pumping to Breit-Rabi state 1, followed by an rf transition to state 2, the on target tensor polarization was found to be $P_{yy} = -1.327\pm0.018$, averaged over the whole seven day run. Comparison with the nuclear tensor mo-ment of the atomic beam obtained by LIF indicates a loss of tensor polarization due to the beam production, acceleration and analyzing system of 21±4%.

For the vector polarization there are no absolute standards. However, from a measured maximum in $A_yP_y$ in the elastic scattering $^4$He($^6$Li,$\alpha$)$^6$Li in a helium polarimeter we determine that $P_y > 0.74\pm0.03$. Assuming a statistical model for the m-state populations, a 21% loss in $P_{yy}$ translates into an 8% loss in $P_y$. Ap-plying this to the LIF measurement of $P_z$ implies an on target vector polarization of $P_y = 0.865\pm0.05$.

Polarized $^6$Li$^{3+}$ from the tandem Van de Graaff has been boosted by the linac, which uses superconducting resonators for acceleration and superconducting solenoids for focusing[11]. No depolarization due to the linac was observed.

## CONCLUSION AND FURTHER WORK

The optically pumped polarized Li source is now in routine operation and has been used for a number of nuclear reaction studies with $^6$Li. The observed loss of polarization between atomic beam and target is in contrast to experience with the Heidelberg source[2] and will be investigated further. An interesting technique for investigating depolarization within the source would be collinear laser induced fluorescence (spectroscopy) of the fast atomic beam leaving the charge exchange cell. Operation with $^7$Li (for which the optical pumping has already been tested[9]) will begin in the near future. The source is sufficiently compact that consideration is being given to installing an optically pumped polarized Li$^+$ source in the terminal of a small Van de Graaff.

## REFERENCES

1. For a review of Nuclear Physics with polarized heavy ions see D. Fick, G. Grawert and I.M. Turkiewicz, Physics Reports 214, 1 (1992).

2. E. Steffens, Nucl. Instr. and Meth. 184, 173 (1981).

3. D. Krämer et al., Nucl. Instr. and Meth. 220, 123 (1984).

4. H. Jänsch et al., Nucl. Instr. and Meth. A254, 7 (1987).

5. H. Reich and H.J. Jänsch, Nucl. Instr. and Meth. A288, 349 (1990).

6. E. Steffens et al., Nucl. Instr. and Meth. 143, 409 (1977).

7. H. Jänsch, E. Koch, W. Dreves and D. Fick, J. Phys. D: Appl. Phys. 17, 231 (1984).

8. W. Dreves, H. Jänsch, E. Koch and D. Fick, Phys. Rev. Lett. 50, 1759 (1983).

9. E.G. Myers, A.J. Mendez, B.G. Schmidt and K.W. Kemper, Nucl. Instr. and Meth. B56/57, 1156 (1991).

10. A.J. Mendez, E.G. Myers, K.W. Kemper, P.L. Kerr, E.L. Reber and B.G. Schmidt, Nucl. Instr. and Meth. (in press).

11. E.G. Myers, A.J. Mendez, K.W. Kemper, P.L. Kerr, E.L. Reber and B.G. Schmidt, Nucl. Instr. and Meth. (in press).

# Round Table Discussions on Optically Pumped Polarized Source

## Chairman: Y. Mori(KEK)

Before starting the discussion on OPPIS, the characteristics and performance of the Atomic Beam Sources(ABS) and the Optically Pumped Sources(OPPIS) were summarized(see Table 1 &2).   Using them, Mori(KEK) compared both types of sources based on beam brightness which was defined in the following equation.

$$B = \frac{I * f}{\varepsilon^2} .$$

(1)

Here, I is the total beam current for the polarized H⁺ beam, e the normalized emittance divided by p ( $\varepsilon = S\beta\gamma/\pi$, S: phase space area, $\beta$ and $\gamma$: relativistic constants) and f the beam fraction occupied by phase space area A.   The results are presented in Fig. 1.   Clegg(TUNL) collected the beam intensities and emittances required for each institute's accelerators and the results are summarized in Table 3.

All of the participants realized that the beam brightness was very important in comparing the performance of polarized ion source.

First of all, it was agreed to focus the round table discussion on how to generate an intense polarized negative hydrogen beam with OPPIS for intermediate and high energy accelerators.   The ultimate goal for the polarized source builder is, of course, that the polarized beam intensity should be the same as that from an ordinary unpolarized source.

## (1)Beam intensity limitation of OPPIS in the present scheme with ECR ion source.

Estimations of the intensity limit on the beam from ordinary OPPIS were presented by Mori.   The beam intensity of OPPIS is determined by the geometrical

220

acceptance of the region from the neutralizer to the ionizer and it is expressed in the following equation.

$$A = \pi \frac{(r_1 + r_2)^2}{2L}.$$

(2)

Here, A is the geometrical acceptance, $r_1$ and $r_2$ the radius of the neutralizer and the ionizer, respectively and the L the distance from the neutralizer entrance to the exit of the ionizer. In the typical configuration of the ordinary OPPIS, $r_1 = 5mm$ and L = 1m (see Fig.2). The radius of the ionizer, $r_2$, affects the output beam emittance. The output beam emittance(normalized by $\beta\gamma$) of OPPIS is determined by the radius(r) of the ionizer cell as follows.

$$\varepsilon = \pi \frac{eBr_2^2}{2m_0 c}.$$

(3)

Here, B is the magnetic field strength at the ionizer and for polarized protons, B>0.15 T. If $\varepsilon$ is set to be less than 1.5 $\pi$mm.mrad( which is acceptable for the post accelerators such as RFQ), r2 should be less than 9mm. Thus, from eq. (1), the acceptance, A, is about 90 $\pi$ mm.mrad. Before neutralization, the typical proton beam intensity extracted from the ECR ion source is about 10-20 mA and the ion temperature of the ECR plasma is about 2-3 eV, therefore, the maximally obtainable intensities for polarized H- and H+ beams from OPPIS are as follows:

polarized H- : 200-300µA
polarized H+ : 2 - 3 mA
emittance: 1.5 $\pi$mm.mrad

The present OPPIS has already achieved almost this level. In order to get more beam current, it is essential to develop a more bright proton ion source than the ECR ion source.

Zelenskii(INR) pointed out that he has already obtained a polarized H⁻ beam of 400 µA with 1πmm.mrad in his scheme, which used a pulse-operated proton ion source (see his contribution for this workshop) developed by Dimov at Novosibirsk for the neutral beam injector of nuclear fusion. He also reported that more than 1mA polarized H⁻ beam in pulsed mode operation could be obtained.    Belov(INR) mentioned that this Dimov type of proton source was very powerful and effective although its lifetime might be somewhat limited(< 1 millions shots or so).

### (2)New scheme of OPPIS with spin-exchange

The possibility of OPPIS with the spin-exchange scheme was discussed. Two types of spin-exchange scheme for making polarized hydrogen beam have been proposed so far.    They are schematically shown in Fig.3(a) and (b), respectively:

[a] Spin-exchange between fast(~keV) hydrogen atoms and optically pumped thermal alkali atoms.

[b]Spin-exchange between thermal hydrogen atoms and optically pumped alkali atoms.

Spin-exchange cross sections for various energies of fast hydrogen atoms measured at KEK were presented by Mori (see his contributed papers in the workshop) and the results showed that the cross sections were somewhat smaller than the theoretical values calculated by Swenson(LAMPF) and Anderson(Wisconsin). Zelenskii showed his recent measurements on the spin-exchange cross sections and his result was almost the same as those predicted by theory.    He also showed Swenson's calculations on the elecron-spin polarization of hydrogen atoms as a function of the target thickness of the optically pumped alkali atoms. He emphasized that this spin-exchange scheme looked very promising to obtain a very intense polarized H⁻/H⁺ beam.    Walker(Wisconsin) noted that the present theoretical calculations are based on a simple model and the difference from the present experimental results might be inevitable .    Everybody agreed that more precise measurements of the spin-exchange cross sections were very important.

Holt(Argonne) mentioned a possibility of spin-exchange scheme with thermal hydrogen atoms for applying to OPPIS.   In this scheme, one of the difficulties is the method of ionizing the polarized hydrogen atoms.    An ECR ionizer might be one of the candidates for ionization as used in ABS, however, contrary to ABS, effusive flows of hydrogen atoms are inevitable in this scheme.    Thus, depolarization caused by wall bouncing might be problem.   On the other hand, once hydrogen atoms are nuclear-spin polarized, it looks rather difficult to depolarize them by wall collisions.   Haeberli(Wisconsin) noted that from his measurements on the storage cell experiments, the depolarization due to the wall collisions was very small even after the hundreds of the wall bounces.    Mori noted that contamination of hydrogen molecules, which was about 30% in the experiment at Argonne, could be a problem for the ECR ionizer.   And he noted that a selective ionization of hydrogen atoms such as a colliding beam scheme might be very useful to generate negative hydrogen ions which has been developed by Haeberli for ABS .    Finally, this round table discussion was finished with the following very encouraging concluding remarks.

*" So far, OPPIS and ABS have been moving independently on each separated trail.   But, from now, they could be unified! "*

Table 1  Characteristics and performance of ABS.

| INSTITUTE | INTENSITY | | | | EMITTANCE ($\pi$mm.mrad,80%) | POLARIZATION | DUTY (%) | IONIZER |
|---|---|---|---|---|---|---|---|---|
| | H- | H+ | D- | D+ | | | | |
| PSI | - | 300 | - | 300 | 1.2 | 0.75(p), 0.8(d) | DC | ECR |
| TUNL | 5 | 35 | 8 | 40 | -(0.6tandem) | 0.8(p), 0.9(d) | DC | ECR |
| IUCF | - | 250 | - | - | 0.8 | - | DC | ECR |
| Sacley | - | 700,2500 | - | 700,2500 | 8, 6 | 0.9 | 0.15, 0.04 | EB |
| | | 150 | | 550 | 1.2, 1.2 | | | |
| Uppsala | - | 40 | - | 40 | 1.5 | (0.9)0.57 | DC | EB |
| COSY | 30 | - | 30 | - | (10) | 0.85 | 1.0 | CB |
| INR | 150 | 6000 | - | - | 1.6 | 0.84 | 0.2(p),0.05(H-) | PI |

EB:Electron Beam Ionizer,   CB:Colliding Beam Ionizer,   PI:Plasma Ionizer¥

Table 2   Characteristics and performance of OPPIS.

| INSTITUTE | INTENSITY | | EMITTANCE | POLARIZATION | DUTY |
|---|---|---|---|---|---|
| | H⁻ | H⁺ | ($\pi$mm.mrad,100%) | | (%) |
| INR | 400 | 4000 | 0.8 | 0.65 | <0.05 |
| KEK | 100 | ~1000 | 2.0 | 0.65 | 0.14 |
| LAMPF | 50, 2 | - | 0.8, <0.1 | 0.58, 0.77 | 10 |
| TRIUMF | ˙ 25 | - | 0.8 | 0.8 | DC |

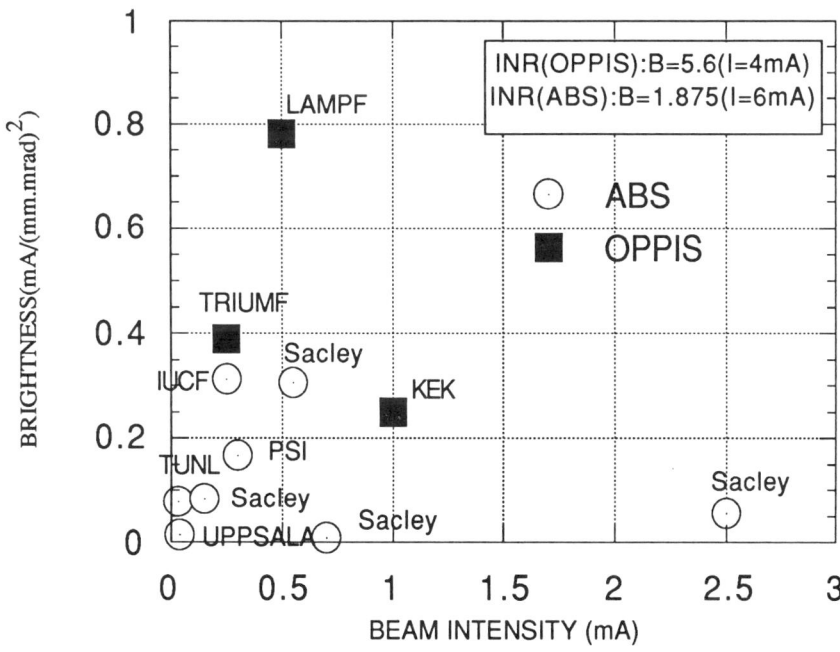

Fig. 1   Beam brightness of ABS and OPPIS.

Table 3    Beam intensities and emittances required for each institute's accelerators.

Unpolarized Beam Current Needs at Major Labs

| Laboratory | Ion Type | Current mA | Pulse Freq./Length | $\varepsilon_n$(mm·mrad) | Comment |
|---|---|---|---|---|---|
| FERMILAB | H+ | 20 – 30 | 15 Hz, 100 μs | 1.5 π | RFQ inj. |
| SSC | H- | 20 | 20 Hz, 100 μs | 1.0 π | RFQ inj. |
| KEK | H- | 10 | 20 Hz, 100 μs | 1.5 π | Inject at 750 keV |
| LAMPF | H+ | 30 | 120 Hz, 800 μs | 0.8 π | Inject at 750 keV |
|  | H- | 16 | " | " |  |
| SACLAY | H+, D+ | 10 | d.c. | 1.0 π | Inject 50 keV |
|  |  | 2 | 1 Hz, 2 ms | 0.6 π | Inject at 750 keV |
| PSI | H+ | 30 | d.c. | ? | Inject at 750 keV For spallation target |
| TRIUMF | H- | 0.3 | d.c. | 0.3 π | Inject at 300 keV |
| IUCF | H+, D+ | 0.25 | d.c. 10 Hz, 2 μsec | 0.6 π | Inject at 600 keV |
| BNL | H- | ? | 18 Hz, 500 μsec | 3 π | Inject at 35 keV |
| TUNL | H+, D+ | 0.2 | d.c. | 0.6 π | 20 – 100 keV |
|  | H-, D- | 0.03 | d.c. |  | Inject at 80 keV |
| COSY | H-, D- | 0.03 | 1 Hz, 10 ms | 10 π | Inject at 4 keV |

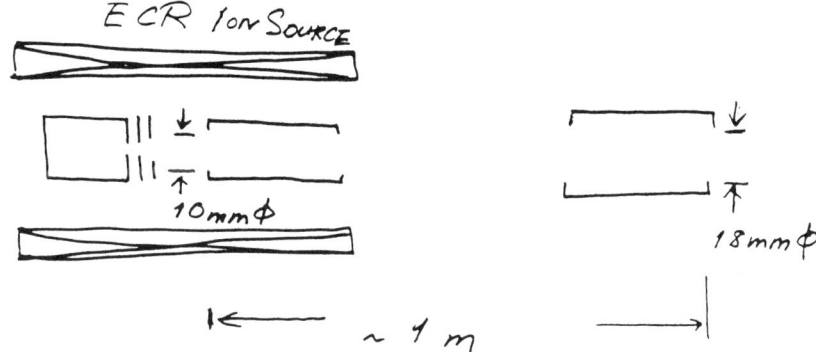

Fig. 2    Configuration of OPPIS

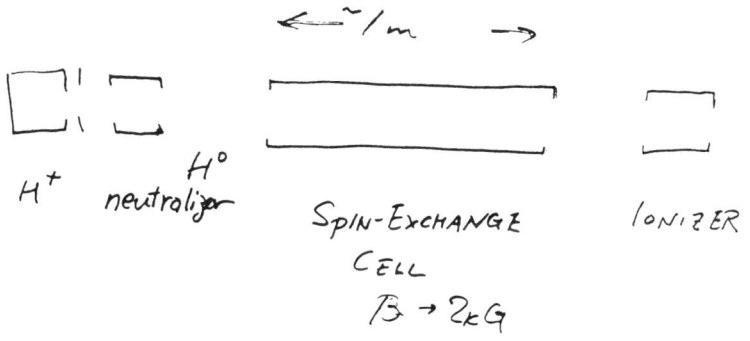

(a)

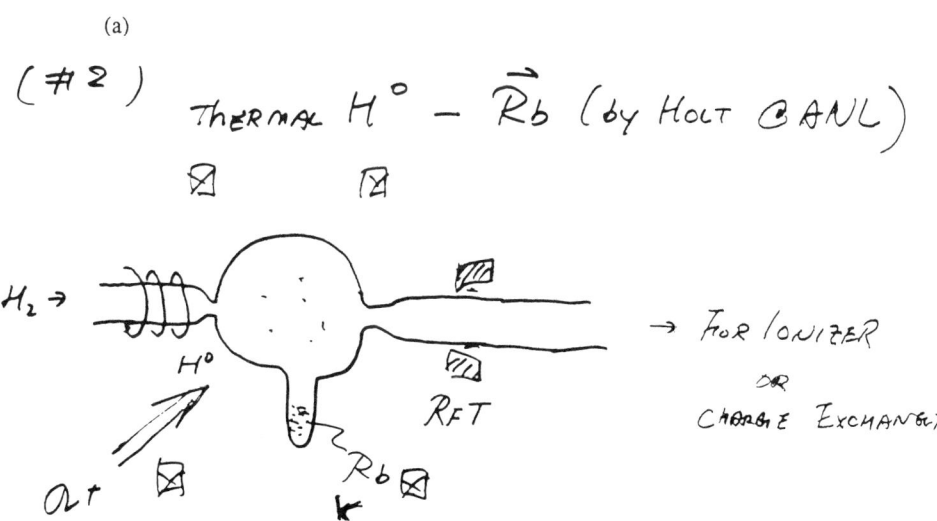

(b)

Fig.3 New scheme of OPPIS with spin-exchnage

# IV. POLARIZED $^3$He TARGETS

# OBTAINING AND MEASURING HIGH NUCLEAR POLARIZATIONS IN OPTICALLY PUMPED He-3

M. Leduc, P.J. Nacher
Laboratoire de Spectroscopie Hertzienne, 24 rue Lhomond,
75231 Paris (Laboratoire de l'E.N.S. et de l'Université, URA 18)
N. Bigelow
University of Rochester, Rochester, NY 14627
C. Larat
LCR, Thomson, Domaine de Corbeville, 91404 Orsay

\*\*\*

## MOTIVATIONS

Nuclear orientation of $^3$He gas through optical pumping of the $2^3S_1$ metastable state is a method to produce large spin polarization using high power lamp pumped LNA lasers operating at 1.08 $\mu$m on the $2^3S$-$2^3P$ transition [1]. Polarized $^3$He gas targets have applications in nuclear and high energy physics. They are also considered as possible spin polarizers for thermal and epithermal neutron beams, using the largely different scattering cross-sections of the two neutron spin states by the $^3$He nucleus [2]. All these experiments require that M, the nuclear $^3$He polarization, be both large and precisely known. In this paper we summarize the ideas underlying an accurate optical measurement of M that we recently performed [3]. We then discuss how its accuracy can be transferred to different experiments and how certain parameters influence both the reachable M values and the build-up time constants.

## ACCURATE OPTICAL MEASUREMENT OF M

Several techniques can be used to measure M. One of them is NMR, which recently gave precise values of M after a very careful check of systematic errors [4]. Our optical method is sketched in figure 1. Absorption signals by the $^3$He atoms in the $2^3S$ state are compared for 2 states of the light polarization. Such signals are 2 combinations of the populations of the magnetic sublevels in the $2^3S_1$, F=3/2 state. In most cases, one can reasonably suppose that these populations have a weighted repartition forced by the nuclear polarization of the ground state atoms through metastability exchange collisions, the dominant process of the pumping kinetics. Values of M can thus be extracted using the model of reference [5]. The sensitivity of the measurement is large at high M values. Note that it might often be more convenient to monitor M just by measuring the degree of circular polarization $\mathcal{P}$ of the 668 nm line emitted by the discharge [6]. We have used our absorption method to calibrate M/$\mathcal{P}$ as a function of pressure (see table 1).

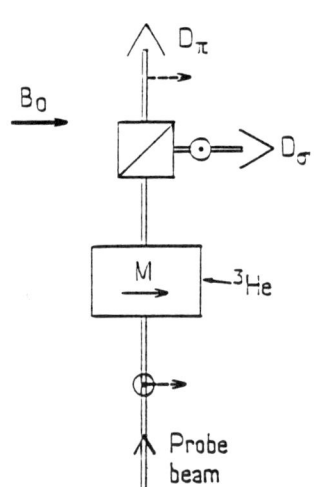

Fig. 1. The weak probe beam is linearly polarized at 45° of the plane of the figure. It is tuned to the $C_9$ transition $(2^3 S_1, F=3/2-2^3 P_0)$. A cube separates this beam into 2 crossed polarizations, $\pi$ and $\sigma$.

| $^3$He pressure (torr) | 0.15 | 0.3 | 0.4 | 0.6 | 0.8 | 1 | 1.6 | 3 | 4.5 | 6.5 |
|---|---|---|---|---|---|---|---|---|---|---|
| $M/\mathcal{P}$ | 6.00 | 6.58 | 6.95 | 7.25 | 7.96 | 8.59 | 10.02 | 11.62 | 14.0 | 17.5 |

Table 1

These $M/\mathcal{P}$ values have been corrected for several systematic errors, as discussed in [3]. They are in good agreement with those of [4]. We suggest that the best way to transfer the accuracy of this measurement to a different experiment would be to attempt to duplicate the conditions of reference [3] and use this to cross calibrate other measurements of M. Note also that the imperfections of the polarimeter optics should be measured separately (see [4] and appendix 2 in [3]).

HIGH POLARIZATION OF He-3

It is pertinent for the design of nuclear targets that both the obtained polarization be high and its build-up time be short compared to relaxation times (leakage time for internal targets, piston cycle time for compressed targets for instance). We recorded polarization signals as a function of several parameters such as discharge level (correlated with nuclear relaxation time $T_1$ in the pumping cell), pressure, quality of polarization of the pump beam and frequency of the pump laser.

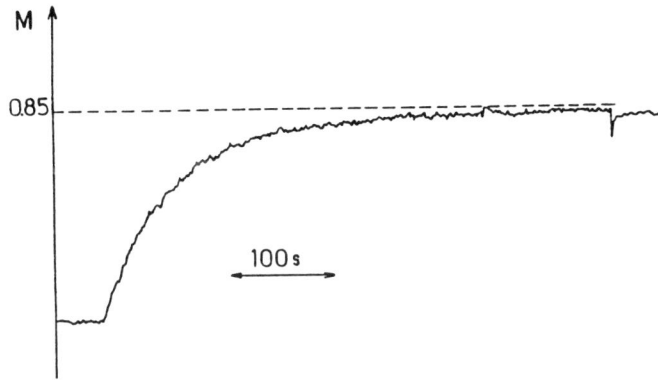

Fig.2. Build-up of the $^3$He nuclear polarization.
Pressure : 0.8 torr; relaxation time $T_1$: 8 min;
discharge frequency : 6MHz; laser power : 1.8w, tuned to
the $C_8$ transition ($2^3 S_1$, F=1/2 - $2^3 P_0$).

Figure 2 shows high M value that we recorded (M=85%). It is
important to note that the build-up of M is non exponential [7].
We thus defined an arbitrary parameter $\Delta_{0.9}$ which is the delay for
M to reach 90% of its maximum value. Figure 3 shows experimental
results obtained for M(t=∞) and $\Delta_{0.9}$ at different discharge levels
(different relaxation rates $1/T_1$). The largest M values are
obtained for the longest $T_1$ achievable, but $\Delta_{0.9}$ decreases faster
than M when the discharge increases. If fast build-up times are
required, a compromise must be found at intermediate discharge
levels.

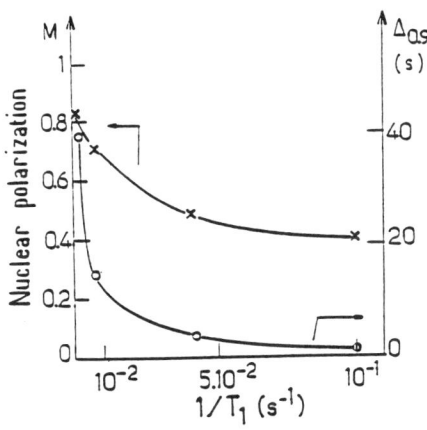

Fig. 3. Nuclear polarization
M(t=∞) and build-up time
parameter $\Delta_{0.9}$ (time for
reaching 0.9M(t=∞)) as a
function of relaxation rate
$1/T_1$ which increases with the
discharge level.

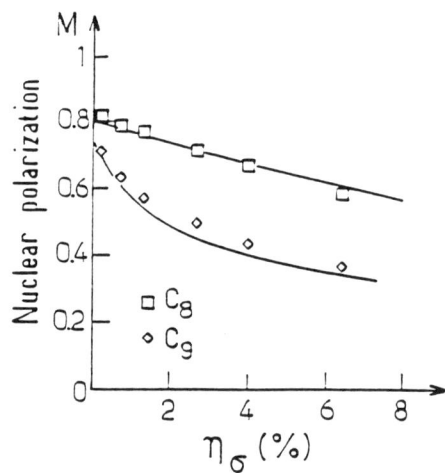

Fig. 4. Observed M(t=∞) values as a function of the admixture $\eta_s$ of $\sigma^-$ light in the $\sigma^+$ pumping beam. The laser is tuned on either $C_8$ or $C_9$ lines. The curves are calculated from [6].

We noticed, as already clearly pointed out in [2], that at high laser power the polarization can be severely reduced by a small admixture of a $\sigma^-$ polarization in the $\sigma^+$ pumping light. Figure 4 shows an experiment in which we intentionally increased the degree $\eta_\sigma$ of admixed $\sigma^-$ light : the drop out of M with $\eta_\sigma$ is larger with the $C_9$ than with the $C_8$ component [5]. Curves in figure 4 are calculated using only one adjustable parameter. Depolarization of the pumping light by the cell windows was shown to be negligible.

For cell pressures above 1 torr, the maximum obtainable M values were observed to decline, which we think is associated to the observed decrease of the relative density of metastables. Similar observations are reported in [8].

We have extended this work to the case of $^3$He-$^4$He mixtures. For the same total pressure mixtures give larger M values than pure $^3$He, both at low laser power and at high laser power and large pressure [9].

REFERENCES

[1] C.G. Aminoff et al, Rev. Phys. Appl. 24 (1989) 827.
[2] K.P. Coulter et al, Nucl. Instr. and Methods, A288 (1990) 1460.
[3] N.P. Bigelow et al, J. Phys. II France (1992) 2159.
[4] W. Lorenzon et al, Phys. Rev. A, 47 (1993) 468.
[5] P.J. Nacher and M. Leduc, Journ. de Physique 46 (1985) 2057.
[6] M. Pinard et al, Can. J. Phys. 52 (1974) 1615.
[7] C. Larat, thèse, Paris VI (1991)
[8] T.R. Gentile and R.D. Mc Keown, Phys. Rev. A 47 (1993) 456.
[9] C. Larat et al, in preparation.

# POLARIZED HE-3 INTERNAL GAS TARGETS

R.G.Milner

*Massachusetts Institute of Technology, Cambridge, MA 02139*

## ABSTRACT

A polarized $^3$He internal gas target has been developed. The target routinely operates at a polarization of 50% and a flow rate of $1 \times 10^{17}$ $^3$He atoms/sec. It has operated with efficiency close to 100% over a period of 1 year in the recently completed CE-25 experiment at the Indiana University Cyclotron Facility. Further, a cryogenically cooled thin-walled storage cell to enhance the target density is under development. A target is under construction for the HERMES experiment at DESY.

## 1. INTRODUCTION

Measurement of electromagnetic spin observables from few body systems is of great current interest. In particular, polarized $^3$He is regarded as an effective neutron target[1,2]. Recently, measurements of spin-dependent electron scattering from polarized $^3$He have been carried out at low $Q^2$ at MIT-Bates[3,4] to probe elastic form-factors of the neutron and at high $Q^2$ at SLAC[5] to probe the deep inelastic spin structure of the neutron. These experiments have been carried out with external beams of polarized electrons and closed cells of $^3$He polarized by laser optical pumping[3,6]. It has long been recognized[7] that the ability to configure a target of polarized atoms in the path of a circulating, polarized beam of a storage ring offers many advantages in the measurement of spin observables. These are:

• isotopically and chemically pure samples of polarized nuclei eliminate the necessity for kinematic deconvolution of the measured asymmetry

• absence of end windows allows scattered particle detection at very forward angles

• rapid reversal ($\sim 1$ second) of the target polarization

• intrinsically low luminosity $\sim 10^{33}$ cm$^{-2}$ s$^{-1}$ allows the use of large acceptance detectors

• a thin walled target cell allows the detection of heavy recoils, e.g. separation[8] of 2 and 3-body breakup channels in $^3\overrightarrow{\text{He}}(\overrightarrow{e}, e'p)$

Experiments have been carried out[9,10] which demonstrate the feasability of the internal target technique. At DESY the HERMES experiment[11] to measure deep inelastic spin dependent scattering from the nucleon is under construction.

## 2. THE PRINCIPLE OF THE HE-3 SOURCE

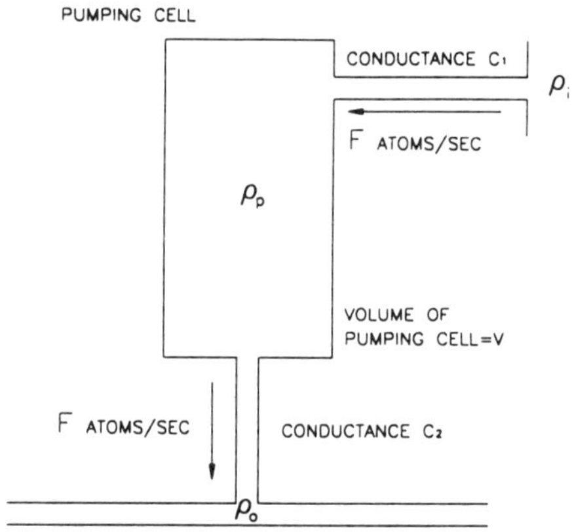

PUMPING CELL

CONDUCTANCE C₁

$\rho_i$

F ATOMS/SEC

$\rho_p$

VOLUME OF
PUMPING CELL=V

F ATOMS/SEC

CONDUCTANCE C₂

$\rho_0$

**Figure 1.** A schematic diagram of the internal target. Typically $\rho_i \approx 5$ mbar, $\rho_p \approx 0.5$ mbar, and $\rho_0 \approx 10^{-3}$ mbar.

The source was developed at the MIT-Bates Laboratory[12]. A schematic diagram is shown in Figure 1. The $^3$He atoms flow through a glass pumping cell at a rate of $F$ atoms/sec. In the pumping cell the atoms are polarized by metastability exchange optical pumping[13]. For a pump-up time ($\sim$ 10 sec) much shorter than the residence time ($\sim$ 80 sec) in the pumping cell, the atoms are polarized to an equilibrium polarization $P_0 > 50\%$. The polarization of the atoms in the pumping cell is measured by detection of the 667 nm circularly polarized light[15]. The target polarization is directed along a magnetic holding field of about 20 gauss.

## 3. THE CE-25 POLARIZED TARGET

Experiment CE-25 was a measurement of spin dependent quasielastic nucleon knockout by a stored polarized proton in the A-section of the IUCF cooler ring[14] in the energy region 200-415 MeV using the polarized $^3$He target and large acceptance non-magnetic detectors. The goal of the experiment was to study both the quasielastic knockout process and the ground state spin structure of polarized $^3$He. Figure 2 shows the internal target. The target polarization was in the vertical direction. The target operated at a polarization of $P_0 = 50\%$ and $F = 1.2 \times 10^{17}$/sec. The storage cell had dimensions of $14 \times 17 \times 400$ mm$^3$ and the resulting target thickness was $1.5 \times 10^{14}$/cm$^2$. Figure 3 shows the polarization of the atoms in the pumping cell as a function of time. The polarization was reversed every 180 sec by reversal of the circular polarization of the optical pumping light. The stored polarized proton beam intensity was up to 100 $\mu$A. At present analysis of the CE-25 quasifree data is in progress.

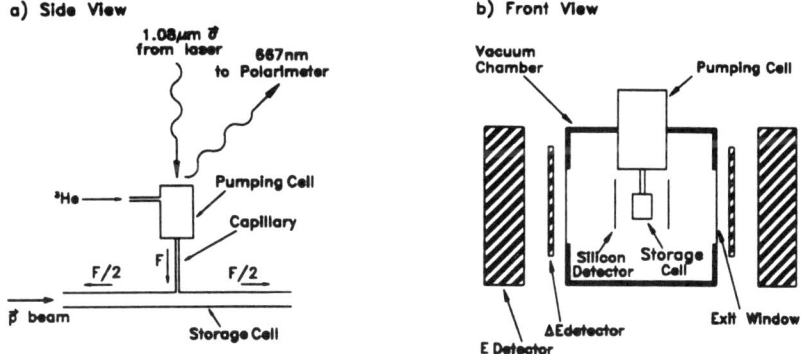

**Figure 2.** A schematic diagram of the CE-25 polarized $^3$He internal gas target.

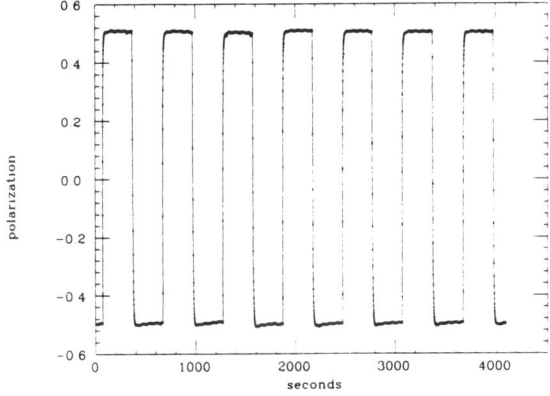

**Figure 3.** The polarization in the pumping cell as a function of time.

## 4. THE CRYOGENICALLY COOLED STORAGE CELL

The velocities of the polarized atoms in the storage cell can be decreased through cooling of the walls to temperature $T$. In this way, the target thickness will increase as $\sqrt{T}$, without any increase in the gas load to the storage ring. Previous data on depolarization of $^3$He atoms on cold surfaces indicate[16] that coatings are not required on non-magnetic cell walls for wall temperatures greater than about 15 $K$. Cooling of the elliptical cross-section HERMES storage cell (dimensions are $25 \times 10 \times 400$ mm$^3$) to 15 $K$ should yield (with $F = 2 \times 10^{17}$ atoms/sec) a target thickness of $10^{15}$ atoms/cm$^2$.

The cell is constructed from two thin sheets of ultra-pure aluminum which are spot-welded together. The cell is then clamped in an aluminum frame, through which liquid helium is passed. Ultra-pure aluminum has relatively high thermal conductivity at low temperatures and cells with wall thickness as low as 50 $\mu$m have been fabricated. A complete design has been carried out[17] and construction is in progress.

## ACKNOWLEDGEMENTS

The polarized $^3$He internal target was developed in collaboration with K. Lee, J-O. Hansen, and J.F.J. van den Brand. I would like to acknowledge the contributions of my colleagues in the CE-25 experiment at IUCF and the HERMES collaboration at DESY. J. Kelsey at MIT is responsible for the mechanical design of both the CE-25 and HERMES $^3$He targets. This research was supported by the Department of Energy under Contract No. DE-AC02-76ER03069 and by the MIT Sloan Fund. The author acknowledges a Presidential Young Investigator Award from the National Science Foundation.

## REFERENCES

1   B.Blankleider and R.M.Woloshyn, Phys. Rev. C **29**, 538 (1984)

2   R.G. Milner, Proceedings of Workshop on Polarized $^3$He Beams and Targets, Princeton NJ, October 1984, p. 186 (AIP Conference Proceedings No. 131)

3   C.E. Jones *et al.*, Phys. Rev. C **47**, 110 (1993)

4   A.K. Thompson et al., Phys. Rev. Lett. **68**, 2901 (1992)

5   P.L. Anthony *et al.*, submitted to Phys. Rev. Lett.

6   T.E. Chupp et al., Phys. Rev. **C36**, 224 (1987)

7   Proceedings of the Ninth International Symposium on High-Energy Spin Physics, Bonn, Germany, 1990, edited by K.-H. Althoff and W. Meyer (Springer-Verlag, Berlin, 1990), Vol. 1

8   J.F.J. van den Brand and R.G. Milner *et al.*, MIT-Bates proposal 89-12, 1989.

9   R. Gilman *et al.*, Phys. Rev. Lett. **65**, 1733 (1990)

10   K. Lee *et al.*, Phys. Rev. Lett. **70**, 738 (1993)

11   HERMES proposal, DESY-PRC-90-01, January 1990

12   K. Lee, J.-O. Hansen, J.F.J. van den Brand, and R.G. Milner, accepted for publication April 1993, Nucl. Instr. and Meth. A

13   F.D.Colegrove, L.D.Schearer, and G.K.Walters, Phys. Rev. **132** 2561 (1963)

14   R.E. Pollock, Ann. Rev. Nucl. Part. Sci. **41**, 357 (1991)

15   W. Lorenzon, T.R. Gentile, H. Gao, and R.D. McKeown, Phys. Rev. A **47**, 468 (1993)

16   C.E. Woodward, Ph.D. thesis Caltech 1991, unpublished

17   'Design and Fabrication Details for the Support Equipment for the Hermes Helium-3 Cryogenic System', June 30 1993, Applied Engineering Technologies Ltd., Woburn, MA

# First polarization experiments at MaMi using a dense polarized ³He gas target

**Collaboration A3:**

Institut für Physik, Universität Mainz - Ecole Normale Superieure, Paris, France - University of Rolla Missouri, USA - Institut für Kernphysik, Universität Mainz - Kelvin Laboratory, Glasgow, UK - Physikalisches Institut, Universität Tübingen

presented by W. Heil

## Introduction and physics motivation

Since 1992 longitudinally polarized electrons in the full energy range of MaMi between 180 and 855 MeV are available [1]. The first two experiments will be devoted to the measurement of the electric formfactor of the neutron $G_{E,n}$ in the range of momentum transfer $Q^2 = 5 - 15 fm^{-2}$ using $^2H(\vec{e}, e'\vec{n})$ and $^3\vec{He}(\vec{e}, e'n)$ [2]. The complete detector set-up for both experiments is shown in fig.(1) and consists of a segmented lead glass calorimeter for electron detection and opposite to it, in direction of $\vec{q}$, the neutron detector with its two scintillator walls which act as a neutron polarimeter in the case of the $^2H(\vec{e}, e'\vec{n})$ experiment. The $^3He$ target cell itself is 20 cm long with V = 100 $cm^3$ and filled with polarized gas at a pressure of 1 bar. In order to discriminate against scattering events from the cell windows ( 20$\mu$m thick), a focusing gas $\hat{C}$erenkov detector [3] is installed between the target and the electron calorimeter which images the center of the target cell ($l_{eff} = 10cm$) onto a photomultiplier. The $^3He$ targetspin $\vec{S}$ can be orientated either parallel or perpendicular to the momentum transfer $\vec{q}$. The measured quantity is then an asymmetry with respect to the sign of the electron helicity given by [4]

$$A_{exp} = P_e \cdot P_n \frac{-2\sqrt{\frac{\tau}{1+\tau}}\tan(\frac{\vartheta}{2})G_E G_M \sin(\theta^*) - 2\tau\sqrt{\frac{1}{1+\tau} + \tan^2(\frac{\vartheta}{2})}\tan(\frac{\vartheta}{2})G_M^2 \cos(\theta^*)}{\frac{G_E^2 + \tau G_M^2}{1+\tau} + 2\tau\tan^2(\frac{\vartheta}{2})G_M^2}$$

(1)

$A_{exp}$ depends on the angle $\theta^*$ between the target spin $\vec{S}$ and $\vec{q}$. With $A_{exp}$ measured in parallel ($\vec{S} \parallel \vec{q}$) and perpendicular ($\vec{S} \perp \vec{q}$) kinematic one can determine the ratio $G_E^n/G_M^n$ given by

$$\frac{G_E}{G_M} = +\sqrt{\tau + \tau(1+\tau)\tan^2(\frac{\vartheta}{2})} \cdot \frac{A_{exp}(\vec{S} \perp \vec{q})}{A_{exp}(\vec{S} \parallel \vec{q})} \quad \text{with} \quad \tau = \frac{Q^2}{4M^2}$$

(2)

DETECTOR SET-UP

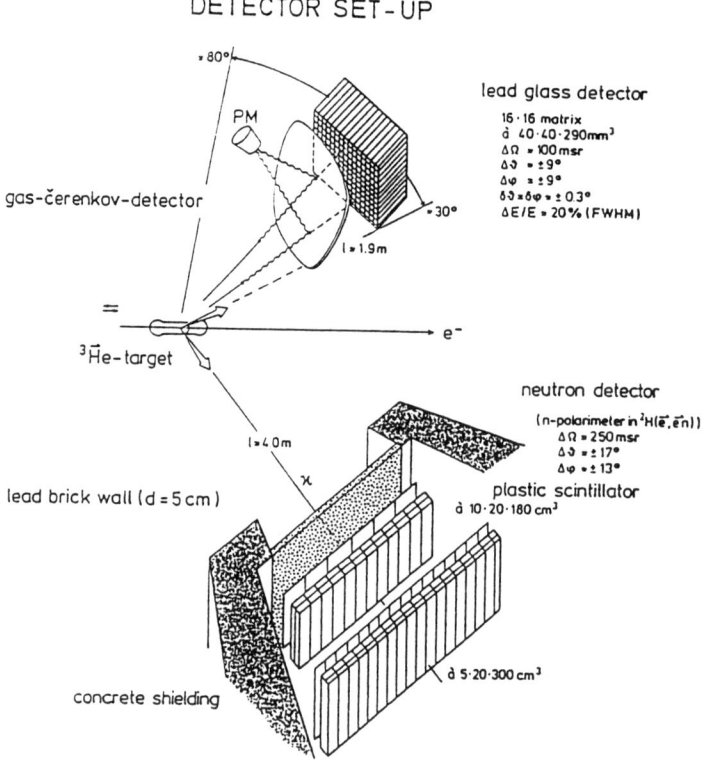

Figure 1: Sketch of the experimental set-up used for the exclusive reactions $^3\vec{H}e(\vec{e}, e'n)$ and $^2H(\vec{e}, e'\vec{n})$ .

First pilot experiments with the polarized $^3$He-target using 20% of the total detector set-up have just been finished at $Q^2 = 8$ fm$^{-2}$ ( $E_o = 855$ MeV, $\vartheta = 43°$ ).

## Dense polarized $^3$He-target

$^3$He gas is spinpolarized by metastability exchange scattering with optically pumped metastable $^3$He atoms in their $1s2s^3S_1$ state [5]. The density at which the pumping is efficient is set by the requirement of a homogeneous weak discharge in the gas and a long lifetime of the metastable atoms. This corresponds to a $^3$He pressure of about 1 Torr [6]. The short spin exchange time constant per $^3$He ground state atom in the order of $T_{ex} \approx 0.3s$ [7] in combination with the availability of intense pump lasers (LNA laser, P$\sim$ 5 W at $\lambda = 1.083\mu$m [8] ) make this pumping scheme so attractive.

In order to reach higher densities - a prerequisite for that kind of experiments - the polarized gas has to be compressed by at least 3 orders of magnitudes. We built two compression systems:

i.) a mercury filled Toepler pump reaching p$\simeq$ 1 bar [9]

ii.) a two stage titanium piston compressor which goes up to p$\simeq$13 bar.

Whereas the piston compressor is still in the testing phase, the Toepler pump is installed in the present experiment and is sketched in fig.(2) .

In the optical pumping cell (OPC) of V =$3l$ and p$\simeq$1 Torr we reach a nuclear spin polarization of $P_o = 60\%$ at a production rate of $\Gamma_{prod} = 1.2 \cdot 10^{15}$ $^3\vec{H}e/sec$ . The mercury piston is activated periodically (27 sec) to compress the polarized gas from the OPC into the target cell. Under stationary operation conditions $\Gamma_{prod}$ is equal to the flowback rate of the gas from the target cell into the OPC whereby it passes through a getter purifier. The getter guarantees a spectroscopic clean discharge in the ÓPC whereas in the target cell a relative nitrogen concentration of $[N_2]/[^3He] \simeq$ $10^{-4}$ has to be maintained in order to quench the formation of molecular ions ($^3He_2^+$) which is the dominant relaxation mode under charged particle beam conditions [10]. By means of a Helmholtz-coil configuration around the target cell ( not shown in fig.(2) ) , the target spin $\vec{S}$ can be rotated away from its direction along the $H_o$ guiding field parallel to $\vec{q}$. In that mode of operation the compression cycle has to be stopped which causes a slight decrease of the target polarization (see inset of fig.(3) ).

## Results

Fig.(3) shows the measured target polarization (NMR) under electron beam conditions at i $= 10\mu$A . The target polarization is fairly constant during 100 h of beam time and reaches 40% . The periodic structure results from the alternate settings of the target spin $\vec{S}$ ( parallel/perpendicular to $\vec{q}$) during the run . The influence of the electron beam on the target polarization can be drawn from the 'beam off' periods in fig.(3); the effect is negligible. From measurements of the wall relaxation time constant $T_1$ before and after the run ( $T_1 = 3.5h$ in both cases), we conclude that the formation of paramagnetic colour centers in the cell tube walls during exposure to the beam did not cause an increased wall relaxation rate. With an average electron beam polarization of $P_e = 30\%$ we accumulated asymmetry runs alternately in 'parallel'(10 min) and 'perpendicular'(60 min) kinematic over measuring periods of 5 hours. Fig.(4) shows the time sequence of asymmetries A($\vec{S} \perp \vec{q}$) (top) and A ($\vec{S} \parallel \vec{q}$) (bottom) where a total of 1030 $\mu$A h of charge was collected. Their weighted

# TOEPLER COMPRESSOR

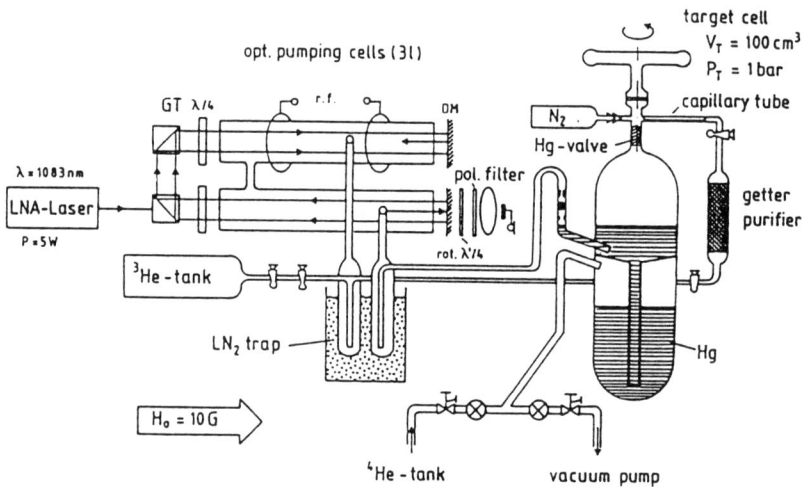

Figure 2:      Sideview of the Toepler compressor

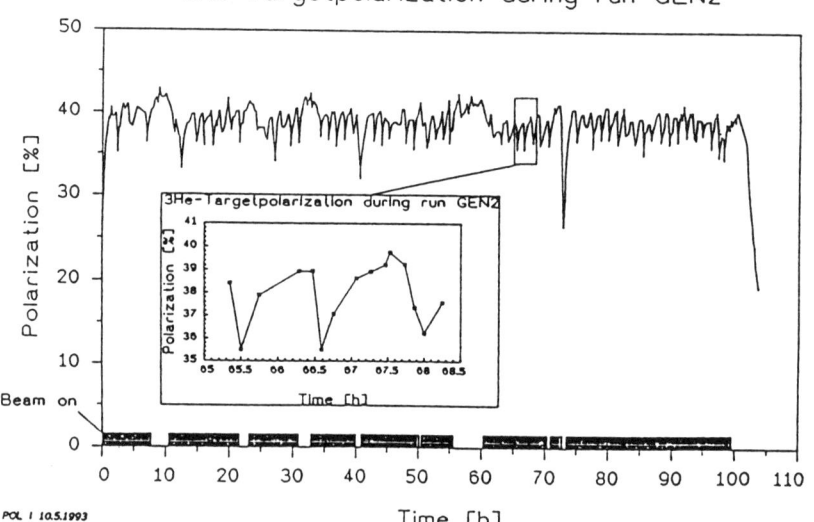

Figure 3: $^3He$ target polaization over the total measuring period of 100 h at i = 10 $\mu$A electron beam current.

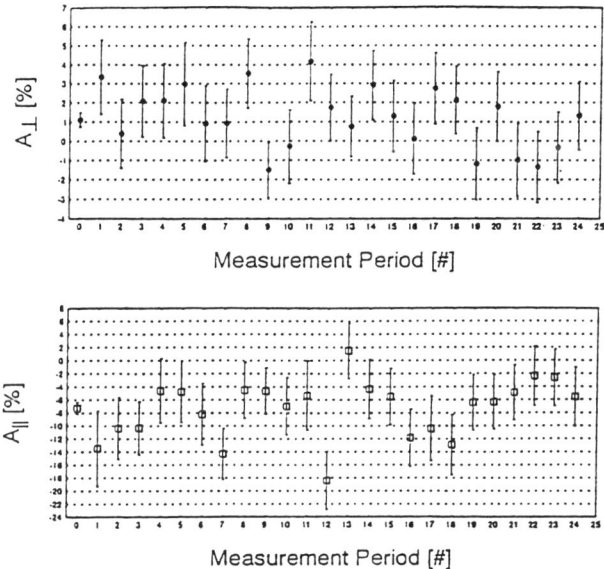

Figure 4: Time sequence of asymmetries $A(\vec{S} \perp \vec{q})$ (top) and $A(\vec{S} \parallel \vec{q})$ (bottom). Weighted averages are listened as leftmost data points.

averages amount to $A_{exp}(\vec{S} \perp \vec{q}) = (1.12 \pm 0.36)\%$ and $A_{exp}(\vec{S} \parallel \vec{q}) = (-7.33 \pm 0.88)\%$ respectively.

# References

[1] K.-H.Steffens et al., Nucl.Instr. and Meth.A325 (1993) 378

[2] A3-Collaboration Mainz, Finanzierungsantrag an die Deutsche Forschungsgemeinschaft, Sonderforschungsbereich 201 (Mainz 1989).

[3] W.Achenbach et al., Nucl.Instr.and Meth. A294 (1990) 234

[4] B.Blankleider and R.M.Woloshyn, Phys.Rev.C29 (1984) 538

[5] F.D.Colegrove, L.D.Schearer and G.K.Walters, Phys.Rev.132 (1963) 2561

[6] R.S.Timsit and J.M.Daniels, Can.J.Phys.49 (1971) 545

[7] J.Dupont-Roc, M.Leduc and F.Laloë, Phys.Rev.Lett.27 (1971) 467

[8] C.G.Aminoff et al.,Rev.Phys.Appl.24 (1989) 827

[9] G.Eckert et al., Nucl.Instr.and Meth. A320 (1992) 53

[10] K.D.Bonin et al., Phys.Rev.A37 (1988) 3270

# The SLAC High-Density $^3$He Target Polarized by Spin-Exchange Optical Pumping

H. Middleton$^c$, G.D. Cates$^c$, T.E. Chupp$^b$, B. Driehuys$^c$, E.W. Hughes$^d$, J.R. Johnson$^e$,
W. Meyer$^d$, N.R. Newbury$^c$, T. Smith$^b$, and A.K. Thompson$^a$

$^a$ Department of Physics, Harvard University, Cambridge, MA 02138
$^b$ Randall Laboratory of Physics, University of Michigan, Ann Arbor, MI 48109
$^c$ Joseph Henry Laboratories of Physics, Princeton University, Princeton, NJ 08544
$^d$ Stanford Linear Accelerator Center, Menlo Park, CA 94025
$^e$ Department of Physics, University of Wisconsin, Madison, WI 53706

Presented by H. Middleton

## ABSTRACT

A new high-density $^3$He target polarized by spin exchange with optically pumped rubidium vapor has recently been used at the Stanford Linear Accelerator in an experiment to measure the longitudinal spin-dependent structure function of the neutron. The $^3$He target operated at a density of $2.3 \times 10^{20}$ atoms/cm$^3$ in a 30 cm long scattering region with polarizations between 30% and 40% measured with NMR techniques. Target cells with several day spin-relaxation times were developed in order to achieve these polarizations.

## INTRODUCTION

Polarized $^3$He has long been recognized as an important nuclear target [1] for studying, among other things, spin-dependent neutron interactions [2,3]. This particular polarized $^3$He target was constructed for use in experiment E142 at SLAC, a measurement of neutron spin-dependent structure functions [4] which involves measuring an asymmetry in the deep inelastic scattering of polarized electrons from polarized $^3$He.

The Pauli exclusion principle provides a conceptual understanding of why polarized $^3$He may be thought of as a polarized neutron, so far as the spin is concerned. Since the two protons spend most of their time in a spatially symmetric S state, their spins must be anti-aligned to satisfy the Pauli principle. Woloshyn has made detailed calculations of the $^3$He nuclear wavefunction and found that the neutrons in a 100% polarized $^3$He sample have an 87% polarization while the protons have only a 2.7% polarization, leaving the neutron spin as the dominant contribution to spin dependent scattering [2].

A more traditional approach is to use polarized deuterium as a polarized neutron target [5,6]. The deuterium nuclear wavefunction is better understood than that of $^3$He, which reduces related systematic errors. The deuteron spin, however, is the result of about equal contributions from neutron and proton spins. Thus, the resulting understanding of the neutron is limited by one's understanding of the proton.

## POLARIZING $^3$He

There are two primary methods of polarizing $^3$He nuclei for target applications, the metastability exchange procedure first demonstrated by Colegrove, Schearer, and Walters [1] and the method of collisional spin exchange with optically pumped alkali-metal vapor introduced by Bouchiat, Carver, and Varnum for the case of $^3$He [7], and developed more generally by Happer [8,9]. Both methods have been developed for use in targets by a number of groups [10–16].

The metastability exchange process involves direct optical pumping of the $1.08\mu$m line in $^3$He. The net spin-exchange rates to the nucleus are reasonably high and the build-up of polarization occurs with time constants of a few seconds. These spin-exchange rates lead to high polarizations of 50% to 70% and make this an excellent system for internal targets in storage rings. The one drawback of this method is the need to produce the metastable $^3$He atoms in an RF discharge at pressures of a few Torr. This constraint has so far limited metastability exchange targets to a few $\times 10^{19}$ atoms/cm$^3$ in fixed targets, although cryogenic and mechanical compression techniques are continually improving the $^3$He density [14,17].

In contrast to metastability exchange, collisional spin exchange with optically pumped alkali-metal vapors can take place at high pressures without sacrificing polarization, but with much slower spin-exchange rates. Spin-exchange optical pumping is a two stage process which begins with optical depopulation pumping of an alkali-metal vapor, in our case, rubidium. The optical pumping is accomplished by illuminating Rb vapor with circularly polarized laser light tuned to the Rb $D_1$ line, which is the transition from the $5S_{1/2}$ ground state to the $5P_{1/2}$ first excited state. The result is a spin polarization of the valence electron. Under typical optical pumping conditions with optically thick Rb vapors, the Rb is nearly 100% polarized since the photon-rubidium spin-exchange rate is $\sim 10^{-6}$s compared to a depolarizing spin-destruction rate of $\sim 10^{-3}$s [18,11,12]. In principle any alkali-metal vapor can be polarized in this manner, but Rb is particularly convenient due to the commercial availability of Ti:Sapphire lasers, which provide several Watts of cw light and are easily tunable to the 795 nm Rb $D_1$ resonance.

Once the Rb vapor is polarized, that polarization is transferred to the $^3$He through spin-exchange collisions [7,19,9]. During any Rb–$^3$He binary collision there is a small probability that the wavefunction of the Rb valence electron will penetrate through the $^3$He atom's electron cloud to the $^3$He nucleus. The hyperfine interaction between the $^3$He nucleus and the Rb valence electron can then induce both species to flip their spins, thereby transferring angular momentum to the $^3$He nucleus from the electron. The cross section for this interaction is very small, $\sim 10^{-24}$cm$^2$ [19]. Consequently, the spin-exchange process is very slow. In targets, typical time constants for the build-up of a $^3$He nuclear polarization are 4 to 40 hours, even though the $^3$He is in constant contact with the $\approx 100\%$ polarized Rb vapor.

## TARGET OVERVIEW

Any spin-exchange optically-pumped polarized $^3$He target will have as its central feature the containment vessel for the Rb and $^3$He. In our case this is a 170 cm$^3$ glass cell containing $\approx 8.4$ atm of $^3$He at 20°C and $\approx 65$ Torr of nitrogen, which serves to increase the efficiency of the optical pumping. The target cell has a double chamber design [12], with the two cylindrical chambers having roughly the same volume and connected by a narrow transfer tube. The lower chamber is the target chamber through which the electron beam passes and has a 30 cm long interaction region. The upper chamber, or pumping chamber, is where the optical pumping occurs and contains a few mg of Rb metal.

The layout of the target system is shown in Fig.1. At the center is the target cell. An oven which encloses the pumping chamber of the cell is used to control the Rb vapor density for optical pumping. The Rb density [Rb] is a few times $10^{14}$ atoms/cm$^3$ in the pumping chamber (160 to 165 °C) and three orders of magnitude less in the colder, $\approx 65$ °C, target chamber. At these temperatures, the pressure in the cell is 11 atm. which corresponds to a $^3$He number density in the target chamber of $2.3 \times 10^{20}$ nuclei/cm$^3$ or a target thickness of $7 \times 10^{21}$ nuclei/cm$^2$.

Five Ti:Sapphire lasers provide the photons for the optical pumping. Each Ti:Sapphire is pumped by a 20 W argon ion laser. This system can routinely produce 20 W at the Rb $D_1$ resonance. The beams pass through focusing/expanding optics and then a quarter-wave plate before being introduced into the pumping chamber of the cell. The five beams enter the chamber through the same window and are arranged to get maximum filling of the chamber's cross section.

A set of 1.4 m Helmholtz coils provides a 20–40 G alignment field for the $^3$He nuclear polarization. This field strength is sufficient to suppress ambient magnetic field inhomogeneities but is still reasonably easy to produce. The $^3$He nuclear polarization is measured with an Adiabatic Fast Passage Nuclear Magnetic Resonance system [20]. This AFP-NMR system uses, in addition to the main field coils, a set of 18 inch Helmholtz RF drive coils and an orthogonal set of smaller pick-up coils located around the target chamber of the cell. A stronger alignment field would lead to larger NMR signals, but this field strength is an acceptable compromise between signal size and equipment costs.

Finally, all of the target equipment except the lasers and the main Helmholtz coils are located inside a vacuum chamber at a few mTorr pressure in order to reduce the background event rates from non-target materials.

## TARGET CELL DESIGN CONSTRAINTS

The primary benefit of using the spin-exchange optical pumping method is the ability of the process to work at high target densities. Ultimately, the target thickness is limited by the use of construction materials which are compatible with high polarizations, in particular, aluminosilicate glass. Tests of our glass cell design indicated a pressure limit of 13 to 15 atmospheres. The cells used during the experiment typically operated at $\approx 11$ atm.

The glass windows of the target chamber where the electron beam enters and exits the target cell are within the acceptance of the spectrometers, and therefore must be made as thin as possible in order to minimize background events. The pressure tests mentioned previously were performed on cells with 1 cm radius convex windows, $100 - 130 \, \mu$m thick over at least 2 mm diameter central region and the cells used in the experiment were also in this range. The background from these windows amounted to 64% of the total events recorded. Hersman has since reported studies on a concave window design where he found a dramatic improvement in strength over the convex design, allowing much thinner windows [21]. In addition to the windows, another 2.8% of the events are from scattering off of the nitrogen in the cell.

Much of the target development effort was directed toward producing high polarizations in the relatively large volume (170 cm$^3$) target cells. Previous work from our group has produced up to 65% polarizations in 8 cm$^3$ spherical cells at 9 atmospheres [10], and work at TRIUMF produced successful 38 cm$^3$ cells [13]. For the SLAC target, these results needed to be achieved in target cells with much larger volumes, more irregular geometries, and larger surface area to volume ratios.

## OPTIMIZING $^3$He NUCLEAR POLARIZATION

The expected $^3$He polarization, calculated from a simple analysis of spin-exchange and $^3$He nuclear relaxation rates, starting from $P_{^3\text{He}} = 0$ at $t = 0$, is

$$P_{^3\text{He}}(t) = \left( \frac{\gamma_{\text{SE}}}{\gamma_{\text{SE}} + \Gamma_{\text{R}}} \right) \langle P_{\text{Rb}} \rangle \left( 1 - e^{-(\gamma_{\text{SE}} + \Gamma_{\text{R}}) t} \right) \tag{1}$$

where $\gamma_{\text{SE}}$ is the spin-exchange rate per $^3$He atom between the Rb and $^3$He, $\Gamma_{\text{R}}$ is the relaxation rate of the $^3$He nuclear polarization through all channels other than spin exchange with Rb, and $\langle P_{\text{Rb}} \rangle$ is the average polarization of the Rb. Maximizing the $^3$He polarization therefore requires making $t$ very long, maximizing $\langle P_{\text{Rb}} \rangle$ and $\gamma_{\text{SE}}$, and minimizing $\Gamma_{\text{R}}$.

The presence of ionizing radiation such as the electron beam is depolarizing to Rb, and therefore can interfere with optical pumping. Furthermore, the radiation tends to darken the glass, reducing laser transmission into the cell. For these reasons, a double chamber cell design was used which allows continuous optical pumping in a chamber spatially separated from the electron beam. The two chambers are connected by a transfer tube, where diffusion times between the chambers are short (tens of minutes) compared to characteristic polarization build-up times, $(\gamma_{\text{SE}} + \Gamma_{\text{R}})^{-1}$. An alternative solution is to use a single chamber cell which is polarized prior to insertion into the electron beam, and then to allow the polarization to decay while taking data [10].

Since $P_{Rb}$ is $\approx 100\%$ wherever the laser light penetrates, $\langle P_{Rb} \rangle$ is maximized by carefully matching the spatial profile of the laser beam to the geometry of the pumping chamber and by adjusting the [Rb] in the chamber so that the absorption length is nearly equal to the length of the chamber. The spin exchange rate $\gamma_{SE}$ is also sensitive to [Rb], and is defined by

$$\gamma_{SE} \equiv \langle \sigma_{SE} \; v \rangle \, [\text{Rb}] \tag{2}$$

where, $\langle \sigma_{SE} \; v \rangle = 1.2 \times 10^{-19}$ cm$^3$/sec is the velocity-averaged spin-exchange cross section for Rb–$^3$He collisions [22,11] and [Rb] should be averaged over the cell. In contrast, the volume of rubidium vapor which can be fully polarized with a given laser intensity will eventually drop off as [Rb] is increased, so $\gamma_{SE}$ cannot be arbitrarily increased without eventually sacrificing $\langle P_{Rb} \rangle$. The optimum Rb density is most easily found by experimentally tuning the pumping chamber temperature to find the highest $^3$He polarization. The target typically operated with $1/\gamma_{SE} \approx 35 - 40$ hours. This is a factor of two larger than for a single chamber cell at the same temperature where all the volume, not just half, is being optically pumped. The factor of two loss in spin-exchange rate is not very important in cases where $\gamma_{SE}$ is somewhat larger than $\Gamma_R$.

## $^3$He NUCLEAR SPIN RELAXATION

One unavoidable limit to the $^3$He relaxation time constant, $\tau_R \equiv 1/\Gamma_R$ is a $^3$He–$^3$He dipolar interaction which occurs during binary collisions in the bulk gas [10]. This interaction couples the $^3$He nuclear spin to the orbital angular momentum of the two $^3$He atoms and will therefore cause depolarization. The relaxation rate $\Gamma_{bulk}$ is proportional to the $^3$He density and varies slightly with temperature, implying a maximum relaxation time constant of 100 hours at the densities and temperatures in the target cells used. This limiting lifetime will be further reduced by relaxation due to collisions with paramagnetic gaseous impurities and cell wall interactions to yield an inherent cell lifetime of

$$\frac{1}{\tau_{cell}} = \frac{1}{\tau_{bulk}} + \frac{1}{\tau_{wall}} + \frac{1}{\tau_{gas}} \tag{3}$$

In addition, there are interactions not inherent to the target cell which further increase the nuclear relaxation rate. Inhomogeneities in the magnetic alignment field induce relaxation in proportion to the diffusion constant for the cell and the square of the gradients transverse to the magnetic alignment field [23]. This effect was very small ($\tau_{\nabla B} > 500$ hours) in the high-density SLAC target, but in experiments where spectrometer magnets are close to the target, the field gradients can be much more important. Nuclear relaxation can also be induced by the presence of ionizing radiation such as an electron beam [24,25]. When a $^3$He atom is ionized, the hyperfine interaction couples the nuclear spin to the unpaired electron spin allowing a transfer of angular momentum if the two initially have opposite spin orientations. Furthermore, electrons from other $^3$He atoms can be transferred to the original ion, creating the potential for depolarizing another atom. This depolarization process continues until the ions are finally neutralized. Under the conditions in our target (high $^3$He density and an admixture of nitrogen), the number of $^3$He nuclear depolarizations per $^3$He ion created is $0.62 \pm 0.08$ according to [25]. Only at the highest currents used was there any effect on the $^3$He polarization. The relaxation time inferred by this drop in $^3$He polarization is greater than 190 hours at our maximum beam current of 3.5 $\mu$A, and the predicted time constant from [25] is 250 hours. The total $^3$He nuclear relaxation rate is given by

$$\frac{1}{\tau_R} = \frac{1}{\tau_{cell}} + \frac{1}{\tau_{beam}} + \frac{1}{\tau_{\nabla B}} \tag{4}$$

when these two external mechanisms are included.

## CELL PRODUCTION

The goal for cell development is to minimize the effect of the walls and gas impurities on the nuclear polarization. The first step in combating these problems is to construct the entire cell from aluminosilicate glass. It is postulated that the extremely small $^3$He relaxation induced by this glass is due to its very low permeability for helium. Since a $^3$He atom does not get trapped for long at the surface during a collision with the wall, interactions will be of short duration, reducing the probability of inducing nuclear spin flips. The $^3$He cells used in [10] and [26] demonstrated that relaxation due to the aluminosilicate glass can be reduced to a negligible level compared to bulk relaxation.

For the SLAC target, we needed to consistently approach the limit of $\tau_{bulk}$ in a cell of more complex geometry and larger volume. For this purpose untreated commercial tubing is completely unsatisfactory. The first step in treating the tubing is to carefully reblow all the glass that forms the target cell by resizing the tubing. For instance, the target chamber glass tubing is initially 12 mm in diameter and is expanded to 21 mm on a glass working lathe. With this procedure the entire cell wall becomes molten so that when it re-forms, it leaves a pristine surface presumably with fewer contaminants and defects. Since aluminosilicate glasses are hard to work, before adopting the resizing method, we tried simply rinsing the glass with nitric acid to remove any surface contaminates. The acid cleaning did improve $\tau_{wall}$ dramatically, increasing it from as little as tens of minutes to a few hundred hours. Unfortunately, $\tau_{cell}$ tended to decay significantly with time in the high-pressure, 5 to 10 atm cells. Low pressure (1 atm) cells do not exhibit this effect, as far as limited tests have shown.

Equally important to the construction methods is the filling process. The cells are attached to a high vacuum system ($\approx 10^{-7}$ to $10^{-8}$ Torr) and baked out under vacuum for 3 to 6 days at 475 °C (Fig.2). The Rb is then distilled into the cell with a hand held torch from a secondary chamber of the vacuum system. Next, a small amount of nitrogen (99.9995% pure) is frozen into the cell. Finally, the initially 99.995% chemically pure $^3$He is introduced into the cell through a trap at liquid $^4$He temperature. This cryogenic trap further purifies the $^3$He by condensing out any contaminants. The cell is also cooled with liquid $^4$He in order to get a high density of $^3$He in the cell while maintaining a pressure of less than an atmosphere. This step is necessary since the cell is permanently sealed by melting closed a constriction in the glass tube where the cell attaches to the vacuum system. Similar cell filling procedures were used for experiments at TRIUMF [13,27] and at LAMPF [10,26].

When this combination of construction technique and filling procedure was followed, out of ten cells produced, all five of those measured had nuclear polarization lifetimes in excess of 30 hours. Three of the five cells, including all those used in the experiment, had measured lifetimes of 50 to 65 hours at room temperature. These numbers, compared to the 95 hour limit of $\tau_{bulk}$ at 20°C, imply that most of the relaxation is caused by the unavoidable $^3$He–$^3$He dipole interaction, although some improvement in $\tau_{cell}$ is still possible. Use of this procedure should ensure production of nearly bulk-limited lifetimes in target cells of any arbitrary geometry and volume so long as extreme care is taken to ensure the cell surfaces are all freshly worked glass, the bakeout is meticulous, the vacuum is good, and filling gases are well cleaned.

The net cell relaxation time is estimated to be about 70 hours when the cell is hot. From (4) and (1), we then find that the $^3$He polarization is approximately 57% to 64% of $\langle P_{Rb} \rangle$, for maximum and minimum electron beam current, respectively. One of the most important limitations is a small $\gamma_{SE}$ caused by the low Rb number densities at our operating temperature of 160 to 165°C. There was sufficient laser power to run at even higher [Rb], but problems with oven materials overheating limited the temperature. Redesigning the oven with high temperature plastics should therefore allow higher [Rb], and a faster $\gamma_{SE}$. With this simple improvement, a doubling of the spin-exchange rate may be possible, leading to a factor of 1.25 improvement in $P_{^3He}$.

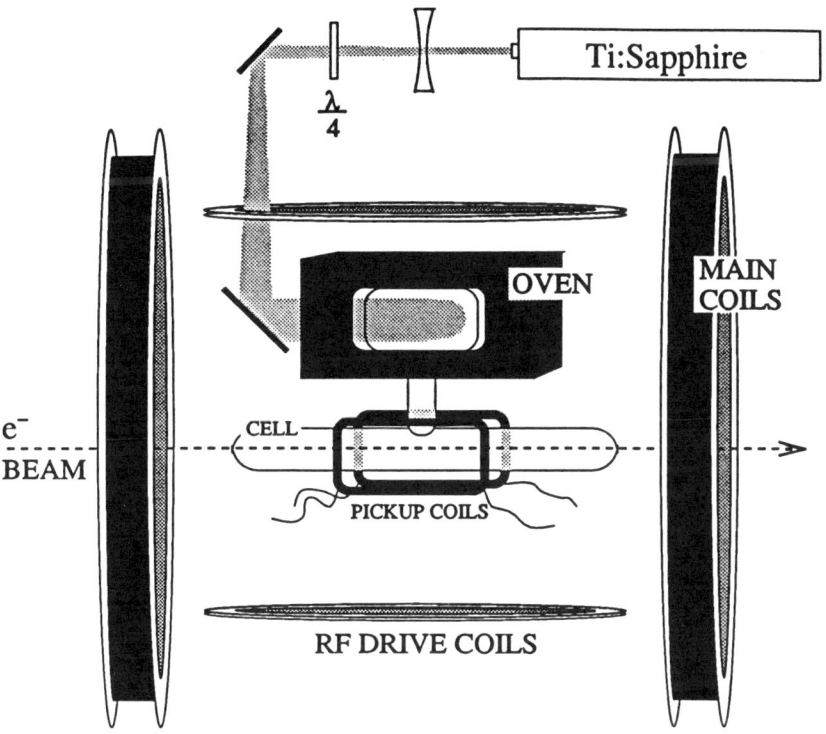

Fig. 1 Diagram of target system.

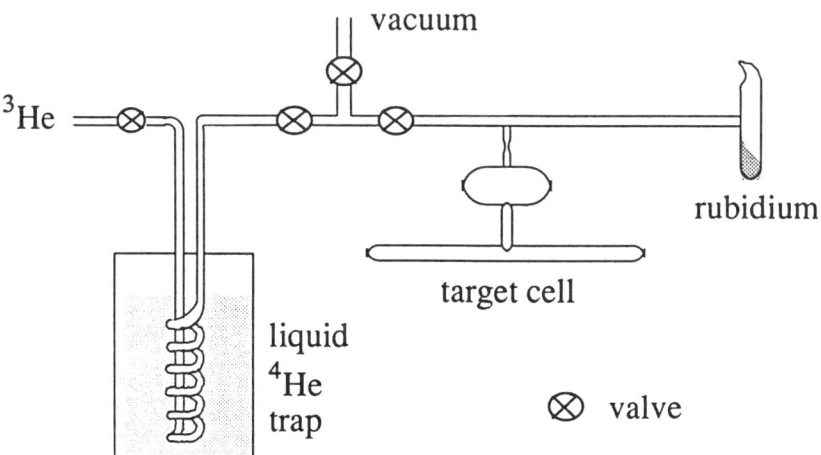

Fig. 2 Schematic of vacuum system.

## I. NMR POLARIMETRY

The polarization measurements were made with the Nuclear Magnetic Resonance technique of Adiabatic Fast Passage (AFP) [20]. The AFP system uses a set of 18 inch Helmholtz drive coils to provide a 92 kHz RF field while the main alignment field is swept through the 29 G $^3$He resonance. At resonance all the $^3$He nuclear spins flip over inducing a signal in an orthogonal set of smaller pick-up coils located around the center of the target chamber of the cell. One possible concern is that the local polarization in the beam may be different from the target chamber's volume-average polarization measured by this system. However, with diffusion times on the order of seconds compared to tens of hours for relaxation and spin-exchange times, it is not possible to maintain any significant polarization gradient in the cell.

The AFP-NMR signal is calibrated with the known thermal equilibrium Boltzmann polarization of the protons in a water sample of the same dimensions. The value of the polarization is easily calculable from Boltzmann statistics and is given by

$$P_{proton} = \tanh\left(\frac{h\nu}{2k_B T}\right) \tag{5}$$

where B=22 G for protons in resonance with the 92 kHz RF. Typical proton signals are $\approx$ 1.8$\mu$V, and the resulting calibration for $^3$He signals is 1.61 $\pm$ .11 % polarization per 10 mV of signal. Figure 3 shows a $^3$He signal and an average of 25 proton signals.

We are also in the process of developing an alternate polarimetry method which involves measuring the shift in frequency of the Rb ESR line due to the magnetic field produced by the polarized $^3$He [10,22]. This technique depends on an atomic calibration constant which at present is known to $\approx$ 3%.

## II. TARGET PERFORMANCE

During the experimental run, the $^3$He polarization was measured every four hours. The results of these measurements, which were taken from November 7 to December 22, 1992, are shown in Fig. 4. The average $^3$He polarization over the entire run is about 36%. During the first three weeks of the experiment, there were a few precipitous drops in the polarization. These problems were caused by materials overheating in the pumping chamber oven, leading to mechanical failure of vacuum seals. The use of high temperature plastics in constructing the oven would easily correct this deficiency. Once further oven problems were avoided by operating at a lower temperature, the target polarization became very stable, running for three weeks with only slow drifts. Toward the end, the drop off in polarization due to an increased beam current is noticeable.

Overall the target required very little maintenance. The laser systems ran for days before requiring brief tuning, collectively producing 16 to 22 Watts. Target helicity reversals were easily done by rotating the alignment field and reversing the laser helicity, requiring only 10 minutes to complete. No other system required any significant attention during the latter half of the run.

Overall, the SLAC target was highly successful, and considering the conservative approach taken in designing this target, there is room for improvement. The target operated at a $^3$He density of 2.3 $\times$ 10$^{20}$ atoms/cm$^3$ in a 30 cm long scattering region. The target thickness of 7 $\times$ 10$^{21}$ atoms/cm$^2$ is the highest yet achieved in a polarized $^3$He target. This high density provides good statistics in spite of a moderately large background event rate, which accounted for about two thirds of the events. Ten to twenty percent increases in $^3$He density along with at least a factor of two reduction in background rates from using thinner windows should be realized in future target cells. The long $^3$He spin-relaxation times of the target cells, estimated to be 70 hours at 65$^\circ$C, led to an average polarization of 36% and a maximum of 42%. The possibility exists for small improvements in target cell lifetimes and a factor of two

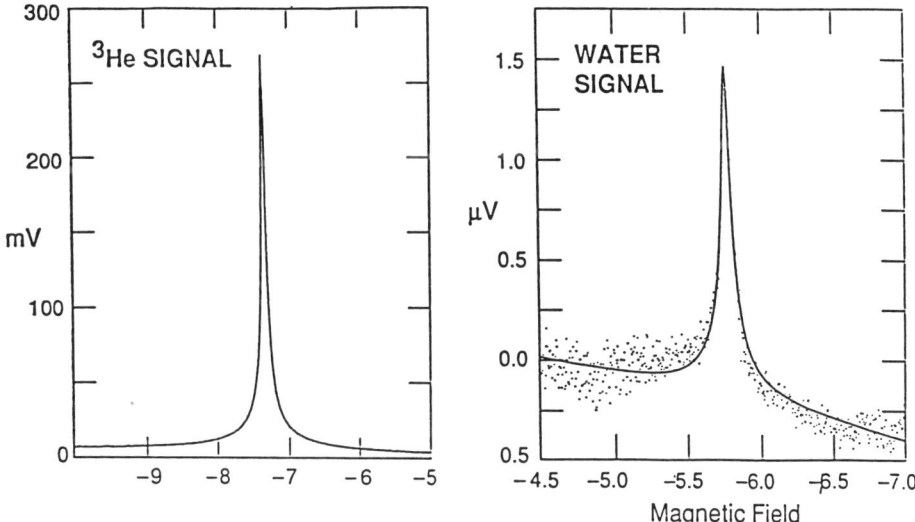

Fig. 3 NMR signals from He cell and water sample.

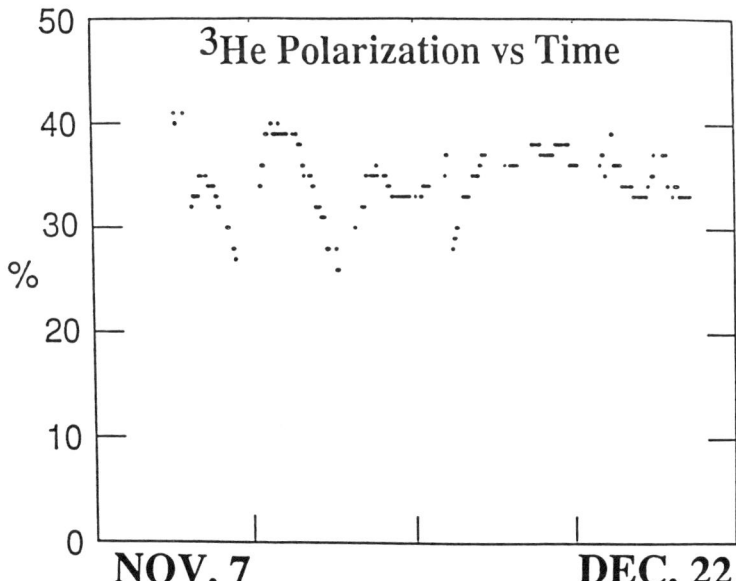

Fig. 4 Evolution of the target polarization during the six week
experimental run.

in spin-exchange time. This combination could bring the $^3$He polarizations well into the 50% range. Even without these improvements, the performance of this $^3$He target was excellent and was central to conducting the most sensitive nuclear spin-dependent structure function measurement to date.

---

[1] F.D. Colegrove, L.D. Schearer and G.K. Walters, Phys. Rev. **132**, 2561 (1963).

[2] B. Blankleilder and R.M. Woloshyn, Phys. Rev. **C29**, 538, (1984); R.M. Woloshyn, Nucl. Phys. **496A**,749 (1989).

[3] R.G. Milner, Proc. Workshop on Polarized $^3$He beams and Targets, ed. R.W. Dunford and F.P. Calaprice, AIP Conf. Proc. **131**, (AIP, New York 1985), p. 186.

[4] P.L. Anthony *et al.*, accepted for publication, Phys. Rev. Lett. **71**, 1993.

[5] V.W. Hughes and J. Kuti, Ann. Rev. Nuc. Sci. **33**, 611 (1983).

[6] B. Adeva *et al.*, Phys. Lett. B **302**, 533 (1993).

[7] M.A. Bouchiat, T.R. Carver and C.M. Varnum, Phys. Rev. Lett. **5**, 373 (1960).

[8] N.D. Bhaskar, W. Happer, and T. McClelland, Phys. Rev. Lett. **49**, 25 (1982).

[9] W. Happer, E. Miron, S. Schaefer, D. Schreiber, W.A. van Wijngaarden, and X. Zeng, Phys. Rev. A **29**, 3092 (1984); X. Zeng, Z. Wu, T. Call, E. Miron, D. Schreiber, and W. Happer, Phys. Rev. A **31**, 260 (1985).

[10] N.R. Newbury *et al.*, Phys. Rev. Lett. **67**, 3219 (1991);N.R. Newbury *et al.*, Phys. Rev. Lett. **69**, 391 (1992).

[11] T.E. Chupp, M.E. Wagshul, K.P. Coulter, A.B. McDonald, and W. Happer, Phys. Rev. C **36**, 2244 (1987).

[12] T.E. Chupp, R.A. Loveman, A.K. Thompson, A.M. Bernstein, and D.R. Tieger, Phys. Rev. C **45** 915 (1992).

[13] B. Larson *et al.* Phys. Rev. Lett. **67**, 3356 (1991).

[14] W. Heil, this proceedings.

[15] M. Leduc *et al.*, Nucl. Sci. Appl. 1, 1 (1983); C.L. Bohler *et al.*, J. Appl. Phys. **63**, 2497 (1988).

[16] R.G. Milner, R.D. McKeown, and C.E. Woodward, Nuc. Inst. Meth. in Phys. Res. **A 274**, 56 (1989); K. Lee *et al.*, Phys. Rev. Lett. **70**, 738 (1993).

[17] L.D. Schearer, private communication.

[18] N.D. Bhaskar, M. Hou, B. Souleman, and W. Happer, Phys. Rev. Lett. **43**, 519 (1979);R.J. Knize, Phys. Rev. A **40**, 6219 (1989).

[19] R.L. Gamblin and T.R. Carver, Phys. Rev. **138**, A946 (1965).

[20] A. Abragam, Principles of Nuclear Magnetism (Oxford University Press, New York, 1961).

[21] W. Hersman, these proceedings.

[22] N.R. Newbury, A.S. Barton, P. Bogorad, G.D. Cates, M. Gatzke, H. Mabuchi, and B. Saam, Phys. Rev. A **48**, 558 (1993); A.S. Barton, N.R. Newbury, G.D. Cates, B. Driehuys, H. Middleton, B. Saam, submitted to Phys. Rev. A, (May 1993).

[23] G.D. Cates, S.R. Schaefer and W. Happer, Phys. Rev. A **37**, 2877 (1988); G.D. Cates, D.J. White, Ting-Ray Chien, S.R. Schaefer and W. Happer, Phys. Rev. A **38**, 5092 (1988).

[24] K.D. Bonin, T.G. Walker, and W. Happer, Phys. Rev. A **37**, 3270 (1988).

[25] K.P. Coulter, A.B. McDonald, G.D. Cates, W. Happer, T.E. Chupp, Nuc. Inst. Meth. in Phys. Res. **A276**, 29 (1989).

[26] N.R. Newbury, A.S. Barton, G.D. Cates, W. Happer, and H. Middleton, submitted to Phys. Rev. A (December, 1992).

[27] B. Larson, O. Häusser, P.P.J. Delheij, D.M. Whittal, and D. Thiessen, Phys. Rev. A **44**, 3108 (1991).

# $^3$HE TARGETS POLARIZED BY SPIN-EXCHANGE OPTICAL PUMPING AT TRIUMF

O. Häusser[1,2], B. Larson[3], E.J. Brash[1], W.J. Cummings[1],
W. Lorenzon[1], and P.P.J. Delheij[2]

[1]Simon Fraser University, Burnaby, B.C., Canada V5A 1S6
[2]TRIUMF, Vancouver, B.C., Canada V6T 2A3
[3]Ohio University, Athens, OH 45701, U.S.A.

## INTRODUCTION

Although $^3$He targets have been polarized by spin-exchange optical pumping for about 30 years [1,2] their use in nuclear and particle physics experiments has been very limited until recently because of low $^3$He density, or low polarization, or both. This situation has changed dramatically with the advent of Ti:sapphire lasers which can produce many Watts of infrared photons suitable for optical pumping of K/Rb/Cs. Other crucial steps were the use of small amounts (50-100 Torr) of nitrogen gas to avoid radiation trapping at high vapor densities [3], the preparation of aluminosilicate glass vessels with very low concentrations of contaminants at the cell walls, and the development of compression methods to achieve high $^3$He densities. At TRIUMF we [4] have produced polarized $^3$He targets of high density (up to 12 atm at 273 K) high polarization (typically 60%), and large volume (initially 17 cm$^3$, later 35 cm$^3$, more recently 70 cm$^3$). Since early 1990 these targets have been used successfully in experiments to measure spin observables for elastic scattering of intermediate energy protons [5], pions [6], and for proton-induced nucleon knockout [7,8]. Charged particle tracking with wire chambers has been used extensively to eliminate background from the walls of the glass vessel. For the weak secondary pion beams at TRIUMF tracking of beam and scattered pions allows full vertex reconstruction. For a pion scattering experiment at LAMPF (E1267) the high pion fluxes and the low duty cycle prohibit the measurement of individual beam pions. Instead, beam collimation and larger diameter (4.5 cm) targets are used to avoid background from the glass side walls. The target volume for 10 atm cells is limited to about 70 cm$^3$ by the available laser power of about 8 Watts at the Rb D1 wavelength (795 nm).

The target polarization is measured by adiabatic fast passage (AFP) NMR, with the NMR signal calibrated by a comparison with signals from a water-filled cell of the same dimension. The $^3$He polarizations measured with the NMR method are in good agreement with results from the $^3$He(p,$\pi^+$)$^4$He reaction for which the identity $A_{nn}=1$ has been exploited to determine the target polarization. A schematic view of the setup used typically for hadronic experiments at TRIUMF is shown in Fig. 1. In the following we describe briefly some of the more recent results and developments at TRIUMF.

## DIPOLAR AND WALL RELAXATION AT 293 K

For targets of the spin–exchange type, the $^3$He polarization attains the asymptotic value $P = P_{Rb}\gamma_{SE}/(\gamma_{SE} + T_1^{-1})$, where $P_{Rb}$ is the average Rb polarization, $\gamma_{SE} = <\sigma_{SE}v>[Rb]$ is the spin exchange rate, and $T_1$ is the longitudinal spin relaxation

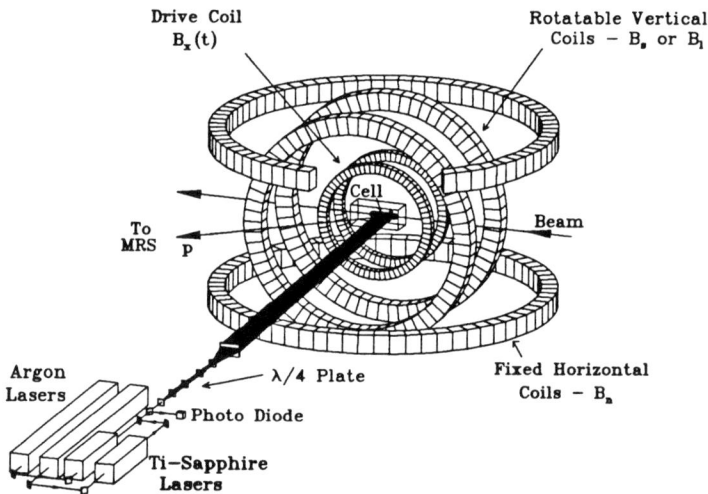

Fig. 1. Schematic view of the polarized ³He target at TRIUMF. The two large coil pairs are part of the adiabatic spin rotator.

time. The relaxation rate

$$T_1^{-1} = T_{1B}^{-1} + T_{1W}^{-1} \approx C_1 \rho + \frac{C_2}{\rho}$$

has contributions from the bulk $(T_{1B})$ and from the walls $(T_{1W})$ which are expected to scale very differently with ³He density $\rho$. The intrinsic dipolar relaxation time $T_{1B}$ depends on orientation dependent molecular forces [9] and sets an ultimate upper limit on the maximum ³He polarization obtainable at very high pressures. We have measured $T_1$ at 293 K, for longitudinal fields of 3 $mT$, and at different ³He pressures. We find that, for a given cell, $T_1$ can sometimes be lengthened by curing the cell at temperatures around 450 K. If we select, from a total of more than 50 cells, those having the longest relaxation times, we find an upper limit for the rate constant $C_1$ by ignoring the wall relaxation contribution in the above equation. Our upper limit (see Table 1), $C_1 \leq 3 \times 10^{-3}$ cm³ g⁻¹ s⁻¹, is about five times smaller than values determined by Chapman [10] at low temperatures using NMR techniques.

Table I $T_1$ relaxation times for ³He gas at 293 K

| volume (cm³) | pressure (atm at 273 K) | $T_1$ (hours) | $(T_1\rho)^{-1}$ ($10^{-3}$ cm³ g⁻¹ s⁻¹) |
|---|---|---|---|
| 2.9 | 3.1 | 220±30 | 3.0±0.4 |
| 70 | 6.09 | 112±12 | 3.0±0.3 |
| 17 | 6.44 | 95±15 | 3.4±0.5 |
| 17 | 8.97 | 74±10 | 3.1±0.4 |
| 35 | 10.5 | 58±8 | 3.4±0.4 |

### AN ADIABATC SPIN ROTATOR FOR POLARIZED $^3$HE TARGETS

The construction of an adiabatic spin rotator (ASR) was motivated by two considerations. First, we expected the polarization losses per spin reversal, known for AFP NMR to be anywhere between 0.1-3% depending on the magnetic environment, to decrease. Because of the rather long time constants for Rb-$^3$He spin exchange (8 hours at Rb concentrations around $4 \times 10^{14}$ cm$^{-3}$) these polarization losses limited the frequency of spin flips during experiments thus increasing the possibility of systematic errors for the target-related spin observables. Second, in nucleon scattering experiments a total of five independent spin correlation parameters are non-zero with the target polarized in the normal, longitudinal or sideways directions.

The ASR [11], shown schematically in Fig. 1, consists of two pairs of orthogonal Helmholtz coils. The first fixed pair provides a vertical magnetic field, whereas the second pair is rotatable and produces a magnetic field in any desired horizontal direction. With suitable controls of the excitation currents of the coil pairs the magnetic field can be rotated adiabatically into any direction in space. For a target cell of known $T_1$ we have performed about 20,000 adiabatic spin reversals, sandwiched between AFP NMR runs. After corrections for $T_1$ and AFP NMR losses the polarization loss per adiabatic spin reversal was measured to be $(9 \pm 3) \times 10^{-6}$. The finite losses, although small, may be caused by nonadiabatic steps in the coil currents.

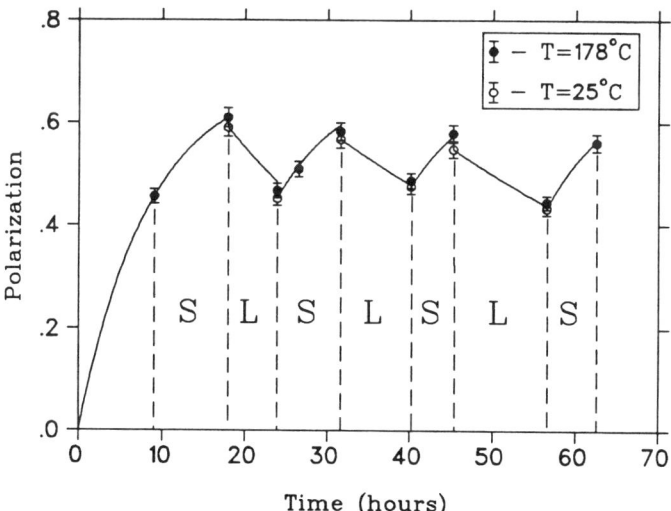

Fig. 2. Polarization cycle used with the adiabatic spin rotator to measure sideways ($S$) and longitudinal ($L$) target observables.

For measurements of elastic proton scattering with longitudinal target polarization we have devised a polarization cycle (see Fig. 2) which required optical pumping in the sideways direction only. Sideways polarization data were taken during optical pumping. After adiabatic rotation of the B-field into the longitudinal direction data were accumulated during ~8-hour periods with the lasers switched off. The target spin was reversed with the ASR every 15 minutes. The target polarization, measured with AFP NMR while the target polarization was in the normal direction, varied between 65% and 45% during the course of the experiment.

## A GAS SCINTILLATION TPC TO MEASURE ASYMMETRIES FROM OPTICALLY PUMPED POLARIZED MUONIC ³HE

Direct spin exchange has been shown [12] to be effective in polarizing muonic ³He at a time scale comparable to the muon lifetime ($\sim 2\mu s$). The triton asymmetry in the $\nu_\mu + t$ muon capture channel can be shown [13] to be very sensitive to the induced pseudoscalar coupling of the semileptonic weak interaction. In an attempt to measure the triton asymmetry relative to the direction of the incident, circularly polarized laser light we have developed a gas scintillation time projection chamber (GSTPC) using UV photons (300-400 nm) emitted by $N_2$ molecules. In addition to observing prompt UV light produced by stopping $\mu^-$ and tritons, electrons are drifted parallel to the laser beam direction into a region of high electric field (light gap) where many secondary UV photons are produced. Using transient digitizers for the secondary photons the origin and direction of the tracks can be deduced. We have carried out extensive studies of the UV yield for different He-$N_2$ mixtures and shown that $N_2$ concentrations compatible with optical pumping of Rb yield useable light signals. We have also obtained stable operation of the GSTPC at elevated temperatures (up to 450 K) necessary to produce sufficiently high Rb densities. In a test run at 293 K we have so far identified tracks from the incoming stopping $\mu^-$ and, separated in time, from the emitted tritons (branching ratio 0.3%). The full experiment will require operation of the GSTPC at 450 K in the presence of Rb vapor, and a fast deflection system for the pump laser.

This work is supported by NSERC (Canada).

## REFERENCES

1. M.A. Bouchiat, T.R. Carver, and C.M. Varnum, Phys. Rev. Lett. 5, 373 (1960).
2. R.L. Gamblin and T.R. Carver, Phys. Rev. 138, 964 (1965).
3. T.E. Chupp et al., Phys. Rev. C36, 2244 (1987).
4. B. Larson et al., Phys. Rev. A44, 3108 (1991).
5. O. Häusser et al., J. Phys. (Paris) Colloq. 51, C6-99 (1990); and to be published.
6. B. Larson et al., Phys. Rev. Lett. 67, 3356 (1991); and to be published.
7. A. Rahav et al., Phys. Lett. B275, 259 (1991); Phys. Rev. C46, 1167 (1992).
8. E.J. Brash et al., Phys. Rev. C47, 2064 (1993).
9. B. Shizgal, J. Chem. Phys. 58, 3424 (1973).
10. R. Chapman and M.G. Richards, Phys. Rev. Lett. 33, 18 (1974).
11. W.J. Cummings et al., to be published.
12. A. Barton et al., Phys. Rev. Lett. 70, 758 (1993).
13. TRIUMF experiment 683, W.J. Cummings and O. Häusser, spokesmen.

# Installation of a polarized $^3$He target in CLAS

F. W. Hersman

Department of Physics, University of New Hampshire, Durham, NH 03824

The CEBAF Large Acceptance Spectrometer (CLAS) provides extensive capability for performing exclusive measurements. It provides momentum analysis of charged particles, trajectory tracking, vertex reconstruction, calorimetry, neutral detection, and Čerenkov and time-of-flight particle identification. The toroidal magnet geometry allows an independent magnetic field at the beam axis for polarized targets. The electron beam of up to 4 GeV energy is available with 75% polarization for currents of up to 1 $\mu$A.

I discuss a new plan for installing a polarized $^3$He target in the CLAS (Fig. 1) for exclusive measurements.[1] (An existing program to install a polarized metastability exchange $^3$He target and emphasizing inclusive measurements [2] is discussed elsewhere.) The key features of the new plan are:

1. Cancellation of field gradients produced by the CLAS inner toroid,

2. Fabrication of a thin, surface-wound magnet for establishing an arbitrary quantization axis,

3. Use of a high density spin-exchange target for minimal sensitivity to remaining magnetic field gradients,

4. Development of thin target windows to eliminate their dominance of the data rate,

5. Achieving maximum CLAS luminosity with the low current, high polarization beam permitted by the high density target.

The CLAS inner toroid sweeps Møller electrons from the vertex drift chamber. The location of the conductors as close as 17 cm from the beam axis leads to gradients within the volume of the pumping cell. Calculations show that a single winding with current flowing in the opposite direction located 4-5 cm closer to the target cancels the leading term in the gradient (Fig. 2).

The standard use of Helmholtz coils to provide the target quantization axis can not be adopted for CLAS without substantial modifications. We have designed a surface wound magnet with a solenoidal winding and a cosine-theta dipolar winding. Independent currents in the two coils will provide a uniform target field in an arbitrary direction. An orthogonal dipolar winding will provide RF perpendicular to the main field for adiabatic fast passage.

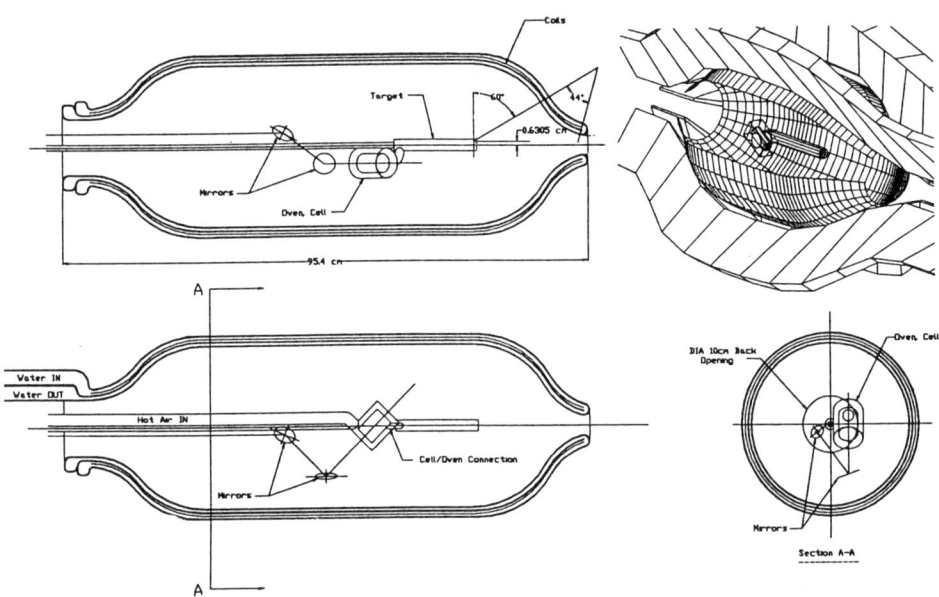

**Fig. 1.** Views of the spin exchange ³He target installed in the CLAS inner toroid, including the surface wound magnet, the target and pumping cells, the optics path, and utilities.

Choice of a high density target offers the best possibility for low sensitivity to magnetic field gradients.[3] The rubidium spin exchange target has operated at pressures of up to 10 bar.[4] A 15 cm long target achieves the limiting CLAS luminosity of $10^{34}$ electrons-nucleons/cm²sec at a beam current of 100 nA. At this current the high polarization strained crystal source offers 75% polarization.

We have developed a glass cell configuration that allows use of thin windows for beam entry and exit from the target cell. It is based on a concave spherical window supported by a hoop on a thicker convex cell. Since the thin window is in compression the full yield modulus of aluminosilicate glass ($\sigma_y$=1000 MPa) can be used in the formula for pressure vessels: $\Delta P_{\max} = 4\sigma_y(t/d)$, where $t/d$ is the ratio of window thickness to diameter. A prototype window 6 mm in diameter with $t$=15 $\mu$m has held 5 bar pressure without rupture (Fig. 3). Although scattering from these windows is significant, it does not dominate the luminosity, and can be removed by tracking in the vertex detector.

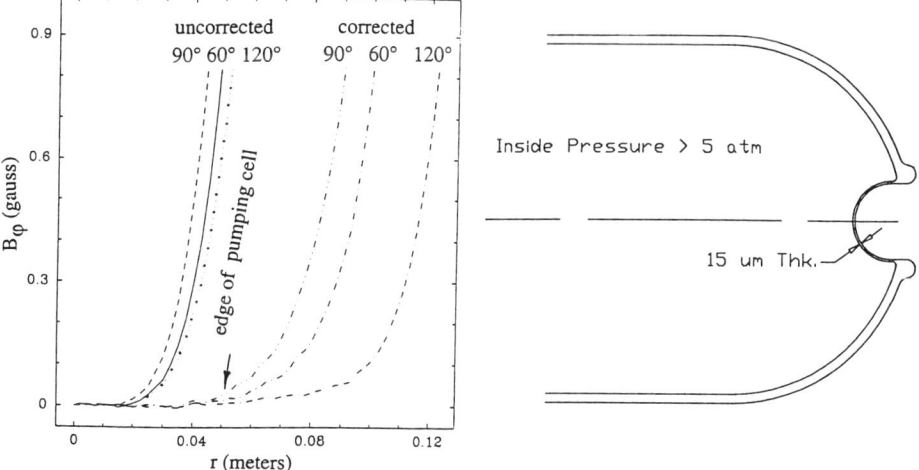

**Fig. 2.**    The azimuthal magnetic field component with and without the cancellation loop.

**Fig. 3.**    A thin aluminosilicate glass window has been successfully blown and tested at UNH.

1. CEBAF Proposal 93-037, F. W. Hersman. Timothy P. Smith, co-spokesmen.

2. CEBAF Proposal 89-007 and updates, R. McKeown, spokesman.

3. T. Chupp, *et al.*, *Phys. Rev.* **C45** (1992) 915, and references therein.

4. H. Middleton, this conference.

## OPTICAL PUMPING OF Helium-3
### Roundtable summary presented by W. Heil

Large samples of highly polarized $^3$He are demanded at present in three fields of physics, namely:

1.  Studying quantum phenomena in the gaseous and fluid phase at low temperature.

2.  Polarizing thermal and epithermal neutron beams by the large spin-dependent n-absorption cross section.

3.  Investigating spin-dependent neutron scattering cross sections, e.g. of the type $^3\overrightarrow{He}$ ($\overrightarrow{e}$, e'n), exploiting the fact that a polarized $^3$He nucleus represents to a good approximation a sample of two paired unpolarized protons and a single polarized neutron.

The polarization of $^3$He may be achieved in different ways, namely: by spin-exchange with a optically pumped Rb-vapour and by meta-stability exchange scattering with optically pumped metastable $^3$He-atoms in their 1s 2s $^3S_1$-state.

Whereas the physics of both methods has been well studied and understood for decades, their technical performance has improved dramatically in recent years mainly due to the development and application of powerful pumping lasers for the desired wavelengths (795nm for Rb-pumping, 1083nm for $^3$He*-pumping).

In the case of $^3$He*-pumping the latest breakthrough came with the development of the so-called LNA-laser (Pioneering work was done by Michele Leduc, ENS Paris).

Its latest version consists of a 79mm long LNA-rod pumped transversely by Kr-arc-lamps in commercial Nd:YAG-laser head. Tuned to resonance and narrowed down to the Doppler width of about 2 GHz by internal etalons it delivers up to 5 W output-power. When circularly polarized, this light beam represents a pumping source of $\dot{L} \approx 10^{20}$ quanta of angular momentum/s.

In the case of the Rb-pumping similar numbers are achieved with the new Ti-sapphire laser, e.g. G. Gates reported from 20 W output power at $\lambda$ =795 nm achieved at SLAC using 5 Ar$^+$-lasers (100 W) as pumping source

for the Ti-sapphire laser. This "brute force" method
certainly is a costly business. Considerably more
advantageous for some applications may be a system
fully based on solid state lasers, e.g. Nd:YAG laser
with high pulse repetition rate $\nu \geq 10$ kHz and conversion into second harmonics at 532 nm as a pump source
for a Ti-sapphire laser (recently reported by Sobel-
mann, Lebedev Physical Institute).

Of utmost importance for the construction of efficient
target- or storage cells of polarized $^3$He gas is the $T_1$
relaxation time constant. H. Middleton reported that at
SLAC wall relaxation time constants between 40 and 100
h can now routinely be obtained, using Corning 1720
alumino-silica-glass cells. A big progress as compared
to earlier results which showed considerably bigger
fluctuations on $T_1$. The influence of a charged particle
beam on the target polarization is negligible at high
pressures, provided that a relatively low concentration
of e.g. $N_2$ impurity gas is maintained in the target
cell. A possible reduction of $T_1$ due to the formation
of paramagnetic colour centers in the cell tube walls
under beam conditions could not be observed.

An important question concerns the efficiency $\eta$ with
which the photon's angular momentum can be transferred
to the number of $^3$He-atoms in the target. $\eta$ differs
significantly for the two pumping schemes mentioned,
since the spin exchange time constants per $^3$He ground
state atom is in the order of $T_{3He - 3He*} \sim 0,3$ s and
$T_{Rb - He} \sim 6000$ s respectively.

The latter, long time constant forms a bottleneck in
the Rb-pumping scheme because it sets a lower limit to
the build-up time of the polarization in the target.

The exploitation of the much faster spin transfer rate
in the $^3$He*-pumping, on the other hand, is handicapped
by the fact that this method functions only in a low
pressure discharge at a level of $p * 1$ Torr. The density
of the polarized gas has to be increased by mechanical
compression, a method which is followed up by the Mainz
group. W. Heil reported from a titanium piston compressor designed for a final pressure around 13 bar.

At present both methods meet the desired order of magnitude in the figure of merit of the polarized $^3$He gas
target.

The last topic dealt with the limitations intrinsic to
the storage of a highly dense polarized $^3$He gas. Due to

dipolar $^3$He-$^3$He-spin-spin interaction one gets a relaxation rate of $\Gamma \sim 1/100$ h at $p \approx 10$ bar. Typical build-up times of the target polarization of $T_p = 30 - 40$ h in case of the Rb-pumping (cell volume $\approx 100$ cm$^3$) lead therefore to a significant reduction of the maximum achievable target polarization.

These limits can certainly be raised , using the $^3$He$^*$-pumping scheme with subsequent compression of the polarized gas. Here one profits from the much higher $^3$He production rates resulting in a faster build-up time of the target polarization (Tp $\sim$ 1-2 h).

# V. MACHINE–TARGET INTERACTIONS

# A STUDY OF STORAGE-CELL INDUCED BACKGROUND

C.W. de Jager
Nationaal Instituut voor Kernfysica en Hoge-Energiefysica (NIKHEF-K),
P.O. Box 41882, 1009 DB  Amsterdam, The Netherlands

## INTRODUCTION

A broad physics program with internal targets has been initiated with the stretcher/storage ring AmPS at NIKHEF-K. The structure of the deuteron ground-state wavefunction is a central question in intermediate energy nuclear physics. Spin-dependent scattering experiments allow access to new and crucial observables through asymmetry measurements which depend on the deuteron orientation parameters. In a first series of experiments[1] electron-induced proton knock-out from polarized deuterium will be studied. In these experiments an atomic beam of deuterium is injected in an open-ended storage cell. Reaction products (electrons and protons) from quasi-free scattering will be measured in coincidence with large-acceptance non-magnetic particle detectors which allows the kinematically complete determination of tensor analyzing powers[2] as a function of missing momentum and energy. The information obtained on the deuteron spin structure will also be of importance for experiments that use polarized deuterium for the study of fundamental aspects of the spin and charge structure of the nucleon. In 1994 a magnetic detector for electrons will be added and finally, in 1995, a stored beam of longitudinally polarized electrons will be made available. In this paper first results are presented of studies of the effect of such a storage cell on beam properties and background.

## INSTRUMENTATION

At NIKHEF-K the existing 500 MeV linear electron accelerator has been upgraded and a stretcher/storage ring (AmPS)[3] added in which an intense (up to 200 mA) beam of up to 900 MeV electrons can be stored. Electrons can be injected into the AmPS ring in three different ways:
1. single-turn on-axis. The stored intensity is limited to the peak current accelerated (80 mA max.), but on injection the betatron oscillations are small so that a fixed storage cell can be used.
2. three-turn off-axis. Now the stored intensity is triple the accelerated current, but the betatron oscillations at the interaction point (IP) are so large that the storage cell has to be opened until the beam has damped.
3. stacking. In order to separate the circulating beam from the injected beam at the injection septum a local bump has to be created on the closed orbit. This again causes such large betatron oscillations at the IP that the storage cell has to be opened during injection. On the other hand, many weak pulses can be injected so that electrons can be accelerated close to the zero-load energy of the accelerator. Stored intensities of up to 200 mA can be achieved.

Polarized deuterium nuclei will be injected into an open-ended storage cell out of an Atomic Beam Source (ABS). The ABS will be operated in a strong ($\approx$ 300 G) field regime, so that maximum tensor polarization values of +1 and -2 can be reached. A flux of 1.6 x $10^{16}$ atoms/s of polarized deuterium into a 15 mm feed tube has been measured. The scattering chamber, containing the storage cell, consists of an aluminium rectangular box with the feed-through to the ABS in the horizontal plane. At the top and bottom thin (0.1 mm stainless steel) windows give passage to the knocked-out proton and the scattered electron, respectively. The complete scattering chamber can be moved vertically over 60 mm in order to allow unrestricted passage of the beam in stretcher mode. Both up- and downstream of the chamber two conductance limiters are mounted.

Very few experimental data[4] exist on the spatial beam distribution of a stored electron beam. After damping the intensity is expected to follow a Gaussian distribution with σ-values of the order of 1 mm or less. On the other hand, disturbing processes, such as rest-gas scattering or possible wakefield effects, will cause the intensity to drop off slower in the tails of the distribution. The contribution of such a halo determines the minimum diameter of a storage cell. Since the effective target thickness of the gas stored in the cell is inversely proportional to the cube of the cell diameter, it is essential to study the beam halo.

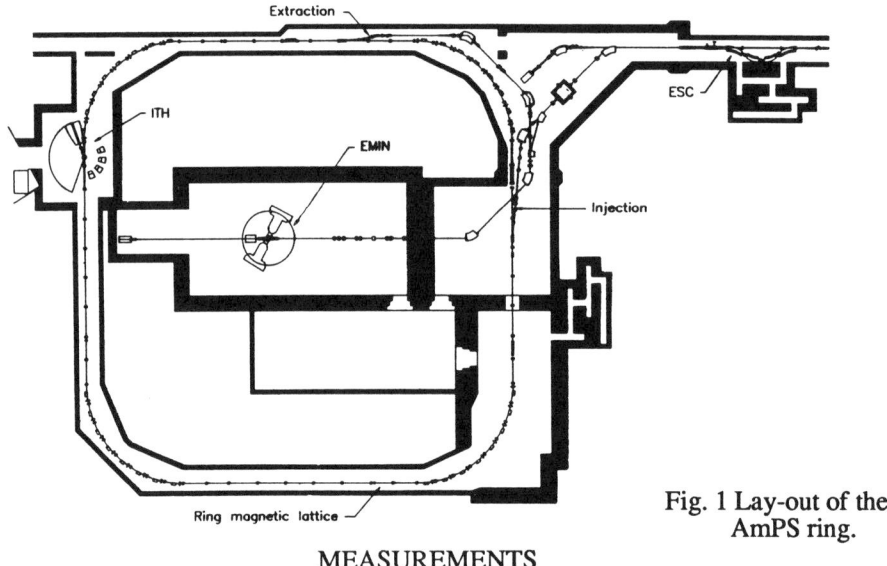

Fig. 1 Lay-out of the AmPS ring.

MEASUREMENTS

Several test measurements have been performed with a stored electron beam at the AmPS-ring. The goal of these tests was to obtain information on the spatial distribution of the stored beam, so that the diameter of the storage cell could be fixed.

At NIKHEF a study of the beam intensity distribution and of the effect of a storage cell was initiated. A set of four independently movable tungsten (9.5 mm thick) scrapers was installed at the IP. Two scintillator telescopes, each consisting of two 2 mm thick plastic scintillators, in the horizontal (vertical) plane at a scattering angle of 20° (38°) and a solid angle of 45 (20) msr, were directed at the IP. The beam intensity was monitored every second with a parametric current transformer. The stored electron beam with an energy of 410 MeV had a lifetime of appr. 100 s.

The countrate was measured in each telescope as a function of the position of the various scrapers of the collimator. In fig. 2 the countrates are shown as a function of the position of all four scrapers. All measurements showed the same behavior, a constant background rate until a position is reached at which the countrate starts to increase drastically until the photomultipliers had to be switched off due to the high rate. The countrate data show that the beam intensity distribution has very long tails, extending over 20 times the predicted value of the gaussian-sigma of the beam. The data also show that the rate is a factor of 10 higher in the horizontal than in the vertical plane This probably confirms that synchrotron radiation losses form a major contribution to the beam lifetime.

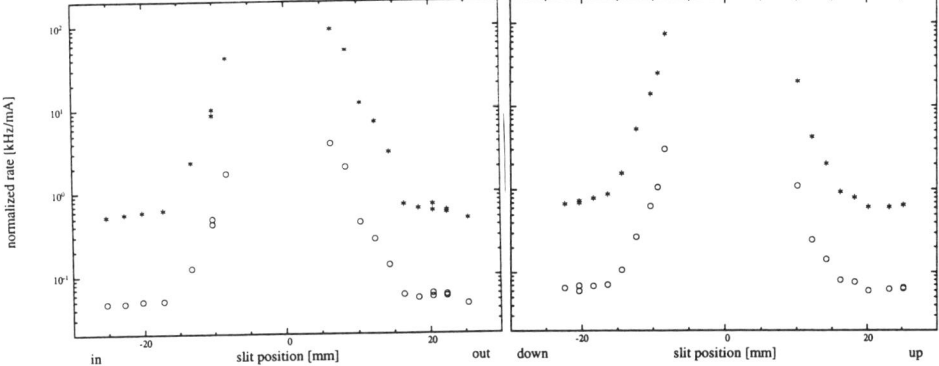

Fig. 2. Count rate [kHz/mA] in the horizontal (stars) and the vertical (circles) scintillator telescope as a function of the horizontal (left figure) and vertical (right figure) scraper position from the centre of the beam.

The beam lifetime was also measured as a function of the position of the scrapers (fig. 3). It can be clearly seen that in the domain where the scintillators were able to measure the countrate, there was no effect whatsoever on the lifetime of the beam. When one scraper starts to intercept a considerable fraction of the beam, the lifetime of the beam starts to drop, until beam can no longer be stored. Assuming a gaussian beam distribution and assuming that a constant fraction of the beam is cut way by the scraper on each revolution, the width of the beam can be calculated for all values of the lifetime. However, since this calculation yields an increasing value of sigma for increasing lifetimes, either the tails of the distribution are non-gaussian or the beam losses can not be described by such a simple model. The average value obtained for the beam width ($\approx$ 0.5 mm) is in agreement with predictions, but the large extension in the vertical plane ($\approx$ 0.8 mm) indicates a strong coupling between horizontal and vertical phase speaces, which is unexpected.

In the second part of the test measurements an empty storage cell, 200 mm long, 25 mm in diameter, made out of 0.1 mm thick aluminium, was installed just downstream of the IP. The beam conditions were the same as for the previous tests.

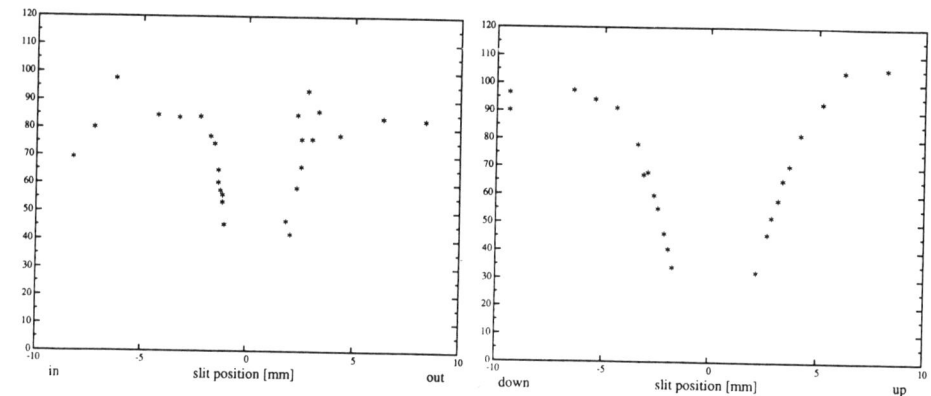

Fig. 3   Beam lifetime [s] as a function of the horizontal (left figure) and vertical (right figure) scraper position from the centre of the beam.

With the local orbit correction the transverse phase space was mapped to find where the beam was stored, respectively stopped by the cell. Even with this small cell it was possible to operate the scintillators, whereas the high voltage of the PM-tubes had to be switched off when the thick scrapers were at 10 mm from the center of the beam. The lifetime of the beam is hardly affected by the cell. The countrates, inside the phase space where beam could be stored,were only a factor of 3 to 5 higher than without the cell.

These studies indicate that internal target experiments at the AmPS ring will certainly be feasible with a storage cell 20 mm in diameter. Further improvements are expected by installing a collimator at a suitably chosen position in the ring.

## ACKNOWLEDGEMENTS

The support of the NIKHEF-K technical staff is gratefully acknowledged. This work is part of the research program of the Nationaal Instituut voor Kernfysica en Hoge-Energie Fysica (NIKHEF-K), supported by the Stichting voor Fundamenteel Onderzoek der Materie (FOM).

## REFERENCES

1. NIKHEF-K proposal 91-12: R. Alarcon, H. Arenhövel, M. Bouwhuis, J.F.J. van den Brand, M. Bucholz, H.J. Bulten, S. Choi, J. Comfort, R. Ent, M. Ferro-Luzzi, W. Haeberli, C.W. de Jager, J. Konijn, J. Lang, W. Leidemann, M. Miller, D. Nikolenko, E. Passchier, S.G. Popov, P. Quinn, I. Rachek, J.J.M. Steijger, O. Unal, H. de Vries, Z.-L. Zhou.

2. H. Arenhövel, W. Leidemann and E. Tomusiak, Phys. Rev. **C46** (1992) 456; W. Fabian and H. Arenhövel, Nucl. Phys. **A314** (1979) 253.

3. G. Luijckx et al, Proc. of the European Particle Accelerator Conference, Nice, 1990.

4. S.G. Popov, 1992, private communication.

# INTERACTIONS BETWEEN A POLARIZED $^3$He TARGET, THE IUCF COOLER AND AN EXPERIMENT

J. Sowinski
Indiana University Cyclotron Facility and Department of Physics
Bloomington, IN 47405

## ABSTRACT

An internal polarized $^3$He target has been in use in the IUCF Cooler Ring with polarized proton beams for about one year. The primary goal has been to investigate the nucleon spin structure of $^3$He via quasi-free scattering. This paper reports on techniques and experience gained in making these first spin correlation measurements using an internal target. We demonstrate that experiments can be performed with these new techniques, with good counting rates and low backgrounds.

## INTRODUCTION

The idea of using a storage cell to increase the thickness of a polarized atomic beam to make an internal target for a storage ring has been around for a long time[1] and has been the topic of a number of workshops such as this one. Even though a demonstration experiment was performed[2] more than 10 years ago, only a very few measurements have been done to this point with internal polarized targets. It is clear from this conference that this technology is leaving the development stage and a number of experiments will be underway shortly.

For the past year we have been performing experiments in the IUCF Cooler Ring with polarized proton beams and a polarized $^3$He internal target developed at MIT.[3] The first spin correlation measurements made with an internal target have been published[4] as a demonstration of the technique. The main goal of this effort is to measure spin observables in quasi-free scattering from $^3$He in order to test models which describe how the spin of $^3$He is carried by the nucleons.[5] Successful production runs for this project were completed recently. The purpose of this talk is to relate some of the details and experience gained in these measurements. The title of this talk is drawn from the fact that the experiment, target and storage ring become an integrated system much more than in conventional techniques and, in particular, I will address some aspects of how they interact. I will also demonstrate that experiments can be performed with these new techniques, with good counting rates and low backgrounds.

## TARGET DESIGN

The optical pumping and performance of the $^3$He target are described elsewhere in these proceedings.[6] The target design must be matched to the vacuum conditions in the ring and the beam dimensions. In addition care must be taken to avoid unnecessary experimental backgrounds from the target structures.

The vacuum requirements in storage rings are much more severe than when traditional nuclear physics techniques are employed. At the IUCF Cooler the ring vacuum is typically $10^{-9}$ Torr. The target chamber must reach this vacuum when not in use so as to not interfere with ring operation for other experiments, hence a careful choice of materials and good vacuum practices are required.

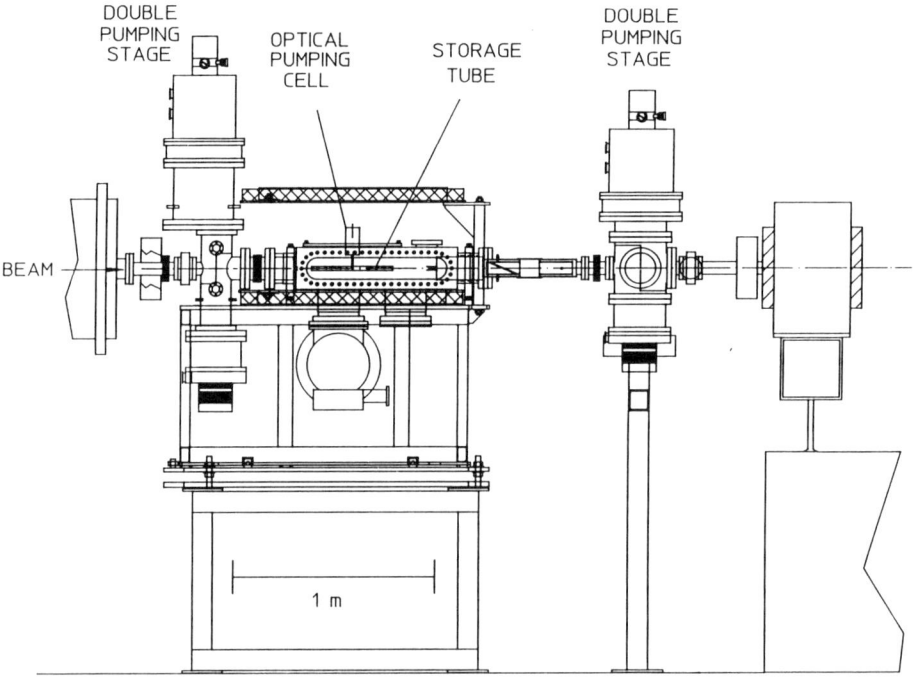

Fig. 1. Target and differential pumping stages.

When the target is in operation a gas flow of $10^{17}$ $^3$He atoms/s must be pumped away in the target region and differential pumping stages used to reach the ambient ring vacuum in as short a distance as possible. We used a 3200 l/s turbo pump in the target region and a smaller turbo pump in intermediate stages on each side, followed by second stages pumped by cryo pumps. Cryo pumps have good pumping speed for $^3$He but a low capacity for this gas. However, the flow is small at the outer stages so a regeneration procedure of 5 minutes is needed only about once per day. For the thin polarized targets the vacuum in the rest of the ring must be good so that the lifetime is dominated by the target. Under our running conditions the target-on lifetime was typically about 1/3 the target-off lifetime.

The storage cell and pumping aperture dimensions were chosen to match the machine aperture calculated from the beta functions using the design acceptance of the ring, $35\pi$mm msr. The machine typically runs at about 1/2 this acceptance so it was felt this would allow the operations staff some freedom in aligning the beam through the target as well as keep beam halo away from the cell and apertures to minimize backgrounds. This resulted in a cell which was $\sim$14mm$\times$16mm$\times$400mm and a target thickness of $\sim$1.5$\times$10$^{14}$atoms/cm$^2$ for the above flow rate.

It is important to minimize the mass of the storage cell structure. In our case we wish to look at quasi-free scattering from $^3$He which would be difficult to separate from quasi-free scattering from the cell. Assuming reasonable beam currents and lifetimes and that the particle losses are distributed evenly about the ring one might expect as many as $10^4$ protons to hit the cell. If the cell was made of 100g of aluminum and presented an area of about 4cm$^2$ to the beam there would be a luminosity on the cell of almost 10% that for the $^3$He target gas. In order to minimize the mass of the cell a design similar to that described in Ref. 7 was used. Thin aluminum sheets were bent into channels which formed the top and bottom of the cell. The sides were covered with 1.7$\mu$m aluminized mylar providing very thin exit foils for low energy recoil nuclei. The total mass of the cell was less than 11g.

## CELL BACKGROUNDS

I would like to convince you now that, with the cell design just described, very low backgrounds are possible. We have investigated this question by looking

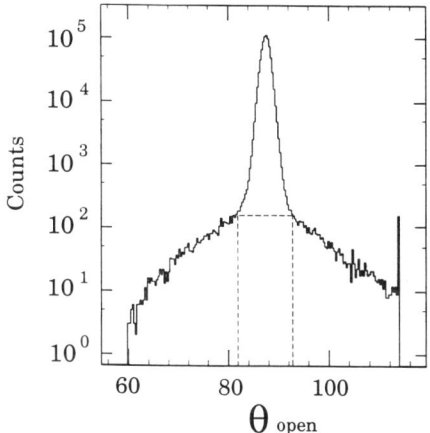

Fig. 2. Angle between protons in p-p elastic scattering at 300 MeV.

at p-p elastic scattering with the target thickness chosen to give the same lifetime as the runs with $^3$He. This provides two charged particles for vertex reconstruction with the angle between the two particles essentially constant. A plot of this opening angle between the two protons is shown in Fig. 2. One sees a peak of about 2° FWHM associated with free scattering on top of a broad distribution that may arise from various background processes. In this spectrum with no cuts the background under the peak is already less than 1% of the peak counts. We are currently reconstructing the vertex with 8mm resolution (FWHM) in height and 14mm (FWHM) transverse to the beam. If we cut to select the regions outside the peak in Fig. 2 we see no preference for these events to be coming from the massive regions of the cell. They seem to be more or less uniformly distributed over a region twice the size of the cell. This would seem to indicate that a large fraction of these events are in fact free scattering from the target where the tracks have been poorly reconstructed due, for example, to large angle scattering of a proton in one of the detectors. In any case it is clear that essentially background free experiments can be performed with storage cells in a ring.

## EFFECTS OF THE TARGET ON THE BEAM

One effect of the target on the beam is to reduce the beam lifetime. For the target thicknesses used in polarized internal targets and the energies of the IUCF Cooler this is dominated by forward angle Coulomb scattering from target nuclei.[8] In addition if the target cell or apertures are made smaller to a size where they cut into the accelerator acceptance (not the case in our experiment) further reductions arise from the decreased acceptance. The effect of such small cells on the beam has been studied previously at IUCF.[7]

A second effect we found important at IUCF was the bending of the beam by the magnetic holding field of the target. We used a field of 10G for most production running but made use of 30G for systematic error studies. Since the $^3$He target required a very uniform field over the optical pumping cell, a large Helmholtz pair was used which produced a significant field over a length along the beam of about 1m. The field integral of this magnet at maximum field is then about 3kG-cm or equal in strength to steerers commonly used in the Cooler. The target field would affect not just the beam locally but the closed orbit around the whole ring. In particular, injection and cooling are very sensitive to small beam misalignments. Without correction the cooling, and hence lifetime and backgrounds, could be different for the two field directions.

To correct the effect of the target field two additional magnets were added to the straight section immediately following the target. The first magnet was used to bend the beam back toward the unperturbed orbit and the second to bend the beam back onto the original orbit. This combination of three magnets then allows for a local motion of the beam without affecting the closed orbit elsewhere in the ring. This can also be done with other ring magnets intervening between the correcting magnets. It is only more complicated to calculate the required strengths. With these magnets deflections of the beam in the cell are less than $\pm 0.1$mm for 10 G.

## LUMINOSITY AND POLARIZATION MONITORING

Monitoring the luminosity, product of beam current and integrated target thickness along the beam, is made difficult by the fact that there are no Faraday cups possible in a ring and that the target thicknesses are very thin. The luminosity can be measured directly through a second set of detectors with good geometry (e.g. left-right symmetric) and a nuclear reaction of known cross section. This is particularly useful with jet targets where the overlap of the beam and jet may vary with time. It can also be possible to measure the beam current and target thickness separately.

In our experiment detectors were installed to observe p-$^3$He elastic scattering. Two left-right symmetric pair of scintillators were placed to detect protons at about $15°$-$20°$ in the lab. Three silicon microstrip detectors were placed along each side of the cell to stop low energy recoil $^3$He nuclei. This gave clean identification of the elastic scattering as well as a good count rate.

The beam and target analyzing powers of the p-$^3$He scattering gave an added benefit in that the luminosity monitors could also monitor the beam and target polarization. We note that these asymmetries cancel out for luminosity determination in left-right symmetric detectors. However there is an added complication that enters when using a luminosity monitor with polarized beam *and* target. In general, the monitor reaction will have a non-zero spin correlation. This is not cancelled out in left-right symmetric detectors. The spin correlation must either be previously measured or calibrated during the experiment. In our experiment we used small angle p-p elastic scattering in a second (unpolarized) target and detector system located in another straight section of the ring. This was operated simultaneously with our experiment during special calibration runs in order to measure the spin correlation in the luminosity monitors. It also gave us a good calibration for the beam analyzing power of p-$^3$He elastic scattering by cross normalizing to p-p scattering.

In addition, we were able to monitor relative luminosity for different spin states by measuring the beam current and target thickness independently. A non-destructive beam current transducer called a parametric current transformer[9] was used to measure the beam current. The gain of this device was determined by periodically turning on a $20\mu A$ current into a wire passing through the device. The zero offset was extracted by killing the beam shortly before the Cooler reset for another fill cycle. In order to measure the relative target thickness the input gas pressure to the target was monitored. Since the flow was in equilibrium the target thickness is proportional to this pressure.

During production runs the ring was filled to $100\mu A$ in 2 min. or less. Higher currents could have been reached but we were limited by the rate of real (p,2p) events into the computer. The beam current decayed with a 1000s-2000s lifetime, depending on energy, over a 10 min. to 15 min. data acquisition period. A typical target thickness of $1.5 \times 10^{14}/cm^2$ resulted in a peak luminosity of $10^{29}/s/cm^2$ and, when averaged over the accelerator cycle, an average luminosity of $5 \times 10^{28}/s/cm^2$. The beam polarization was typically 75% and the target polarization typically 50%.

## CONCLUSION

I am happy to report that very successful runs with polarized proton beams in the IUCF Cooler and a polarized $^3$He target have taken place over the past year. These experiments demonstrate that internal targets can be used for competitive measurements and that the promise of a very clean background free environment can be attained.

I gratefully acknowledge the efforts of my collaborators. I wish there were room to list them all, but instead I must refer you to the author list of Ref. 4. I also wish to thank the technical staff at IUCF and the other collaborating institutions for their work on this project.

## REFERENCES

[1] W. Haeberli, Experientia, Supple. **12**, 64 (1966).
[2] M.D. Barker et al., in Polarization Phenomena in Nuclear Physics-1980, eds. GG. Ohlsen et al., (American Institute of Physics, New York, 1981), p. 931.
[3] K. Lee et al., submitted to Nucl. Inst. Meth.
[4] K. Lee et al., Phys. Rev. Lett. **70**, 738 (1993).
[5] See talk of J. van den Brand at this conference for details.
[6] See talk of R. Milner at this conference for details.
[7] M.A. Ross et al., Nucl. Inst. Meth., A326, 424 (1993).
[8] R.E. Pollock et al., submitted to Nucl. Inst. Meth.
[9] K. Unser, IEEE Trans. Nucl. NS-28, 2344 (1981).

# OPTIMUM LUMINOSITY OF POLARIZED GAS TARGETS IN STORAGE RINGS

B. v.Przewoski

*Indiana University Cyclotron Facility, Bloomington, IN 47405*

In conjuction with stored beams polarized targets that utilize an atomic beam source have become feasible. Storage cells are already in use at facilities such as the Indiana Cooler, the Test Storage Ring in Heidelberg or the electron ring in Novosibirsk, VEPP$-3$. They serve to enhace the target thickness that would be achieved by simply crossing the stored beam with a polarized, atomic beam by $1-2$ orders of magnitude.

In designing a target cell, several contradicting criteria have to be taken into account. For a cylindrical cell with diameter D, the target thickness is proportional to $1/D^3$ thus making a small cross section most desirable. The beam lifetime, however, decreases linearly with the acceptance. As D decreases, the latter will eventually be determined by the small cell. Therefore, both beam current and target thickness depend on the cell dimensions. Also, background due to scattering from the cell walls increases with decreasing size of the cell. Obviously, a cell geometry that maximizes the luminosity must exist.

We have performed a series of measurements to optimize the geometry for the storage cell that will be used in conjunction with the Wisconsin Atomic Beam Sorce [Ref.1]. A detailed description of the measurements and their interpretation can be found in Ref.2.

The experimental setup consists of a set of four tantalum slit jaws located in the A$-$section of the Indiana Cooler. The slits could be moved individually with a precision of $\pm 0.1$mm and were used to mock up cells of different size. The A$-$section was chosen, because it has no dispersion and the smallest beta functions in the ring, and is thus the ideal location for the future polarized hydrogen target. At the location of the slits the beta functions are $\beta_x = 1.3$m and $\beta_y = 1.8$m. For the test, a diffuse hydrogen gas target was introduced in the adjacent, downstream section (G$-$section) of the ring.

The goal of the first measurement was to understand the dependence of the beam lifetime on the ring acceptance. At a given location along the ring the maximum coordinates for a particle can be calculated from the acceptance A and the beta function $\beta_\xi$ at that location, where $\xi$ is x or y. Assuming zero dispersion and mixed x and y phase space, a particle remains stored if its coordinates are less than $\xi_{max} = \sqrt{A\beta_\xi}$ and $\theta_{\xi,max} = \sqrt{A/\beta_\xi}$. If

© 1994 American Institute of Physics

one then assumes that Rutherford scattering is the dominant loss mechanism, one can calculate the probability that a stored ion is lost by integrating the Rutherford cross section from $\theta_{\xi,max}$ to $180°$. From the $\theta_{\xi,max}^{-2}$ –dependence of the integral follows that the square root of the beam lifetime is proportional to the slit position, as long as the slit edge cuts into the machine acceptance. Once the slit is positioned at $\xi \geq \xi_{max}$, the mesured liftime should become independent of the slit position. The results of a measurement where three of the slit jaws were kept fixed and one jaw was moved are shown in Fig.1 together with the theoretical expectation.

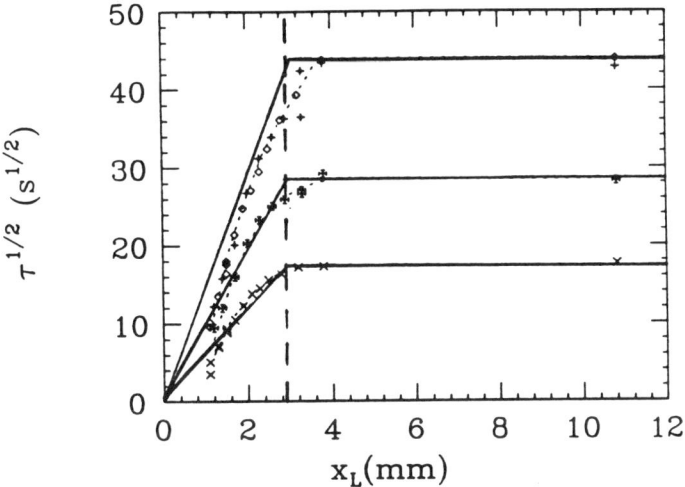

Fig.1: Beamlifetime as a function of slit position

As can be seen from the figure the data deviate from the calculation. The acceptance does not appear to have a sharp boundary. Instead, a particle that is still inside the acceptance can be lost or a particle from outside the acceptance can still circulate around the ring. Also, the location of the transition from linear to constant $\sqrt{\tau}$ does not occur at the same position for different target thickness. At present we have no explanation for the disagreement of the measurements with the theoretical expectation.

In order to investigate the characteristics of the beam as a function of cell size the positions of all four slit jaws were varied simultaneously. The beam lifetime as a function of target thickness was then measured for four different geometries. The results are shown in Fig.2 together with calculations that have been obtained as described in the following. The geometries were (from top to bottom) 4.6 by 6.3 mm$^2$, 5.6 by 7.5 mm$^2$, 6.6 by 8.7 mm$^2$ and 7.6 by 9.9 mm$^2$

The time dependent luminosity L(t) is proportional to the target thickness $d_t$, the revolution frequency $f_R$ and the lifetime $\tau$ of the stored beam. In general, the lifetime itself depends on the target thickness, a fact that has to be taken into account to optimize the time−averaged luminosity. However, in the limit of thin targets, i.e. in the regime where the beam lifetime is dominated by the residual gas in the ring, the lifetime no longer depends on the target thickness. For thin targets the effects of the restgas and the target are independent. Then

$$\frac{1}{\tau} = \frac{1}{\tau_t} + \frac{1}{\tau_0} \tag{1}$$

is valid with $\tau_t$ being the lifetime due to the target and $\tau_0$ the lifetime due to the restgas. For thin targets the time averaged luminosity is largest when $\Phi(d_t)=\tau \cdot d_t \cdot f_R$ is largest [Ref.2]. Subsituting $\tau$ from equ.1 we get

$$\Phi(d_t) = \frac{(\tau_o \cdot f_R \cdot d_t)}{(1 + \tau_0/\tau_t)}. \tag{2}$$

In the limit of thin targets $\tau_t$ becomes infinite and $\Phi$ is simply proportional to $d_t$. As the target thickness increases $\tau_t$ becomes more important. In the limit of zero emittance the lifetime is inversely proportional to the target thickness. Then the target related loss cross section

$$\sigma_t = (\tau_t \cdot f_R \cdot d_t)^{-1} \tag{3}$$

is independent of $d_t$. The loss cross section $\sigma_t$ follows from the Rutherford cross section integrated over angles larger than the acceptance angle at the target under the assumption of a well cooled beam. In a separate measurement [Ref.2] we found that the singles rate as a function of scattering angle follows indeed the $\theta$−dependence of Rutherford scattering. The data in Fig.2 have been fitted substituting equ.3 into equ.1 and treating the target loss cross section $\sigma_t$ as free parameter; the beam lifetime due to the residual gas in the ring was measured for each slit geometry. From Fig.2 it is evident that the calculation describes the measured data very well.

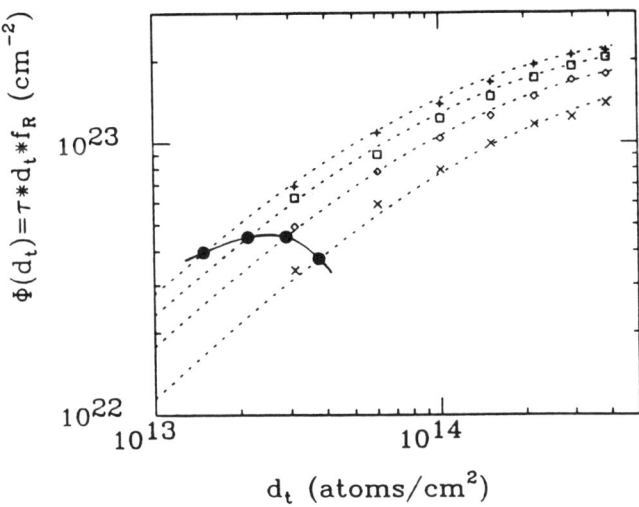

Fig.2: $\Phi$ (proportional to the time averaged luminosity) as a function of target thickness.

So far, we have not taken into account that the flow of hydrogen atoms out of an atomic beam source into a storage cell is limited. For a given flow of atoms into the cell the total target thickness depends linearly on the length of the cell. The conductances of the halves of the cell upstream and downstream of the feedtube determine the gas density $\rho_o$ at the feedtube into the cell. The conductance of the cell depends on the cell dimensions and thus a working point for each cell geometry must exist in the $\Phi,d_t-$diagram. The black dots in Fig.2 correspond to such working points. It can be seen that a maximum for $\Phi$ and thus the luminosity exists for a typical fixed flow of $10^{16}$ atoms per second. Obviously, for too large cells the target is not thick enough and for too small cells the machine acceptance is too much restricted.

In summary we have studied the influence of a restriction to the acceptance in the presence of gas, such as a storage cell would present to a stored ion beam. The cell geometry can be optimized for a given gas flow to yield maximum luminosity.

REFERENCES

1.  A.D. Roberts, T. Wise and W. Haeberli (in these Proceedings)
2.  M.A. Ross et al., Nucl. Instr. and Meth A **326**, 424(1993).

# Beam-Induced Target Depolarization

Edward R. Kinney

*University of Colorado, Boulder, CO 80309, USA*

ABSTRACT: *The interaction between the magnetic fields induced by intense electron pulses in storage rings and polarized hydrogen and deuterium atoms stored internally by a cell has been investigated. A brief review of the principle depolarization mechanisms is given, along with a study of a single depolarizing resonance. The results of full simulations for longitudinally polarized targets in the HERA ring are presented.*

## Introduction

A number of experiments have been planned which use polarized atoms passing through or trapped near a circulating particle beam to perform measurements of scattering from polarized nuclei. The extreme thinness of the target is combined with the large intensity typical of stored beams giving quite practicable luminosities. However, the current densities of the circulating bunches in the ring can be of the order of $10^5$ A/cm$^2$, resulting in magnetic fields on the order of several kilogauss near the beam. Such fields will cause the precession of the electron and nuclear spins during the duration of the bunches. This can result in nuclear depolarization through either a large precession of the nucleus or, more strongly, through depolarization of any uncoupled electron spin followed by hyperfine interaction between the electron and nuclear magnetic moments. A strong external magnetic field maintained in the vicinity of the internal target can greatly ameliorate these effects by decoupling the much larger electron moment from that of the nucleus. Care must be taken to avoid values of the holding field at which the atomic state transition energies are equal to integral multiples of the circulating frequency of the bunches. In this case, the atoms are depolarized resonantly.

### Direct and Resonant Depolarization

As an example of how one may estimate the general size of the non-resonant effects, consider the storage ring VEPP-3 at the Institute of Nuclear Physics, Novosibirsk, with average current 200 mA, circulating frequency of approximately 4 MHz, and a bunch of 30 cm length and roughly circular cross section of 1 mm radius. More detail can be found in Ref. 1. These beam paramters correspond to a peak current of 50 A, and if one ignores the finite length of the pulse, a peak magnetic field $B$ of 100 Gauss. The classical precession for a moment $\mu$ is given by $\phi = \mu B \Delta t$ which yields 0.5 mrad for a nuclear magneton, and 0.9 rad for a Bohr magneton; clearly depolarization mechanisms involving the electron will be predominant. Holding (static) magnetic fields of similar strength to the bunch field are required to overcome the precession of the electron.

The values of the holding field at which resonant depolarization occurs can be easily estimated by dividing the transition frequencies by the circulating frequency of the bunch and noting when the result is a whole number. Figure 1 displays a graph of such results for deuterium transitions involving only nuclear spinflip in the VEPP-3 storage ring; the positions of the resonances are indicated

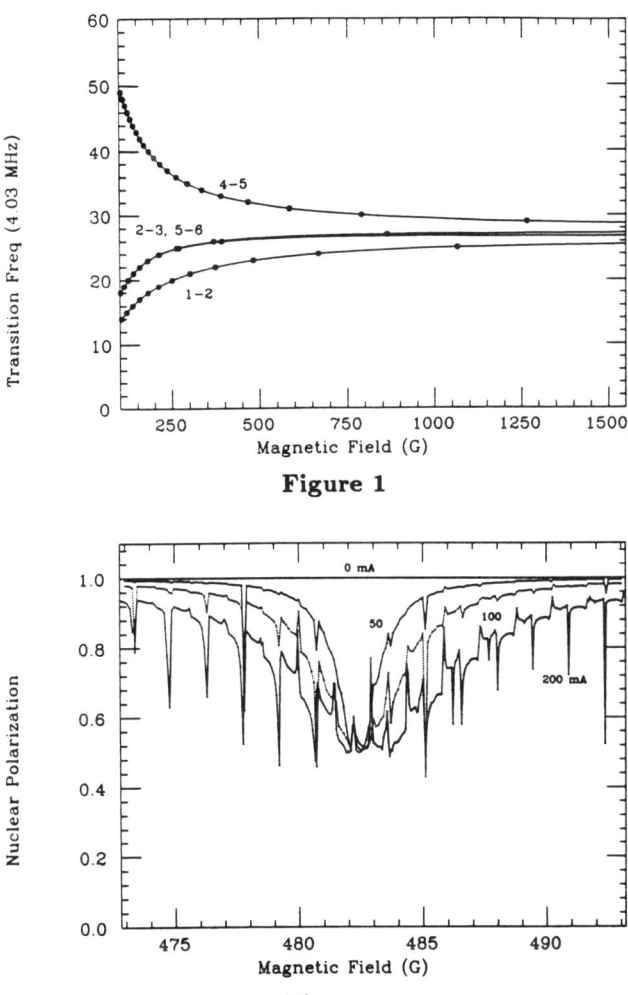

Figure 1

Figure 2

as well. The $1 \leftrightarrow 2$ resonance near 500 G was investigated in detail using a quantum mechanical calculation similar to that described below, for a fixed deuterium atom as function of holding field. The average vector polarization of the deuteron, initially in state 1 ($m_d = +1$) is shown in Fig. 2 for several different values of the average electron current in the ring. The resonance occurs at the simply predicted field value and exhibits a variation in width as a function of the driving current. The sharp spikes are in fact the electron spinflip resonances. In a real storage cell, the resonance width is further modified by the motion of the atoms in the holding field which will always have some level of non-uniformity.

A simulation of the target polarization has been developed to provide reliable estimates of the average polarization one can expect for a given atomic species, beam structure and experimental setup. The atoms are propagated through space classically, while the spin wavefunction is propagated quantum mechanically by numerical integration of the Schrödinger equation. The effect of wall collisions on the spin wavefunction is not included for lack of a quantum-mechanical model of the interaction. The field strength from the pulse is approximated by a gaussian in the longitudinal direction and an inverse variation with radius in the transverse direction; the field direction is taken to be azimuthal only. The static field is allowed non-uniformity in strength and direction. The particular computational difficulty of this simulation is the need to perform a numerical integration of a system of 6 x 6 coupled equations (of a complex function) with a sub-picosecond step size for a reasonable number of atoms, most of which remain in the storage cell for several milliseconds. This task has required the use of a CRAY to date, but recently developed RISC workstations may provide sufficient computing power. Codes for $^1$H, $^2$H, and $^3$He have been developed, and applied to the conditions at VEPP-3, HERA, the Bates and NIKHEF Stretcher Rings.

At present the only available data on these bunch-field effects comes from studies at VEPP-3, where the depolarization of a tensor-polarized atomic deuterium beam[2] was measured as a function of average electron current. The experimental results are in reasonable agreement with the predictions of the calculation,[1] giving some confidence in the simulation of the depolarization.

**Predictions for the HERMES Target**

A large number of studies have been performed for the internal targets of the HERMES experiment,[3] which will be used to measure the spin-dependent structure functions of the neutron and proton at DESY. The electron beam structure at HERA is much more severe, as the bunches are extremely small in space, while the average current and circulation frequency are comparable to VEPP-3; the peak magnetic fields are here of the order of several kilogauss. Furthermore, a hydrogen target is planned which has a hyperfine structure requiring much larger fields to decouple the electron from the proton. In the case of longitudinally polarized nuclei, the static holding field has little effect on the electron beam, hence one can choose a sufficiently strong field to ensure good target polarization. The results[4] for this case are summarized in Fig. 3, which displays the average nuclear polarizations along the length of the storage cell for $^1$H, $^2$H, and $^3$He. The hydrogen isotopes are in a holding field of over 3 kG, while the helium target only needs 20 G.

The case of transverse holding fields is complicated by their bending on the electron beam which requires not only the use of compensating magnets, but also results in significant release of energy in the form of synchrotron radiation. It appears that one can operate at significantly lower holding fields if one uses the beam only after the current in HERA has dropped to 30–40 mA and the holding field is extremely uniform ($\pm$ 1%). Finding a practical solution is under intense study presently.

In conclusion, one finds that although the bunch-field depolarization must be considered, a practical solution to minimize the effects can usually be found. In the future, more refined calculations are planned which will incorporate the

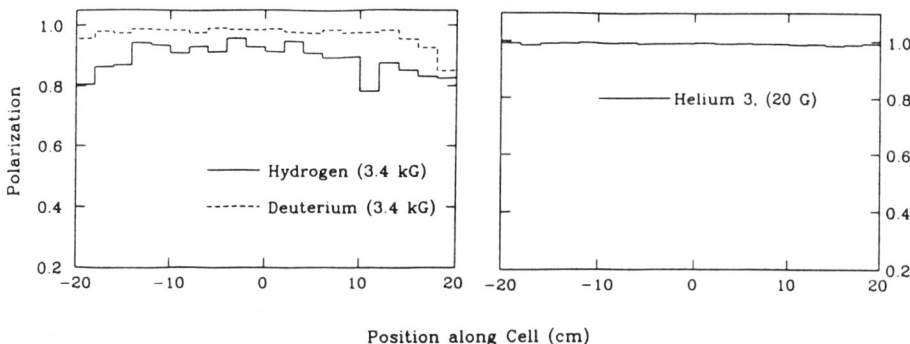

Position along Cell (cm)

**Figure 3**

effects of wall collisions. Finally, one needs additional measurements to check the calculated results in detail, as several large experiments are being designed according to the predictions of these simulations.

**Acknowledgements**

This work was supported by the U.S. Dept. of Energy, Nuclear Physics Division under grant DE-FG02-86ER-40269. The calculations reported here were made possible by a grant of time on the Cray computers at the National Energy Research Computation Center, Livermore, CA.

**References**

1. R. Gilman et al., Nucl. Instr. and Meth. A327 (1993) 277.

2. A. V. Evstigneev et al., Nucl. Instr. and Meth. A238 (1983) 12.

3. "A Proposal to Measure the Spin-Dependent Structure Functions of the Neutron and Proton at HERA," the HERMES collaboration, DESY-PRC-90-01 (1990).

4. "Simulations of the Atomic Polarization and Density in the HERMES Polarized Internal Target," E. R. Kinney, HERMES Internal Report 3/90, (1991).

Workshop on polarized Ion Sources and Polarized Gas Targets

May 23-27, 1993, Madison, Wisconsin

ROUND TABLE DISCUSSION:

# Machine-Target Interaction

H.O. Meyer

Dept. of Physics, Indiana University, Bloomington, IN

This workshop is one in a series. The participants are traditionally scientists interested in the technology, and atomic physicists interested in the principles, but in ever growing numbers also the "users" interested simply in the use of a polarized target for their experiments. The users are mainly concerned with making the target apparatus compatible with their beams and detector setups. It is clear that this process will involve compromises and careful advance planning. For this reason, an afternoon session of this workshop was devoted to the aspects of the interaction between targets and experimental environments with the intent to catalog possible problems and to ask ourselves what possible solutions might be. This marks the first time in this workshop series when an entire half day was invested in the topic of matching of tools to the final task at hand, which is rather timely, as the first nuclear physics experiments with polarized gas targets are about to start, and the first experimental facts that bear on this topic are available.

Polarized gas targets are inherently thin: with the use of a buffer cell, thicknesses of $10^{13}$ to $10^{14}$ atoms/cm$^2$ can be achieved. Their only attractive use is with intense, accumulated beams in storage rings. Since the introduction of a target in a recirculated beam is a much more serious intrusion than would be the case for a single-pass environment, the problems of target-machine interactions are quite serious.

During the session on target-machine interactions talks were presented by representatives of storage ring facilities where buffer cell targets have actually been operated. The talks were followed by a round-table discussion with K. de Jager, D. Torpokov and K. Zapfe as pannel members in which the following topics were addressed.

1. The machine <u>acceptance is limited</u> by the buffer cell

Beam lifetimes as a function of machine acceptance have been measured with movable slit systems for stored protons as well as electrons. The observed relationship is qualitatively understood from basic accelerator theory. This is sufficient as a basis for optimizing the cell dimensions. Some observations are not yet fully explained. One of them is that the acceptance boundary seems to be blury, or, in other words, the life expectancy of an individual particle varies smoothly across the acceptance limit. Another unexplained effect is that the beam lifetime depends on the beam current, as if there is a heating mechanism that depends on the number of particles present (intra-beam scattering is not large enough to explain this).

2. Need for <u>dynamic cells</u>?

It is possible to move the cell out of the way during the filling of the ring. For the target operation the cell is then assembled. An arrangement similar to a clam shell has been used for this purpose in Novosibirsk. There is clear evidence (from an electron machine) that this technique results in higher stored currents.

3. <u>Image charges</u> on the buffer cell walls

The peak current is the important parameter, thus rings for electrons are more susceptible to this than those for protons. It is known that abrupt changes in the diameter of the beam pipe should be avoided. Clearly, a buffer cell presents a potential problem.

4. Material close to the beam causes <u>experimental background</u>

Unwanted reactions of stored particles near the edge of the acceptance with the wall of the cell could, depending on the experiment, lead to unacceptable background or singles rates in some detectors. Singles rates have been investigated with an electron beam and movable slits at NIKHEF. Attempts to use slits at some ring location other than the target to define an acceptance that is smaller than the one wich is given by the cell have not been successful at IUCF.

It seems that the generation of background is hard to understand. In addition, the exact nature of the problem depends on the specific experiment. In the discussion, we heard about collimators in front of the cell, after the cell or somewhere else in the ring. The collimators could be massive (providing some shielding), or flimsy (to remove the intercepting particles during the next orbit). They could even be active, i.e., consist of a scintillator that is used to produce a veto signal.

Another important question is whether the background halo is always distributed around the ring or could, in principle, be localized (at a place other than the location of the experiment).

Clearly, a lot more experimental evidence is needed before this important aspect of experimentation is moved out of the domain of wild speculation.

# AUTHOR INDEX

## A

Alexander, N., 50
Altmeier, M., 92
Anderson, L. W., 138, 142, 146, 208
Arima, H., 88

## B

Baumann, R., 76
Belov, A. S., 102
Bigelow, N., 231
Boyd, I. D., 44
Brash, E. J., 253
Brown, R., 72
Buchholz, M., 131

## C

Cates, G. D., 244
Chupp, T. E., 244
Clausnitzer, G., 76, 80
Clegg, T. B., 115, 119
Coulter, K. P., 27, 125, 131
Crosson, E. R., 115
Cummings, W. J., 253

## D

de Jager, C. W., 27, 265
de Vries Nikhef, H., 27
Delheij, P. P. J., 253
Derenchuk, V., 72
DeZarn, W. A., 14
Doskow, J., 14
Driehuys, B., 244
Dudnikov, V. G., 102
Dulick, M., 200
Dutto, G., 173

## E

Eggert, M., 92, 97
Eversheim, P. D., 92

## F

Fan, Q., 50
Felden, O., 92
Feuerstein, B., 76
Flierl, H. P., 76
Frolov, V. V., 27
Fujita, J., 84
Fujiwara, M., 204

## G

Gebel, R., 92
Gilman, R., 27
Glombik, A., 92
Goto, A., 84

## H

Haeberli, W., 10, 14
Hatanaka, K., 84
Häusser, O., 253
Heil, W., 239, 260
Hersman, F. W., 257
Hill, K. M., 40
Hirose, T., 88
Holt, R. J., 27, 125, 131
Honig, A., 50
Horoi, M., 76
Hughes, E. W., 244

## I

Ikegami, K., 84
Inabe, N., 84
Ishida, S., 84

# AIP Conference Proceedings